Underwater Acoustic Reflection from Surfaces and Shapes

Werner G. Neubauer

Author/Editor

PENINSULA PUBLISHING

Underwater Acoustic Reflection from Surfaces and Shapes

Originally published in 1986 by the United States
Department of the Navy's Naval Research Laboratory,
Washington, DC, under the title "Acoustic Reflection from
Surfaces and Shapes." For clarity, retitled in this reprint
edition as: "Underwater Acoustic Reflection from Surfaces
and Shapes."

Published by:
Peninsula Publishing
1630 Post Road East, Unit 312
Westport, Connecticut, 06880 USA

E-mail: sales@PeninsulaPublishing.com
Telephone: 203-292-5621
Website: http://www.PeninsulaPublishing.com

Library of Congress Catalog Number: 2016943025

ISBN: 978-0-932146-32-8

Printed in the United States of America

FOREWORD

This book collects, under single cover, the acoustic reflection and scattering work, theoretical and experimental, that has been carried out on acoustic reflection and scattering by the group that was first started in the Sound Division at the Naval Research Laboratory by the author under Dr. Raymond Steinberger in 1958.

The funds to produce this book were provided cooperatively by three codes in the Office of the Naval Research: Dr. Nicolas Basdekas, Code 432, Dr. Logan Hargrove, Code 421, and Dr. Peter Rogers, Code 425UA and the Acoustics Division of the Naval Research Laboratory. Their support and encouragement is gratefully acknowledged.

I wish to thank my co-workers, past and present, and co-authors of the papers that are the foundation of this book for their contributions. The long association, co-authorship and many conversations about the book and its contents with Dr. Louis R. Dragonette have been a rewarding association for me. Also, work with Dr. Lawrence Flax has been stimulating, provocative, and a great pleasure. In various ways and to varying degrees the contributions of the following are recognized and deeply appreciated:

> Dr. C. M. Davis
> Dr. Herbert Überall
> Richard Vogt
> Dr. Susan Numrich
> Dr. Peter Uginčius
> Luise Schuetz
> Leonard Burns
> Janet Mason

Without the encouragement of the Superintendent of the Acoustics Division, Naval Research Laboratory, Dr. John C. Munson, this volume would probably not exist. The production of a book has editorial, mechanical, emotional, financial, and a seemingly unending list of considerations and difficulties. It is not possible to mention all those who had some hand in some aspect of producing these pages, but their willing and expert help is nevertheless greatly appreciated.

Typing, proofing, and coordinating many aspects of the production of this book are the result of unceasing always cheerful efforts of Mrs. Roberta Hopkins. I wish to thank her for her great help and patience. The text was first typed on a word processor, transmitted by a communications link to the Technical Information Division, Computerized Technical Composition Section at the Naval Research Laboratory. That group, headed by Mrs. Dora Wilbanks, composed the book and prepared it for printing. This was primarily the patient and capable effort of Judy Kogok who made it all come together in coherent fashion. Her contribution and the willing and cheerful assistance of Mrs. Jean Moon is gratefully acknowledged.

Werner G. Neubauer

PREFACE

The intent of this book is to make accessible selected reflection and reflection-related work of a group that has been conducting such research since approximately 1958. I have myself carried out or been involved in or responsible for the work described in each of the chapter in this book. In some cases, I have played the part of no more than editor, with only minor modifications introduced from the original publication. In all cases, when a single published paper is an entire or major content of a chapter it is referred to at the title of the chapter. For many chapters my role was more appropriately that of editor than author. Time and budgetary considerations prevented extensive rewriting. I regret that, as a result, notation and mathematical conventions are not necessarily consistent among all of the chapters. Various degrees of new material are included in several chapters.

The material is intended to be collective rather than encyclopedic. Therefore, regretfully, this book does not include the work of other groups and individuals who have contributed significantly to the field of acoustic reflection and scattering.

Werner G. Neubauer

INTRODUCTION

Underwater Acoustic Reflection from Surfaces and Shapes is a magnificent collection of mini-studies of *underwater* acoustic reflection and scattering, theoretical and experimental, from varied shapes and surfaces. These studies were carried out by the Acoustics Division at the Naval Research Laboratory during the period from 1958 through 1978 under the leadership of Dr. Werner G. Neubauer, with the assistance and guidance of Dr. Louis R. Dragonette. The driving force for this research, of course, addressed the improvement of understanding of sound interacting with submarine targets with the ultimate goal being to improve detection of submarines by submarine and surface ship sonars or to improve the invisibility of submarines *from* detection. Accordingly, the varied shape and surface targets addressed in the studies were related to those of a submarine or sections thereof: cylinders, spheres, spheroids, ellipsoids and flat surfaces.

The book has 23 chapters, each a different study. Depending on the purpose of an individual study, the shapes and surfaces addressed were made of various materials; possessed various degrees of elasticity and rigidity; and had various internal volumes, from completely solid to various degrees of internal space. Echo reduction effectiveness of reflection-reduction coatings and absorbing and non-absorbing surfaces was also measured. And the forecited surface and shape targets were "painted" by steady state or transient acoustic signals as determined by the purpose of each individual study.

The seminal studies in this book can provide underwater acousticians and modelers with valuable qualitative and quantitative insights on the reflected signals, reverberation and noise returned from submarines that will eventually lead to feeding the sonar equation.

Charles Wiseman
Publisher

CONTENTS

Chapter   Page

1   APPROXIMATE FORMULATION OF REFLECTION
    BY THE SUMMATION FORMULA   1

2   REFLECTION FROM A FINITE PLANE
    AND EXPERIMENTAL MEASUREMENTS   17

3   THEORY AND DEMONSTRATION OF
    CREEPING WAVES   35

4   CIRCUMFERENTIAL WAVE INTERPRETATIONS IN
    CYLINDER REFLECTIONS   55

5   LAYERED ELASTIC ABSORPTIVE CYLINDERS   125

6   NONABSORBING AND ABSORBING
    CYLINDERS   139

7   A CYLINDRICAL CAVITY IN AN
    ABSORPTIVE MEDIUM   151

8   ELASTIC SPHERES — STEADY-STATE SIGNALS   161

9   ELASTIC SPHERES AND RIGID SPHERES AND
    SPHEROIDS — TRANSIENT SIGNALS   177

10 ABSORBING SPHERES   195

11 LONGITUDINAL WAVES INCIDENT ON AN
    ELASTIC SPHERICAL OBSTACLE IN
    ANOTHER ELASTIC MATERIAL   205

12 RUBBER CYLINDERS AND SPHERES   225

13 ELASTIC CYLINDERS AND SPHERES AT HIGH ka   237

14   REFLECTION AND VIBRATIONAL MODES
OF ELASTIC SPHERES 241

15   RESONANCE EXCITATION AND
SOUND SCATTERING 257

16   CALIBRATION OF ACOUSTIC SCATTERING
MEASUREMENTS USING SPHERE REFLECTIONS . 279

17   EVALUATION OF A REFLECTION-REDUCTION
COATING 289

18   REFLECTION OF A BOUNDED BEAM 297

19   RAYLEIGH AND SHEAR SPEED DETERMINATION
USING SCHLIEREN VISUALIZATION 315

20   SCHLIEREN VISUALIZATION OF "HEADWAVES"
ON PLATES 325

21   SCHLIEREN VISUALIZATION TO DETECT FLAWS
ON PLATES 347

22   EXPERIMENTAL FACILITIES 357

23   SPEED OF SOUND IN FRESH WATER 373

INDEX 397

Chapter 1

APPROXIMATE FORMULATION OF REFLECTION BY THE SUMMATION FORMULA*

INTRODUCTION

Approximate solutions to reflection and scattering problems, beyond their usefulness in computations over limited ranges of parameters, have a value in clarifying physical understanding. This is true of the Kirchhoff approximation. The fundamental approximation attributed to Kirchhoff has basic limitations; but, also, the description of it leads to a basic understanding of reflection by finite bodies. There seems to be no direct application of such an approximation to reflection by Kirchhoff himself. Meecham [1], however, has considered its use for reflection problems. The attribution of the approximation stems from its use in a problem that is similar to the reflection problem, that of propagation of a wave through a circular aperture in a plane screen. In words, it demands that the wave in the aperture is everywhere what it would be if the screen were absent. Thus, there is a restriction of the aperture size relative to the wavelength and, also, of the accuracy of the resultant formulation describing a field value which depends on the location of the field point relative to the aperture. It can be shown [2] that for an aperture size greater than the wavelength a good approximation is achieved. Similarly, an angular limitation to the main lobe between the first two nulls, caused by the diffraction, is adequately described. The angular region for reasonable accuracy by the Kirchhoff approximation can exceed the main lobe as will be shown in Chapter 2.

The reflection from a body, whose dimensions are large compared to the acoustic wavelength, at a great distance from it, can be approximately described by the use of Fresnel zones on the surface of the reflector. In a way similar to the Kirchhoff approximation, as the reflector dimensions approach the acoustic wavelength, Fresnel zone analysis gives poor results. The Fresnel formula can be interpreted to describe the far field acoustic pressure as [3]

$$p_r = \frac{p_{r1}}{2} + \frac{p_{rn}}{2},\qquad(1)$$

*This development was first presented in: Werner G. Neubauer; J. Acoust. Soc. Am. **35**, 279-285 (1963).

where p_{r1} and p_{rn} are contributions to the total reflection from the first and n^{th} quarter-wave (Fresnel) zones respectively. On most regular curved convex bodies, such as spheres and cylinders, the last (n^{th}) zone is almost normal to the direction of incident wave propagation, especially if the body is much larger than an acoustic wavelength. Therefore, the last zone contribution to total reflection is negligible and the reflection by the body is substantially one half that from the first (central) zone. An example of the application of the Kirchhoff approximation will be given in which Fresnel zones are subdivided. It has been called the summation formula method. By this means, the contribution to the elements of acoustic reflection by bodies will be brought out.

It is convenient to express the solution to the wave equation in the velocity potential ϕ by

$$\phi = \frac{A}{r} \exp i\omega \left(t - \frac{r}{c} \right). \tag{2}$$

This assumes a lossless infinite homogeneous diverging acoustic wave traveling with a speed c. The potential is at any point in the field at a distance r from a point source. That source has a strength $A \exp (i\omega t)$ at unit distance, oscillating harmonically with angular frequency ω. A smooth, rigid finite reflector insonified by such a wave results in a velocity potential in the field which depends on the reflector size, shape, and orientation. A continuous source results in a velocity potential which is the summation of the potential due to that source and the potential due to the reflection by the scatterer. Here the interaction between the source and reflected wave will be ignored and only the velocity potential resulting from interaction of an incident wave with the reflector will be considered. For simplicity, consider the far field reflection in the backscattered direction, or that which is monostatic reflector.

CONTINUOUS SURFACE

Since $\omega = 2\pi f$ and $\lambda = c/f$, Eq. (2) may be written

$$\phi = \frac{A}{r} \exp \left[2\pi \left(ft - \frac{r}{\lambda} \right) \right]. \tag{3}$$

The far field demand on the problem forces r_i, the distance to the point on the reflector nearest the source, to be considered sufficiently large that $r_i \gg \lambda$. If, in addition, each dimension of the reflector is also very small compared to r_i, a portion of the spherical wave incident on the reflector may be considered plane to any desired degree. We will consider the space between the normal plane (Fig. 1) and the reflector to contain a wave that is sufficiently plane so that spherical divergence may be neglected there. Now the velocity potential at the reflector surface may be written

$$\phi_i \approx B \exp\left[-i2\pi \frac{r_i + \Delta r}{\lambda}\right],\tag{4}$$

where $r_i + \Delta r$ is the distance from the source to an elemental area da on the reflector. As a consequence of our previous assumptions, $r_i \gg \Delta r$ in Eq. (4) b is a constant factor containing the periodicity of the source and the source strength as well as the spherical divergence causing diminution of velocity potential over the distance r_i, so $B = (A/r_i) \exp(i2\pi ft)$. In the exponent of Eq. (4) the ratio r_i/λ may be divided into an integral and a fractional part, since $\exp(-i2\pi n) = 1$ and $\exp(-i2\pi a) = \alpha = $ constant where n is an integer and a is a number less than unity. Now

$$\phi_i = \alpha B \exp\left[-i2\pi \frac{\Delta r}{\lambda}\right].\tag{5}$$

As a part of the Kirchhoff approximation, and to satisfy the boundary condition on particle velocity at the rigid surface, the reflecting surface is regarded as an infinite number of elemental areas da, each of which operates as a simple source in an infinite baffle and radiating through half space. This construction is known to limit the applicability of the forthcoming analysis with respect to the curvature, extent, and perimeter of the surface.

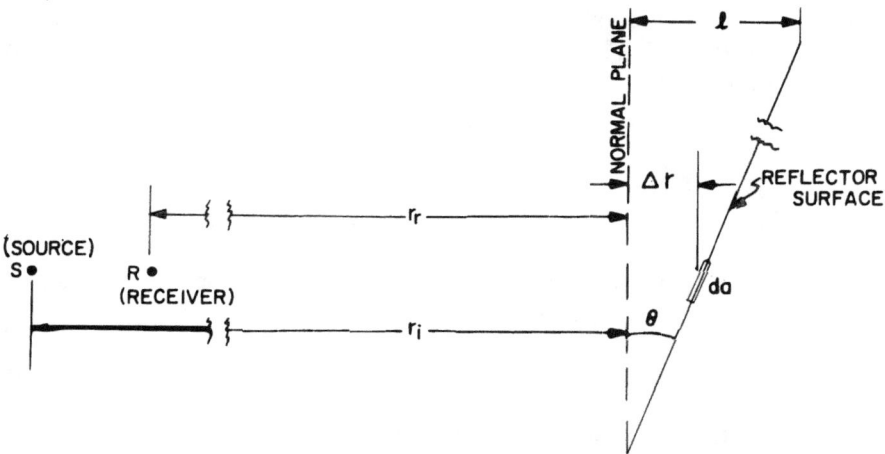

Fig. 1 — Geometry for monostatic reflection

The incident particle velocity is

$$\xi_i = -\frac{\partial \phi_i}{\partial} r_i = \frac{i2\pi}{\lambda} \phi_i + \frac{\phi_i}{r_i} = ik\phi_i\left[1 - i\frac{\lambda}{2\pi r_i}\right] \qquad (6)$$

where $k = 2\pi/\lambda$. The imaginary term in Eq. (6) is not significant since $r_i \gg \lambda$, so $\xi_i \cong ik\phi_i$.

The strength Q of the elemental area da as a source whose normal particle velocity is equal and opposite to that of the normal component of the incident particle velocity is $Q = -\xi_n da = -ik\phi_i \cos\theta\, da$, where θ is the angle between the normal to da, and the direction r_i of the incident wave. See Fig. 2. The velocity potential at R, the position of the receiver, due to the area da acting as a source is

$$d\phi_r = \frac{Q}{2\pi(r_r + \Delta r)}$$

$$\exp\left[-i2\pi\frac{r_r + \Delta r}{\lambda}\right] \approx -\frac{i\alpha' k}{2\pi r_r} \phi_i \exp\left[-i2\pi\frac{\Delta r}{\lambda}\right] \cos\theta\, da. \qquad (7)$$

We can simplify, as we did for the incident phase, and determine that $\exp(-i2\pi r_r/\lambda) = \alpha'$, where α' is a constant for a fixed value of r_r just as α was a constant for a fixed value of r_i. Substituting for the incident field ϕ_i from Eq. (5), Eq. (7) becomes

$$d\phi_r = -\frac{i\alpha\alpha' k}{2\pi r_r} B \exp\left[-i4\pi\frac{\Delta r}{\lambda}\right] \cos\theta\, da. \qquad (8)$$

The exponent $-i4\pi\Delta r/\lambda$ may be written $-i2\pi\Delta r/(1/2\lambda)$. The factor $\Delta r/(1/2\lambda)$ will take values which may be expressed again as an integral part $m = 0, 1, 2, \ldots$ and b, a number less than unity, as before in deriving the expressions in treating α and α' so that

$$\exp[-i2\pi(m + b)] = \exp(-i2\pi b),$$

and

$$d\phi_r \approx \frac{i\alpha\alpha' k}{2\pi r_r} B \exp(-i2\pi b) \cos\theta\, da. \qquad (9)$$

The field at r resulting from the entire reflecting surface is the integral of the contributions $d\phi_r$ from the elemental areas da and is given by

$$\phi_r = -\frac{i\alpha\alpha' B}{2\pi r_r} \int_s \cos\theta \exp(-i2\pi b)\, da, \qquad (10)$$

where s is the reflector surface. Geometrically, $da \cos\theta$ is the projection da' of the element da onto the normal plane (Fig. 1) and, therefore, the integral may be taken over the projection of the entire surface

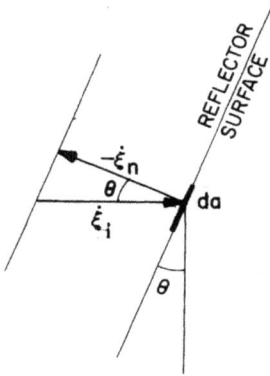

Fig. 2 — Particle velocity at the reflecting element

onto the normal plane instead of the surface itself. So Eq. (10) may be written as

$$\phi_r = -\frac{i\alpha\alpha'kB}{2\pi r_r}\int_{s'} \exp\left(-i2\pi b\right)da'. \tag{11}$$

Now we have an expression for an acoustical quantity, the velocity potential, from which may be derived other quantities in which one might be interested, in terms of constants derived from measurable distances or areas on the reflector or on a projection of the reflector onto the normal plane that can be measured. In most descriptions of the acoustical field, quantities that are required to describe the reflected field are acoustical quantities such as pressure or particle displacement measured on or near the body itself. Equations (10) and (11) require only area and distance measurements, but are encumbered by the limitations of the Kirchhoff approximation.

RIGID, FINITE PLANE

As an example of the application of the formulas which have been derived, let us consider a rectangular reflecting plane of width d and height h. Let the plane be located in an orthogonal coordinate frame with the height h parallel to the Y-axis, the X-axis dividing the normal plane in half and let Z be the direction of propagation. The positive Z direction is the direction of the incident wave considered plane by the time it reaches the reflector. See Fig. 3. Since $b + m = \Delta r/(1/2\lambda)$, $b = (2x \tan \theta/\lambda) - m$ and $da' = dxdy$, Eq. (11) becomes

$$\phi_r = -\frac{i\alpha\alpha'B}{\lambda r_r}\int_{-1/2h}^{1/2h} dy \int_0^{d\cos\theta} \exp\left[-\frac{i4\pi\lambda}{\lambda}\tan\theta\right]dx. \tag{12}$$

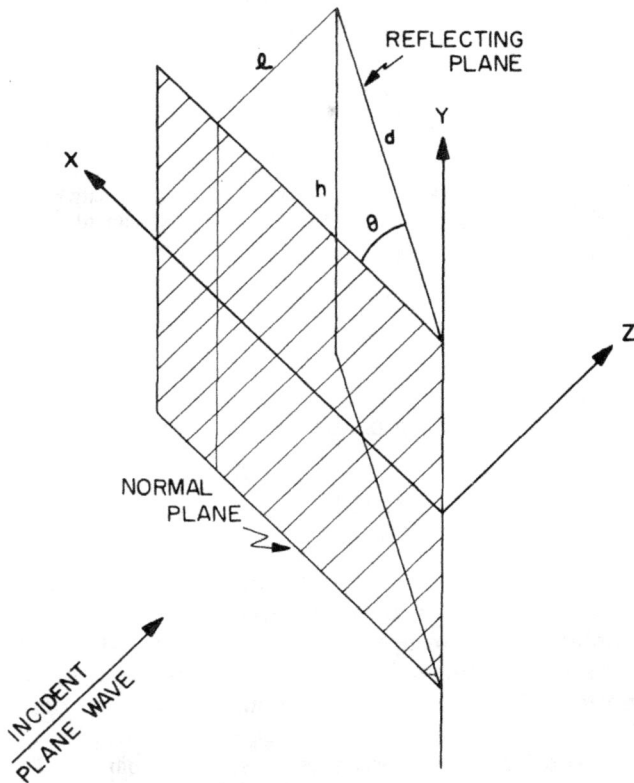

Fig. 3 — The orientation of the plane reflector
in Cartesian coordinates

Equation (12) can be written approximately as a finite sum as

$$\phi_r \approx - \frac{i\alpha\alpha'B}{\lambda r_r} \sum_{j=1}^{q} \sum_{p=1}^{t} \Delta x_p \Delta y_j \exp\left[-\frac{i4\pi}{\lambda}x_p \tan\theta\right]. \qquad (13)$$

This equation may be considered the velocity potential from the normal plane itself if it were composed of areas $\Delta x_p \Delta x_j$ qt, in number. Each area is considered to be vibrating with phase $(4\pi/\lambda)$ $x_p \tan\theta$ where $x_p = p\Delta x_p$. Because of the choice of the orientation of the plane in the coordinate system $\sum_{j=1}^{q} \Delta y_j = h$ and Eq. (13) may be written

$$\phi_r = - \frac{i\alpha\alpha'B}{\lambda r_r} \sum_{p=1}^{t} \Delta x_p \exp\left[-\frac{i4\pi}{\lambda}\lambda_p \tan\theta\right]. \qquad (14)$$

Consider the reflecting plane divided into smaller planes spaced at integral multiples of $\lambda/2$ from the normal plane, and parallel to it as shown in Fig. 4, where the geometry is being viewed looking down along the Y-axis. These subplanes will intersect the reflecting surface to define on it what may be considered half-wavelength zones. Subdivide the distances between these subplanes with even more closely spaced planes, again parallel to the normal plane, equally subdividing the $\lambda/2$ distances into an arbitrary number ν. Let the smallest planes intersect areas having widths on the reflector surface small compared to a wavelength. In Fig. 4 the construction is viewed in the $X-Z$ plane for $\nu = 6$. For any orientation parallel to the Y-axis, the projections of the subdivided surface areas, all of length h, are proportional to the line length $\Delta x_1 = \Delta x_2 = \cdots = \Delta x_p$. These are the magnitudes of the vectors in the complex plane in Fig. 5. The constant phase increment is the angle between successive vectors, and each time the factor $[\Delta x_p/(\lambda/2) \tan \theta]$ becomes an integral value i.e., when $p = \nu$, the sum becomes zero. In Fig. 4, only two and a fraction $(1/\Delta)$ subzones are taken; in the second, $\lambda/2$ zone. The resultant relative velocity potential at R is proportional to the vector Λ. It must be remembered that a constant factor h, which was a coefficient for the sum, accounts for the height of the plane and must be taken into account as a scale factor for Λ.

The graphical solution of Eq. (12) can now readily be seen to be a circle tangent to the real axis in the complex plane if $\Delta x_p \to 0$ as $p \to \infty$.

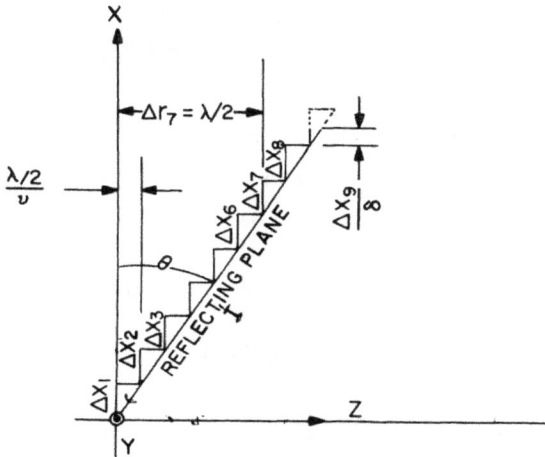

Fig. 4 — A graphical composition of the plane reflecting surfaces as finite normally oriented areas

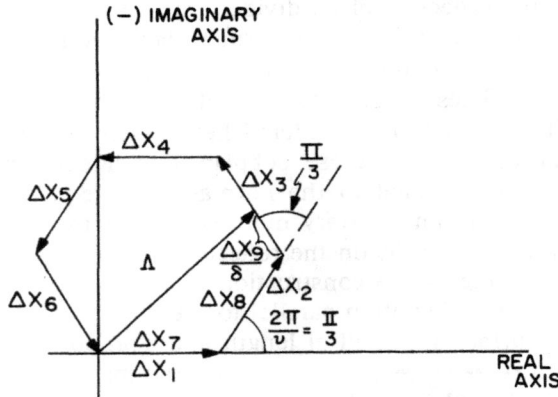

Fig. 5 — The plot in the complex plane of ϕ_r,
approximated by the sum in Eq. 14

FINITE WEDGE

A wedge, as depicted in Fig. 6, can be considered by the previous method if it is regarded as two finite planes rigidly fixed with respect to each other. Under the same conditions as in the previous section, the velocity potential at a distance R will get the summation in phase and amplitude of the velocity potential of each plane separately. In integral form this may be expressed as

$$\phi_r = -\frac{i\alpha\alpha'B}{\lambda r_r}\left[\int_{-h/2}^{h/2} dy \int_0^{d_1\cos\theta} \exp\left(-\frac{i4\pi}{\lambda}\tan\theta\right)dx\right.$$

$$\left. +\int_{-h/2}^{h/2} dy \int_0^{d_2\cos\gamma} \exp\left(-\frac{i4\pi x}{\lambda}\tan\gamma\right)dx\right] \qquad (15)$$

which upon integration yields

$$\phi_r = \frac{\alpha\alpha'hB}{4\pi r_r}\left[\frac{\exp-[(i4\pi d_1/\lambda)\sin\theta]}{\tan\theta} - \frac{\exp-[(i4\pi d_2/\lambda)\sin(\theta+\beta)]}{\tan(\theta+\beta)}\right.$$

$$\left. -\cot\theta + \cot(\theta+\beta)\right]. \qquad (16)$$

For a cube, in which case $\beta = \pi/2$ and $d_1 = d_2 = d$, Eq. (16) becomes

$$\phi_r = \frac{\alpha\alpha'hB}{4\pi r_r}\left[\frac{\exp[-(i4\pi d/\lambda)\sin\theta]}{\tan\theta} + \frac{\exp[-(i4\pi d/\lambda)\cos\theta]}{\cot\theta} - 2\csc 2\theta\right].$$

$$(17)$$

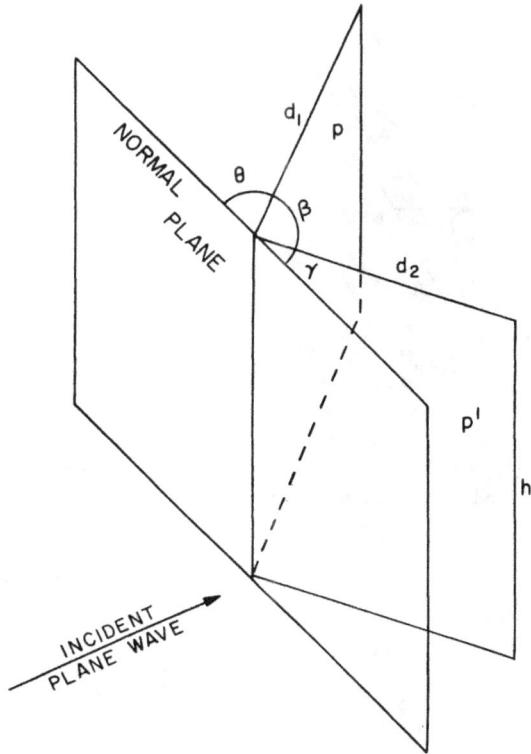

Fig. 6 — The wedge reflector

Taking finite elements along a plane, as shown in Fig. 5, Eq. (15) may be expressed as a sum similar to Eq. (14) for the plane as,

$$\phi_r = \frac{i\alpha\alpha'hB}{\lambda r_r}\left[\sum_{p=1}^{t}\Delta x_p \exp\left(-\frac{i4\pi}{\lambda}\lambda_p \tan\theta\right)\right.$$

$$\left. + \sum_{p'=1}^{t'}\Delta x_p' \exp\left(-\frac{i4\pi}{\lambda}\lambda p'\tan\gamma\right)\right]. \qquad (18)$$

Figure 7 shows two means of arriving at the complex resultant Λ_r. In Fig. 7(a), the resultant obtained from the separate consideration of each face by adding Λ and Λ' associated with faces P and P' respectively in Fig. 6. The same resultant Λ_R is obtained if Δx_p and $\Delta x_{p'}$ are first added and then summed over p, as shown in Fig. 7(b), in which $p = t > p' = t'$. Here the definition is made that for $t > t'$, $\Delta x_{p'}$ is $\equiv$ 0. Since $|x_p \tan\theta| = |x_{p'} \tan\gamma|$ (Fig. 8), Eq. (18) reduces to

$$\theta_r = -\frac{i\alpha\alpha'hB}{\lambda r_r}\sum_{p=1}^{t}(\Delta x_p + \Delta x_{p'})\exp\left(-\frac{i4\pi}{\lambda}x_p \tan\theta\right). \qquad (19)$$

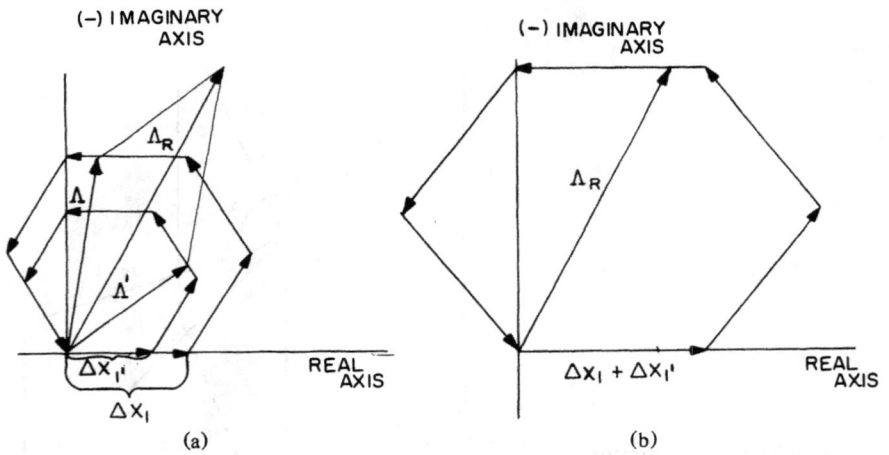

(a)   (b)

Fig. 7 — Plots in the complex plane for reflection from a wedge: (a) each taken separately and their resultants added, and (b) both planes taken as contributing to the same zone

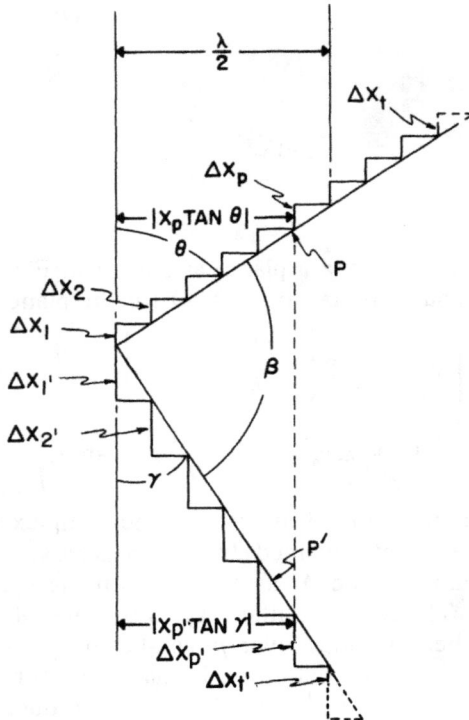

Fig. 8 — The graphical composition of the reflecting wedge surfaces as finite normally oriented areas

RIGID, DOUBLE CURVED SURFACE

If the same construction and analysis already applied to a plane and a wedge is applied to a curved surface, the finite moduli Δx_p of Eq. (19) would, in general, not be equal for equal phase increments as $p \rightarrow t$ (see Fig. 9). In addition, a restriction must be imposed on the principal radii of curvature of the surface. It is sufficient to demand that at all points of the reflector the principal radii of curvature be greater than the incident sound wavelength so that the small but finite areas Δa_p may be considered essentially plane. The general expression for the velocity potential ϕ_r as an integral over the illuminated surface of the reflector is

$$\phi_r = - \frac{i\alpha\alpha'kB}{2\pi r_r} \int_s [\exp(-i2k\Delta r)] \cos\theta \, da, \qquad (20)$$

where $k2\pi/\lambda$. The equivalent expression over the projections da' of the subzones onto the normal planes is

$$\phi_r = - \frac{i\alpha\alpha'B}{2\pi r_r} \int_{s'} \exp(-i2\pi k\Delta r) \, da. \qquad (21)$$

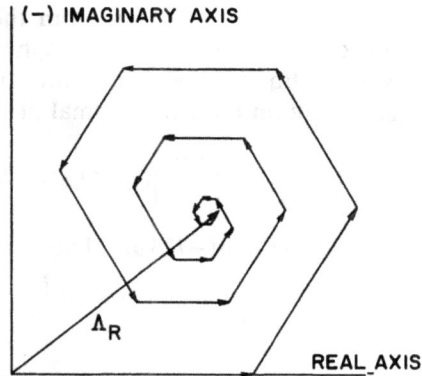

Fig. 9 — A plot in the complex plane for reflection from a general shape as approximated by the sum in Eq. (24)

The distance from the normal plane, along the Z-axis, is considered to be l and is divided into μ equal parts, the integrals in Eqs. (20) and (21) respectively can be approximated by sums

$$\phi_r = \frac{-i\alpha\alpha'kB}{2\pi r_r} \sum_{p=1}^{\mu} \Delta a_p \cos\theta \exp\left(-i2k\frac{pl}{\mu}\right) \qquad (22)$$

$$\phi_r = - \frac{l\alpha\alpha'kB}{2\pi r_r} \sum_{p=1}^{\mu} \Delta a_{p'} \exp\left(-12k\frac{pl}{\mu}\right). \qquad (23)$$

As shown in Fig. 10, $\Delta a_{p'}$ is the projection onto the normal plane of that portion of the reflector surface between the pth and the $(p + 1)$th plane subdividing l. The complex plane is shown for such a sum. Equation (22) can be evaluated by measuring the areas on the reflecting body between successive cutting planes or alternatively the areas of their projections onto the normal plane. The associated value of pl/μ to the area measured on the surface of the reflector or on the normal plane must, of course, also be known. The magnitude of the velocity potential at R then for a given orientation of the reflector may be expressed approximately as

$$|\phi_r| = -\frac{k|B|}{2\pi r_r} \left[\sum_{p=1}^{\mu} \Delta a_{p'} \sin\left(2k\frac{pl}{\mu}\right) \right]^2 + \left[\sum_{p=1}^{\mu} \Delta a_{p'} \cos\left(2k\frac{pl}{\mu}\right) \right] \quad (24)$$

Application of this formula result in formulas for the reflection from rigid quadric surfaces which have been published [5] and a table for various spheroidal shapes is reproduced here in Table 1.

It would be useful to know how much of an approximation such an approximate formula employing the Kirchhoff approximation and then an approximate summation to an integral will cause. For this, let us consider the reflection from the sphere. Figure 11 shows the solution for the reflection from a sphere using the Kirchhoff approximation given by Eq. (24) which results from the integration of Eq. (11) over the projection onto the normal plane of a sphere radius a. The result is

$$\phi_r = -\frac{\alpha\alpha'\alpha''B}{2r_r} \left[\cos 2ka - \frac{\sin 2ka}{2ka} + i\left(\sin 2ka - \frac{\sin^2 ka}{ka} \right) \right], (25)$$

where $\alpha'' = \exp(-i2ka)$. The magnitude of Eq. (25) is

$$|\phi_r| = -\frac{a|B|}{2r_r} \left[1 - \frac{\sin 2ka}{ka} + \frac{\sin^2 ka}{(ka)^2} \right]^{1/2}. \quad (26)$$

The square-root expression is what is plotted against ka in Fig. 11. The exact solution, which will be given later, is also shown. It can be seen that the periodicity of oscillation of the solution with increasing ka is significantly different and the magnitude in the region of 2 can be very much different. However, as ka increases, these oscillations in both cases decrease and the difference in the solutions tend to be no more than plus or minus 10% or approximately plus or minus 1 dB. In addition, an approximation is introduced by approximating the integral by a finite sum. The integral is approached to a desired degree by the choice of the ratio $1/\lambda\mu$ or the choice of μ/p for a given ka. For ka greater than 6, the difference between the two curves is never greater than 16%, or approximately 1.5 dB.

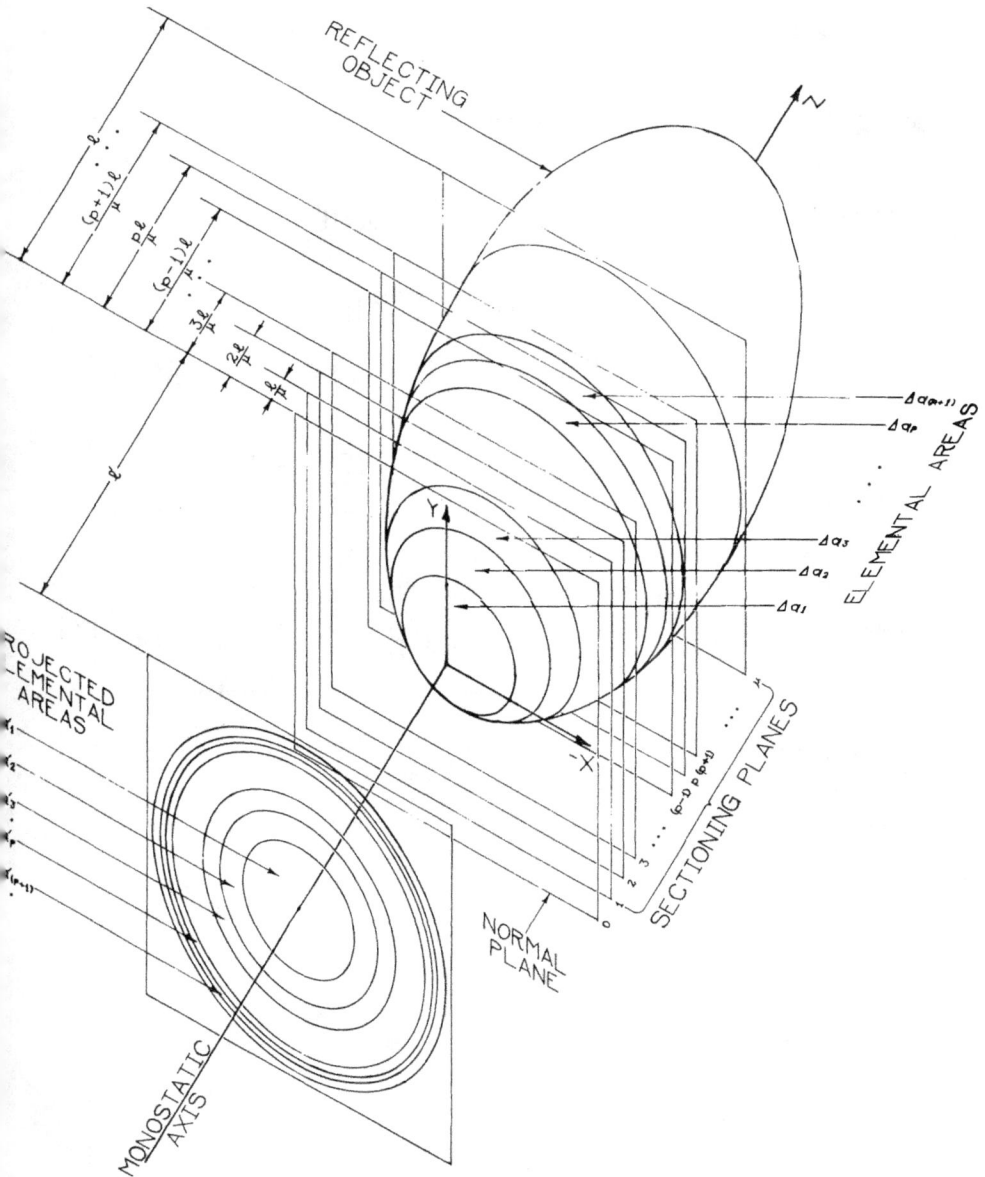

Fig. 10 — The geometrical quantities involved in the derivation of the reflection summation formula. For clarity, the contours defined on the reflecting object are not projected onto the normal plane, but onto a plane parallel to the normal plane at a constant distance from it. The family of projected contours on this paralel plane is identical to the family of contours which would be obtained if the projection were made onto the normal plane.

Table 1 — Reflection Characteristics of the Rigid Ellipsoid and Its Special Cases

Reflecting Object	Aspect	Geometrical Conditions	General Elemental Area $\Delta a_{p'}$	Magnitude of the Reflected Velocity Potential $|\phi_r|$
Ellipsoid	Along a Principal Axis (l_1, l_2, l_3)	$l_1 \neq l_2 \neq l_3$	$\dfrac{\pi l_2 l_3}{\mu^2}(2\mu - 2p + 1)$	$\dfrac{|B|}{r_r}\left(\dfrac{l_2 l_3}{l_1^2}\right)\left(\dfrac{l_1}{2}\right)\left(\dfrac{\sin^2(kl\,s)}{(kl\,s)^2} - \dfrac{\sin(2kl\,s)}{kl\,s} + 1\right)^{1/2}$
Sphere	Any	$l_1 \neq l_2 \neq l_3$	$\dfrac{\pi l_1^2}{\mu^2}(2\mu - 2p + 1)$	$\dfrac{|B|}{r_r}\left(\dfrac{l_1}{2}\right)\left(\dfrac{\sin^2(kl\,s)}{(kl\,s)^2} - \dfrac{\sin(2kl\,s)}{kl\,s} + 1\right)^{1/2}$
Prolate Spheroid	End-On	$l_1 \neq l_2 \neq l_3$	$\dfrac{\pi l_2^2}{\mu^2}(2\mu - 2p + 1)$	$\dfrac{|B|}{r_r}\left(\dfrac{l_2 l_3}{l_1^2}\right)\left(\dfrac{l_1}{2}\right)\left(\dfrac{\sin^2(kl\,s)}{(kl\,s)^2} - \dfrac{\sin(2kl\,s)}{kl\,s} + 1\right)^{1/2}$
Prolate Spheroid	Beam	$l_1 \neq l_2 \neq l_3$	$\dfrac{\pi l_1 l_3}{\mu^2}(2\mu - 2p + 1)$	$\dfrac{|B|}{r_r}\left(\dfrac{l_2 l_3}{l_1^2}\right)\left(\dfrac{l_1}{2}\right)\left(\dfrac{\sin^2(kl\,s)}{(kl\,s)^2} - \dfrac{\sin(2kl\,s)}{kl\,s} + 1\right)^{1/2}$
Oblate Spheroid	End-On	$l_1 \neq l_2 \neq _3$	$\dfrac{\pi l_2^2}{\mu^2}(2\mu - 2p + 1)$	$\dfrac{|B|}{r_r}\left(\dfrac{l_2}{l_1}\right)\left(\dfrac{l_1}{2}\right)\left(\dfrac{\sin^2(kl\,s)}{(kl\,s)^2} - \dfrac{\sin(2kl\,s)}{kl\,s} + 1\right)^{1/2}$
Oblate Spheroid	Beam	$l_1 \neq l_2 \neq l_3$	$\dfrac{\pi l_1 l_3}{\mu^2}(2\mu - 2p + 1)$	$\dfrac{|B|}{r_r}\left(\dfrac{l_3}{l_1^2}\right)\left(\dfrac{l_1}{2}\right)\left(\dfrac{\sin^2(kl\,s)}{(kl\,s)^2} - \dfrac{\sin(2kl\,s)}{kl\,s} + 1\right)^{1/2}$

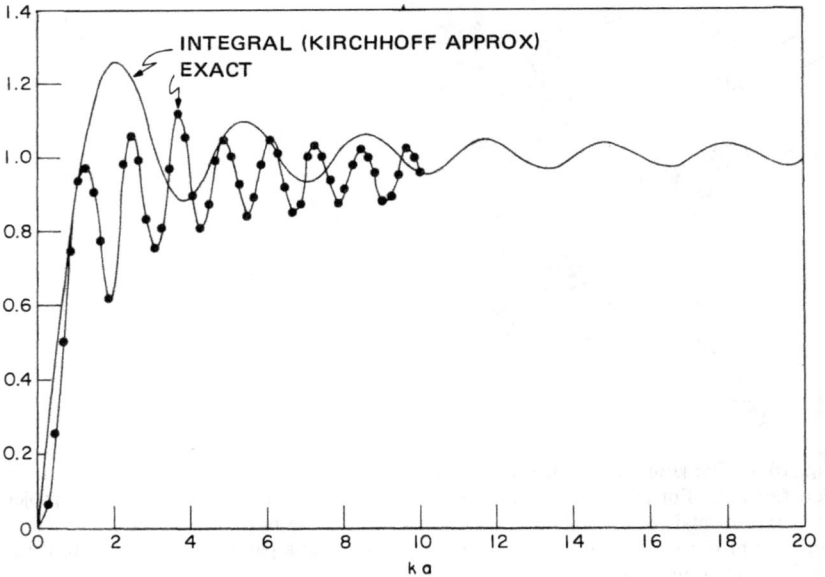

Fig. 11 — The reflection from a rigid sphere

The error introduced by use of the summation formula related to the choice of μ for a sphere is given in Table 2. These error values were obtained by evaluating $\Delta a_{p'}$ for a sphere and algebraically summing the series thus obtained. Passing to the limit by allowing $\mu \to \infty$ as $\Delta a_{p'} \to 0$ results in Eq. (26). Of course, this error is not directly applicable to what one would expect for an arbitrary shape, since the error introduced would definitely be a function of the local curvature of the reflector. Evaluations such as this can however serve as a guide in the application of the formula summation.

Table 2 — The Error in
Approximation of the Integral by a
Finite Sum for a Chosen μ.

μ	Error (%)
8	11
16	2
24	1
48	0.3

CONCLUSION

The summation formula can be employed to utilize measured areas on the target to directly arrive at an acoustical field quantity representative of the reflection, say the velocity potential. The formula is only useful in those cases where an insignificant amount of energy can be considered to penetrate the target and cause elastic vibration of the target itself. As will be seen later, there are regions even for very practical shapes and dimensions that will permit application of the assumption of very little penetration into the target and, therefore, reasonable approximation by a formulation assuming a rigid boundary condition.

REFERENCES

1.   W.C. Meecham, "On the Use of Kirchhoff Approximation for the Solution of Reflection Problems," J. Rational Mech. Anal. 2, 323 (1956).

2.   S. Hanish, *A Treatise on Acoustic Radiation*, book published by the Naval Research Laboratory (1981).

3.   F.A. Jenkins and H.E. White, *Fundamentals of Optics*, McGraw-Hill Book Company, Inc. (1950).

4.   A.J. Rudgers, "Calculation of the Monostatically Reflected Velocity Potential in the Far Field of Certain Finite Rigid Bodies," NRL Report 6285, December 1965.

Chapter 2

REFLECTION FROM A FINITE PLANE AND EXPERIMENTAL MEASUREMENTS*

INTRODUCTION

A further application of the Kirchhoff approximation to the problem of the reflection from a rectangular plane considered to be rigid will allow a comparison with experimental results. Theoretical description will be accomplished by approximate evaluation of the retarded potential in a rectangular coordinate system.

THEORY

The velocity potential ϕ at a distance r_a from an elemental reflecting area da is

$$d\phi_r = (dQ/2\pi r_a) \exp - (ikr_a), \tag{1}$$

where dQ is the strength of the elemental source aa on the reflector, radiating into half space. This essentially constitutes the Kirchhoff approximation. In terms of the incident particle velocity ξ_i,

$$dQ = - \xi_i \cos \psi \cos \theta \, da = - ik\phi_i' \cos \psi \cos \theta \, da, \tag{2}$$

where ϕ_i' is the incident velocity potential in the absence of the reflecting plane, k is the wave number of the incident wave and the angles ψ and θ are those defined by Fig. 1. For a simple harmonic spherically diverging wave emitted from a source at a distance r_i

$$\phi_i' = (A/r_i) \exp [ik(ct - r_i)]. \tag{3}$$

At a sufficiently large r_i, i.e., $r_i \gg \lambda$, where λ is the wavelength, a plane wave approximation over a limited region may be achieved. Assuming the reflecting plane to have a wave incident on it under these conditions,

$$d\phi_r = - (i\phi_i/r_a\lambda) \cos \psi \cos \theta \exp [- i2\pi (r_a + r_i')/\lambda] \, da, \tag{4}$$

*The theory and experiments described here first approved in: Werner G. Neubauer and Louis R. Dragonette, J. Acoust. Soc. Am. **41**, 656-661 (1967)

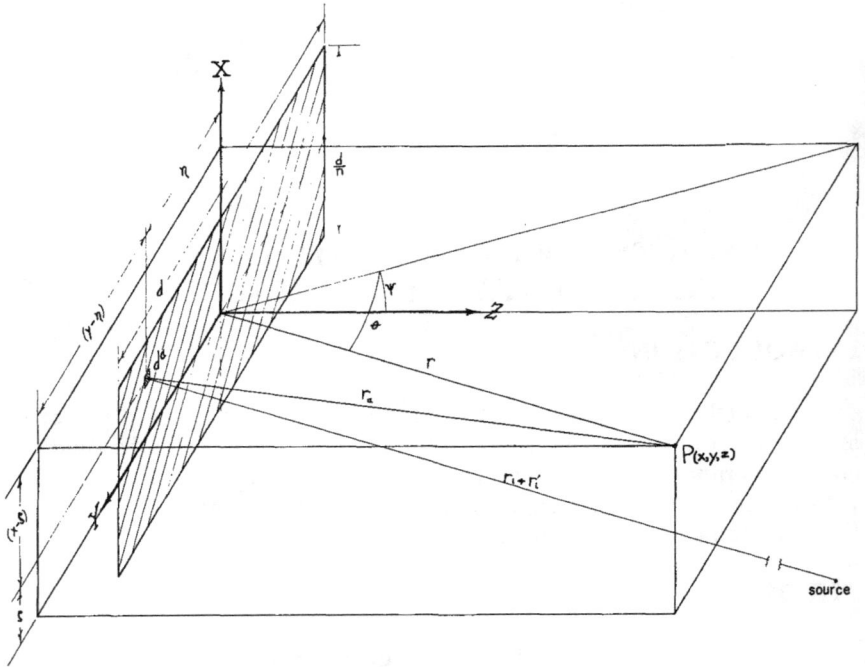

Fig. 1 — The geometry which defines r_i and r_a

where ϕ_i represents the incident velocity potential on the normal plane which just touches the corner of the reflecting plane nearest the source. The plane wave approximation permits $\phi_i' \cong \phi_i$. The quantity $(r_a + r_i')/\lambda$ represents the phase retardation that occurs as the wave transits from the normal plane to the reflector and back to the field point p. The distance r_i' is the distance from the normal plane to the element da on the reflecting plane. Expressed as an integral over the reflecting plane

$$\phi_r = -\left(\frac{i\phi_i \cos \psi \cos \theta}{\lambda r}\right) \int_S \exp\left[-\left(i2\pi \frac{r_a + r_i'}{\lambda}\right)\right] da, \qquad (5)$$

where r_a has been approximated by r, which means that the divergence from each elemental area da to a reflected point p is the same for all elements on the reflecting plane. From the geometry in Fig. 1, da equals $d\zeta \, d\eta$ and it is possible to derive the relationship

$$r_a^2 = (x - \zeta)^2 + (y - \eta)^2 + z^2 = r^2 + \zeta^2 + \eta^2 - 2x\zeta - 2y\eta. \qquad (6)$$

It can also be shown that

$$r_i' = [(d/2n) - \zeta] \sin \psi \cos \theta + [(d/2) - \eta] \sin \theta, \qquad (7)$$

where d and n are defined in Fig. 1. The quantity r_a can be taken to be the first two terms of the binomial expansion

$$r_a \cong r + (\zeta^2 - 2x\zeta + \eta^2 - 2y\eta)/2r, \tag{8}$$

and the exponential in Eq. (5) can be cast into the form of the Fresnel integral so

$$2\pi(r_a + r_i')/\lambda = (\pi/2)[4(r_a + r_i')/\lambda] \cong \pi/2(\alpha' + u'^2 + v'^2), \tag{9}$$

where

$$\alpha' = (4r/\lambda) + (2d/n\lambda) \sin\psi \cos\theta + (2d/\lambda) \sin\theta \tag{9a}$$

$$u'^2 = (2/\lambda r)\zeta^2 - [4/\lambda][(x/r) + \sin\psi \cos\theta]\zeta,$$

and

$$v'^2 = (2/\lambda r)\eta^2 - [4/\lambda][(y/r) + \sin\theta]\eta. \tag{9b}$$

Solving Eq. (9a) for ζ and substituting

$$u'^2 = u^2 - (2/\lambda r)(2x)^2,$$

it follows that from Eq. (9a)

$$u = (2/\lambda r)^{1/2}(\zeta - 2x).$$

Similarly, Eq. (9b) results in

$$v = (2/\lambda r)^{1/2}(\eta - 2y),$$

where $x = r \sin\psi \cos\theta$ and $y = r \sin\theta$. Now Eq. (5) becomes

$$\phi_r = -\left(\frac{i\phi_i \cos\psi \cos\theta}{2}\right) \exp\left(\frac{-\alpha i\pi}{\lambda}\right) \tag{10}$$

$$\times \int_{u_1}^{u_2} \exp\left(\frac{-i\pi u^2}{2}\right) du \int_{v_1}^{v_2} \exp\left(\frac{-i\pi v^2}{2}\right) dv,$$

where

$$\alpha = 2r + (xd/nr) + (yd/r) - 4(x^2/r) - 4(y^2/r)$$

and

$$u_{1,2} = (2/\lambda r)^{1/2}[\mp(d/2) - 2x],$$
$$v_{1,2} = (2/\lambda r)^{1/2}[\mp(d/2n) - 2y].$$

Also,

$$|\phi_r/\phi_i| = \left|\frac{p_r}{p_0}\right| = \left|\frac{\cos\psi\ \cos\theta}{2}\right| \left|\int_{u_1}^{u_2} \exp\left(\frac{-i\pi u^2}{2}\right) du\right|$$

$$\times \left|\int_{v_1}^{v_2} \exp\left(\frac{-i\pi v^2}{2}\right) dv\right|, \tag{11}$$

where p_0 and p_r are the incident and reflected pressures, respectively. Since this result was obtained by expanding r_a in a binomial series and retaining the first two terms, it is subject to the limitation that

$$[(\zeta^2 - 2x\zeta + \eta^2 - 2y\eta)r^2]^2 \leqslant 1.$$

In geometric terms this means that the region inside of a hemispherical volume containing the reflector is to be excluded from consideration. The integrals in Eq. (10) can be expressed in terms of the complex Fresnel integral $F(u_1)$ given by

$$F(u_1) = \int_0^{u_1} \exp\left(\frac{-i\pi u^2}{2}\right) du, \tag{12}$$

which by application of "Euler's" formula, can be expressed as

$$F(u_1) = \int_0^{u_1} \cos\left(\frac{\pi u^2}{2}\right) du - i \int_0^{u_1} \sin\left(\frac{\pi u^2}{2}\right) du. \tag{13}$$

The conventional definitions are

$$C(u_1) \equiv \int_0^{u_1} \cos\left(\frac{\pi u^2}{2}\right) du$$

and   (14)

$$S(u_1) \equiv \int_0^{u_1} \sin\left(\frac{\pi u^2}{2}\right) du.$$

A similar definition can be made for a second integral $F(u_2)$ and by manipulation of the integrals it may be shown that

$$F(u_1, u_2) \equiv F(u_2) - F(u_1) = \int_{u_1}^{u_2} \exp\left(\frac{-i\pi u^2}{2}\right) du$$

$$= [C(u_2) - C(u_1)] - i[S(u_2) - S(u_1)] \tag{15}$$

whose magnitude is

$$|F(u_1, u_2)| = \{[C(u_2) - C(u_1)]^2 + [S(u_2) - S(u_1)]^2\}^{1/2}. \tag{16}$$

This magnitude is the quantity required to compute the reflected pressure and is the line length from u_1 to u_2 in the complex $C(u)$, $S(u)$ plane. The associated phase for this magnitude may be taken at the angle that the directed line $u_1 u_2$ makes with the positive $C(u)$ axis. The functions $S(u)$ and $C(u)$ have been tabulated [1] and their appropriate expansions have been evaluated by computer [2].

EXPERIMENTS

The pressure reflected monostatically from the faces of solid metal blocks with parallel sides was experimentally measured. The purpose of experiments was to determine the degree of agreement that could be achieved between an experiment using a real elastic reflecting material in water and the theory which assumes the reflector to be rigid and is based on the Kirchhoff approximation. The reflectors were suspended in the field so that no reflections could be measured from suspension wires. The blocks were placed in the far field of spherical-wave sources that subtended an aperture over which the incident wave was essentially plane. Sufficiently long pulses were used so that their flat portions could be assumed to be steady-state insonification of the block, which justified comparison of the results with the steady-state theory. Figure 2 gives examples of two reflected pulses indicating how nearly flat the envelopes of the pulses were. The slight decrease in amplitude toward the end of the pulse in Fig. 2b would indicate a degree of penetration of acoustic energy into the block and subsequent interaction at the front face to cause a reduction in amplitude. An experimental determination of dilatational wave speed, density, and acoustic impedance are tabulated in Table 1 for three materials that were used.

Reflected pressures were referred to the reference pressure p_0 measured at the position later occupied by the reflecting plane before measuring the monostatic reflection by the plane. The receiver was a probe hydrophone which occupied a small portion of the field. In all cases, even in the near field measurements, the variation of the incident pressure over the area of the reflecting plane was never greater than 3%. A listing of all reflectors is given in Table 2 along with the differences from theory for a rigid reflector at normal incidence. Figure 3 shows the reflected axial pressure amplitude for brass blocks of different thicknesses. Since the pressure can be either higher or lower than what one would expect from a rigid plane, the indication is that internal energy is a significant contributor to that degree. These differences from rigid reflection were not possible to account for by considering a simple three-medium problem i.e., water, brass, water

(a)

(b)

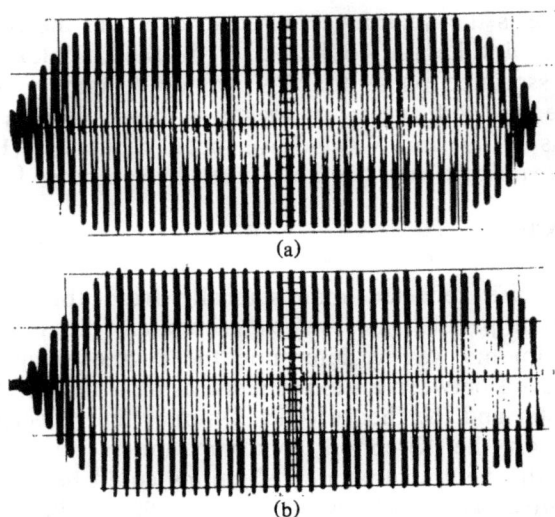

Fig. 2 — Reflected pulses from 4.5λ square faces
of (a) K-8, and (b) brass, at normal incidence

Table 1 — Materials and Their Properties

Material	c Dilatational sound speed (m/sec)	ρ Density (kg/cm)	ρc Acoustic impedance (kg/sec · m²)
K-8	6 901	1.49×10^3	102.8×10^6
Brass	4 445	8.41×10^3	37.38×10^6
Aluminum oxide	10 696	3.91×10^3	41.82×10^6

Table 2 — Reflection from Rectangular Areas.
(Dimensions are given in units of wavelength in water.)

Face size in wavelengths	Depth in wavelength	Material	Difference from rigid theory at normal incidence (%)
2.8 × 2.8	2.1	K-8	1.2
2.8 × 2.8	2.1	Brass	17
2.9 × 3.8	3.8	K-8	2
2.9 × 3.8	3.8	Brass	9
2.9 × 3.9	3.9	K-8	0.8
3.6 × 3.6	2.7	K-8	1.9
4.5 × 4.5	3.4	K-8	<2[a]
4.5 × 4.5	3.4	Al_2O_2	10
4.5 × 4.5	3.4	Brass	12

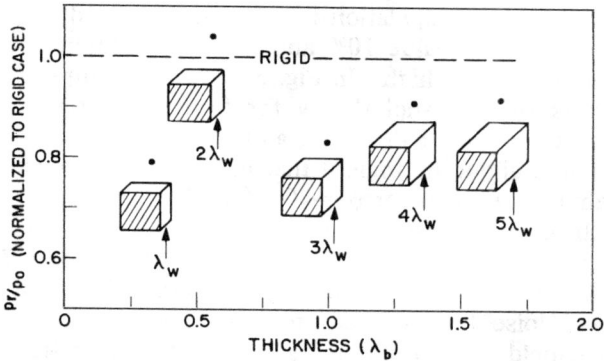

Fig. 3 — The monostatically reflected axial pressure from $3\lambda_w$ square face for different thicknesses of brass blocks (λ_w, wavelength in water; λ_b, wavelength in brass). The reflecting plane is shaded.

and a subsequent reflection to the first water medium. The way axial pressure can be expected to change is shown in Fig. 4 for a Kennametal [3] K-8 block. It can be seen that the Fresnel theory continues to give reasonable quantitative results, even in the near field when $r = 15\lambda$. Where Fresnel theory is only 6% higher than the actual experimental

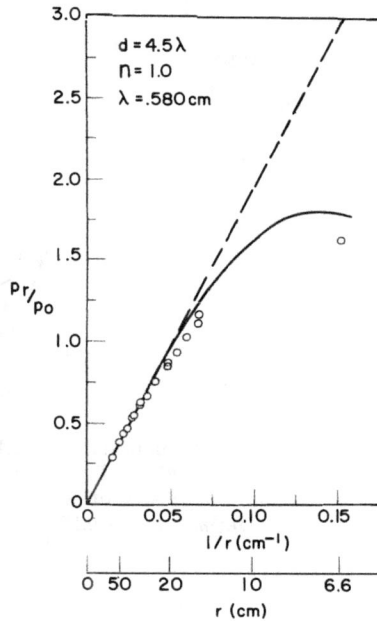

Fig. 4 — The monostatically reflected axial presure from the 4.55λ square face of a K-8 block vs $1/r$ for $1/r$ extrapolation of the far field (---); Fresnel theory (—); experimental measurement (O). $d = 4.55\lambda$; $n = 1$; $\lambda = 0.580$ cm.

value, while the $1/r$ extrapolation is 20% too high; and at $r = 6.6\lambda$ the Fresnel theory gives a value 10% above the experiment, while the $1/r$ extrapolation is 65% too high. In Fig. 5 there is a comparison of experimental results with Fresnel theory for far-field monostatic reflection from otherwise identical K-8, brass, and aluminum oxide blocks whose faces are rotated about an angle θ that includes the first side lobe. All measurements were taken for $\psi = 0$. (See Fig.1.) In this experiment a wavelength of incident sound was 0.584 cm. Similar comparisons for identical brass and K-8 blocks are shown in Figs. 6 and 7. In these experiments, the wavelengths of sound were 0.952 and 0.692 cm, respectively. Noise limited measurement at angles larger than those shown. Near-field measurements using a 4.55λ square-face K-8 block at a distance 15λ from the receiver are shown in Fig. 8. A similar comparison for a block 6.6λ from the receiver is shown in Fig. 9.

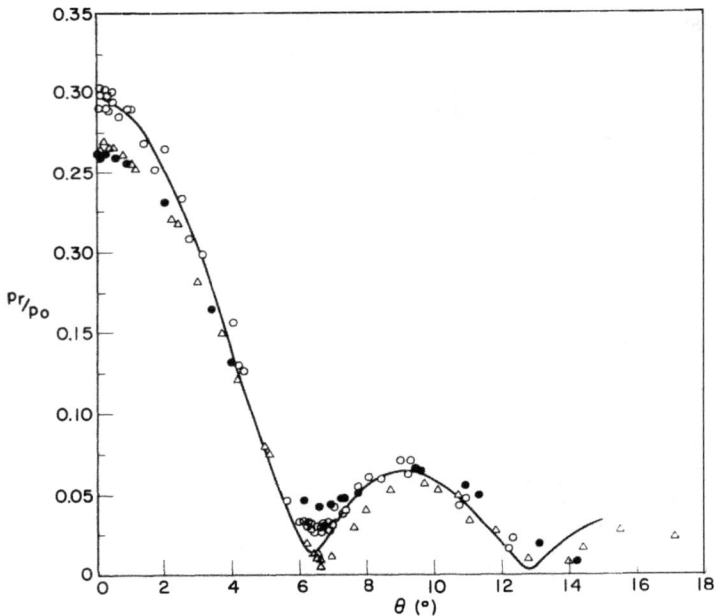

Fig. 5 — The monostatically reflected axial pressure vs reflecting plane rotation angle (θ) for Fresnel theory (—); and experimental measurement of faces of blocks of K-8 (O); brass (●); and aluminum oxide (Δ). $d = 4.5\lambda$; $n = 1$; $\lambda = 0.584$ cm; $r = 68.5\lambda$.

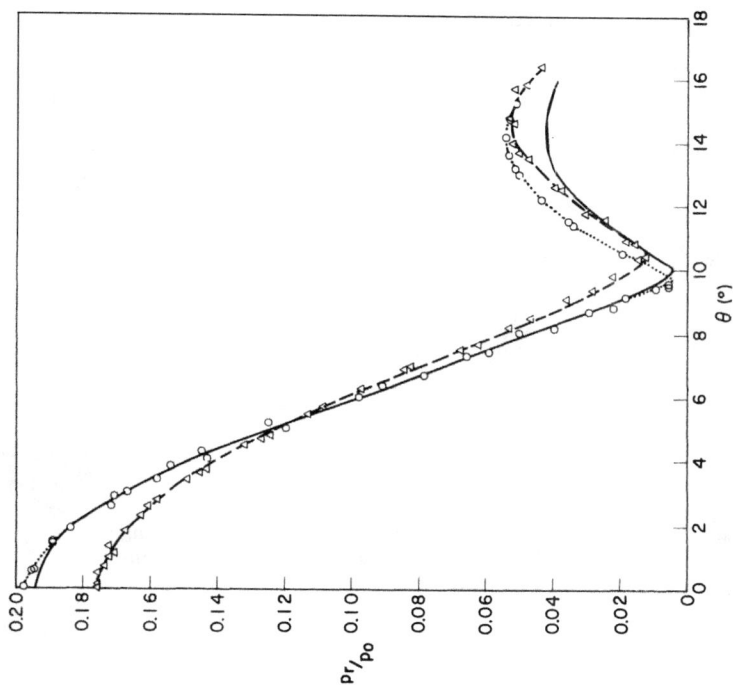

Fig. 7 — The monostatically reflected axial pressure *vs* reflecting plane rotation angle (θ) for Fresnel theory (—), and experimental measurement of faces of blocks of K-8 ($\cdot\cdot O\cdot\cdot$); and brass ($-\Delta-$). $d = 2.9\lambda$; $n = 0.75$; $\lambda = 0.692$ cm; $r = 56.11\lambda$.

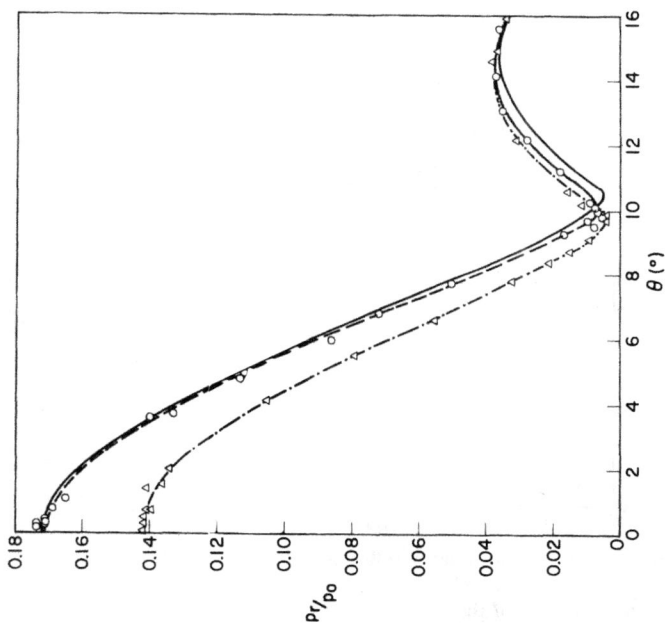

Fig. 6 — The monostatically reflected axial pressure *vs* reflecting-plane rotation angle (θ) for Fresnel theory (—), and experimental measurement of faces of blocks of K-8 ($-O-$); and brass ($-\cdot\Delta-\cdot$), $d = 2.8\lambda$; $n = 1$; $\lambda = 0.952$ cm; $r = 44.83\lambda$.

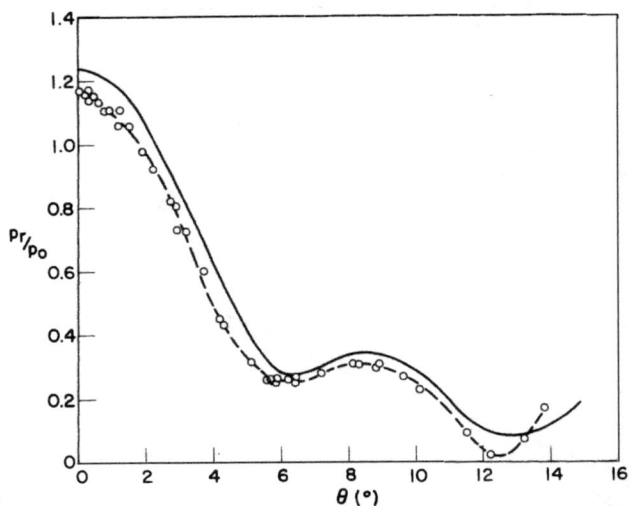

Fig. 8 — The monostatically reflected axial pressure vs reflecting plane rotation angle (θ) for Fresnel theory (—), and experimental measurement of the face of a K-8 block (—O—). $d = 4.55\lambda$; $n = 1$; $\lambda = 0.580$; $r = 15\lambda$.

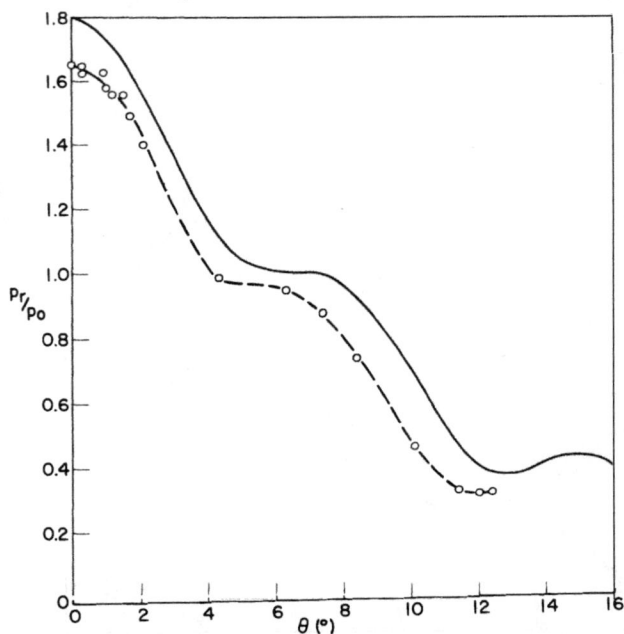

Fig. 9 — The monostatically reflected axial pressure vs reflecting plane rotation angle (θ) for Fresnel theory (—), and experimental measurement of the face of a K-8 block (—O—). $d = 4.55\lambda$; $n = 1$; $\lambda = 0.580$; $r = 6.6\lambda$.

CONCLUSION

The comparison is favorable between the Kirchhoff theory and experimental measurements using real materials for relatively small finite planes whose edge length is approximately four wavelengths. With only slight reduced accuracy, the theory even allows description of the near field in the region between one and two edge lengths distant from the plane. It appears that the differences found between approximate reflection theory for a reflecting plane and measurements of reflection from planes of relatively high-impedance materials can be similar to those imposed by the approximate theory itself.

For high acoustic impedance materials, quantitative agreement at normal incidence if found to be within 2% for reflecting planes of real material and for lower specific acoustic impedance material agreement is somewhat poorer. For instance, for brass it was still found to be within approximately 25% or 2.5 dB.

Appendix

RADIATED FIELD OF A RECTANGULAR PISTON*

By procedures similar to those in this chapter an acoustic approximation to the acoustic field radiated by a rectangular piston can be calculated. A rectangular-plane piston with dimensions d by d/n is located in a plane infinite, rigid baffle. Assume a rectangular-coordinate system with the origin at the center of the piston, as shown in Fig. A-1. The radiation into half-space, in terms of the velocity potential resulting from the simple-harmonic motion normal to the surface of the piston, may be expressed [1] as

$$\phi = - (A/2\pi)\exp(i\omega t) \int\int (1/r_a)\exp(-ikr_a)\,da, \qquad (A-1)$$

where A is the uniform piston-velocity amplitude, r_a is the radial distance from the contributing-piston element, k is the wavenumber $2\pi/\lambda$, where λ is the wavelength, and the integral is taken over the radiating area.

Approximating r_a in the phase, by the first two terms of an expansion about $r_a = r$,

$$r_a \simeq r + [(\zeta^2 - 2x\zeta + \eta^2 - 2y\eta)/r^2], \qquad (A-2)$$

which is subject to the condition that

$$[(\zeta^2 - 2x\zeta + \eta^2 - 2y\eta)/r^2]^2 \leqslant 1. \qquad (A-3)$$

This limitation is satisfied for all field points on or outside of the hemisphere that circumscribes the piston.

A similar approximation was previously given by Freedman [2] in which a more limiting phase expansion was made about $(r = z)$, resulting in a greatly restricted region of validity for the solution.

---

*This development first appeared in: Werner G. Neubauer, J. Acoust. Soc. Am **38**, 671-672 (1975).

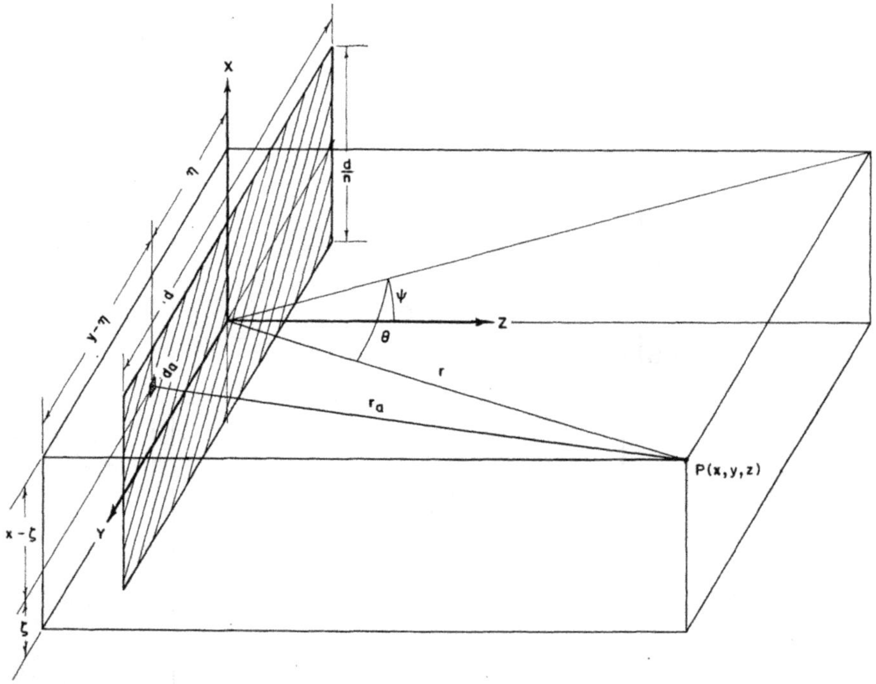

Fig. A-1 — Geometrical orientation for the radiation from a rectangular piston

Manipulation and transformation of Eq. (A-1), yields [3]

$$\phi = - (A/2k)\exp\{[i\omega t - (i\pi/\lambda)][2r - (x^2/r) - (y^2/r)]\}$$

$$\times \int_{u_1}^{u_2} \exp(-i\pi u^2/2)\, du \int_{v_1}^{v_2} \exp(-i\pi v^2/2)\, dv, \qquad \text{(A-4)}$$

where $u = (2/\lambda)^{1/2}(\zeta - x)$, $v = (2/\lambda r)^{1/2}(\zeta - y)$, and

$$u_{1/2} = - (2/\lambda r)^{1/2}[x \pm (d/2n)], \quad v_{1/2} = - (2/\lambda r)^{1/2}[y \pm (d/2)]$$

$$\text{(A-5)}$$

The integrals in Eq. (A-4) can be expressed in terms of the complex Fresnel integral [4].

The scalar field described by evaluating Eq. (A-4) is shown in Figs. (A2-A5). Comparison is made with a numerical integration of Eq. (A-1), as well as recomputed results, using the formulation of Freedman in Figs. (A2-A4). Essential agreement between the present approximation and the numerical integration is seen to exist even relatively close to the piston face and far from the piston axis. In addition, expected agreement is found with the results after Freedman on and near the central axis.

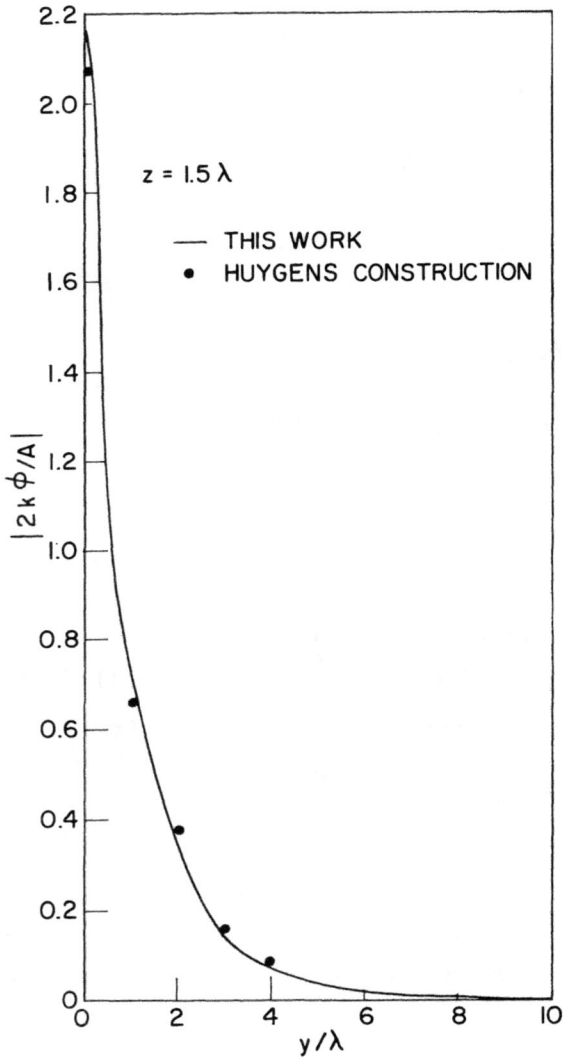

Fig. A-2 — Plot of $|2k\phi/A|$ computed in a rectangular-coordinate system over a range of y/λ for $x = 0.05\lambda$ and $z = 1.5\lambda$, for a piston for which $d = 2\lambda$ and $n = 2$ (i.e., a $\lambda \times 2\lambda$ piston). Similar results of a numerical integration (Huygens construction) are plotted for comparison. ————: This work. ●: Huygens construction.

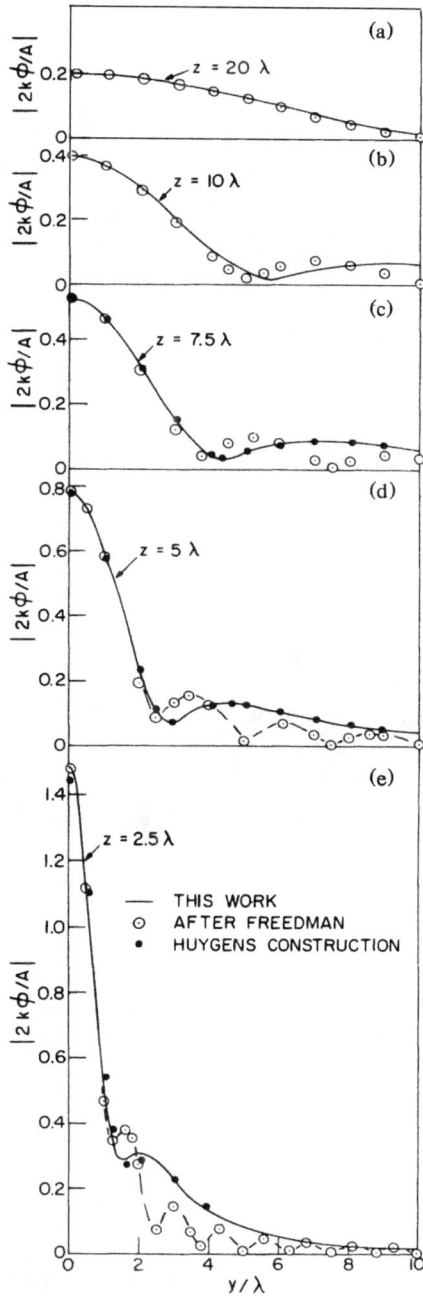

Fig. A-3 — Plots of $|2k\phi/A|$ computed in a rectangular-coordinate system over a range of y/λ for $x = 0.05\lambda$ and (a) $z = 2.5\lambda$, (b) $z = 5\lambda$, (c) $z = 7.5\lambda$, (d) $z = 10\lambda$, (e) $z = 20\lambda$, for a piston for which $d = 2\lambda$ and $n = 2$ (i.e., a $\lambda \times 2\lambda$ piston). Similar results of an analysis following Freedman and a numerical integration (Huygens construction) are plotted for comparison. ————: This work. —⊙—: After Freedman. ●: Huygens construction.

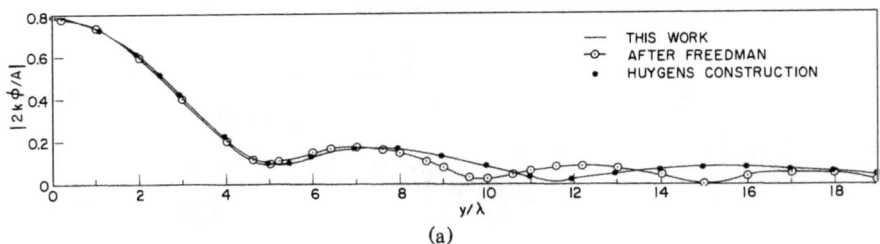

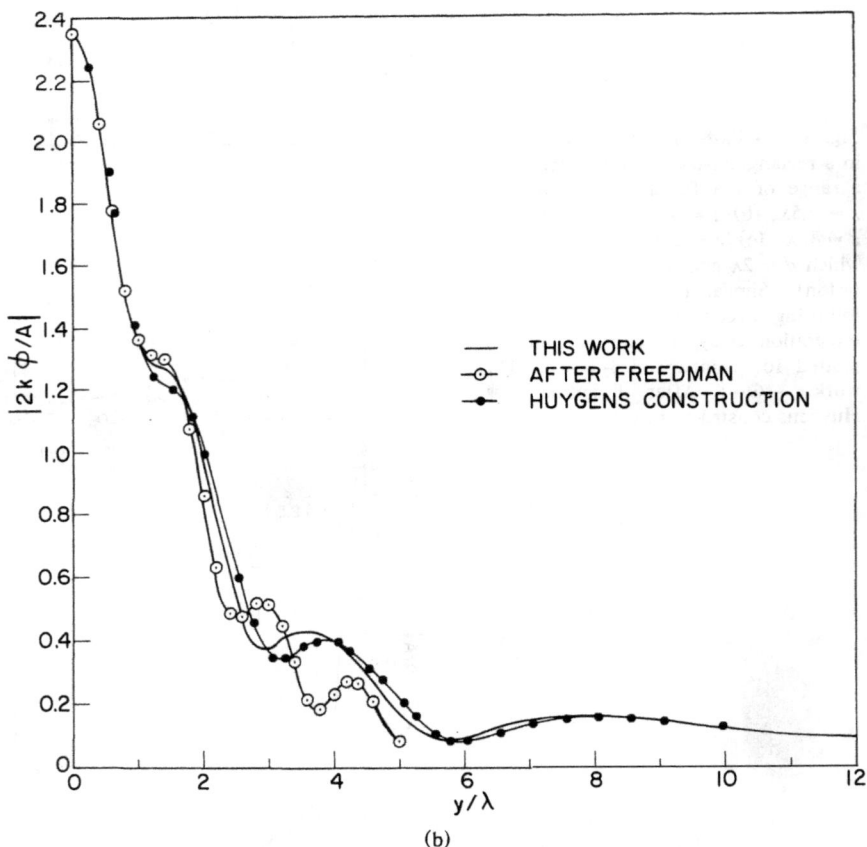

Fig. A-4 — Plot of $|2k\phi/A|$ computed in a rectangular-coordinate system over a range of y/λ for $x = 0.05\lambda$ and (a) $z = 5\lambda$ and (b) $z = 20\lambda$ for a piston for which $d = 4\lambda$ and $n = 2$, (i.e., a $4\lambda \times 2\lambda$ piston). Similar results of an analysis following Freedman and a numerical integration (Huygens construction) are plotted for comparison. ———: This work. —⊙—: After Freedman. ●: Huygens construction.

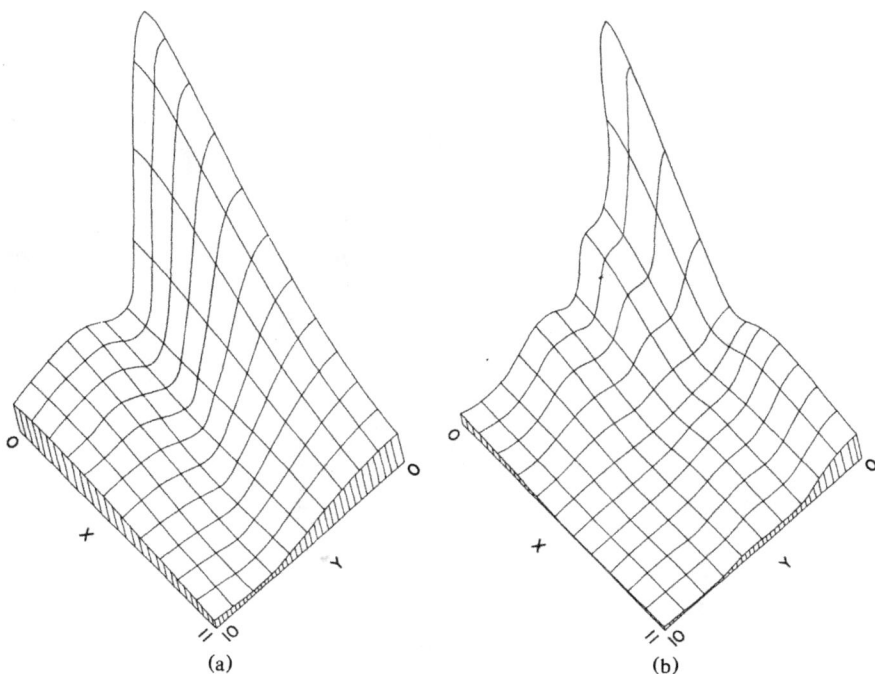

Fig. A-5 — Isometric plots of one quadrant of the field in a plane parallel to the piston face at a distance $z = 7.5\lambda$ for (a) a $2\lambda \times \lambda$ piston and (b) a $4\lambda \times 2\lambda$ piston.

REFERENCES

1. J. W. Strutt Lord Rayleigh, *Theory of Sound* (MacMillian Co. Ltd., London, 1940), Vol. 2, p. 107.

2. A. Freedman, "Sound Field of a Rectangular Piston," J. Acoust. Soc. Am. **32**, 197-209 (1960).

3. A means of obtaining graphical results as well as a more detailed discussion may be found in W. G. Neubauer, NRL Rept. 6286 (1965).

4. A treatment of Fresnel integrals is given in S. A. Schelkunoff, *Applied Mathematics for Engineers and Science* (D. Van Nostrand Co., Inc., Princeton, N.J., 1965), p. 385.

Chapter 3
THEORY AND DEMONSTRATION OF
CREEPING WAVES*

INTRODUCTION

Among the multitude of classical problems which A. Sommerfeld [1] considered is the problem of propagation of radio waves around the earth. He solved it by using a mathematical method which is now known as the "Sommerfeld-Watson transformation." Also van der Pol and Bremmer [2] addressed themselves to the same problem, and Franz [3-7] advanced the theory considerably by applying it to diffraction of electromagnetic waves around conducting cylinders and spheres. It was he who coined the term "creeping wave," (Kriechwelle), to describe the phenomenon of circumferentially propagating waves.

The same creeping-wave theory furnishes a most useful description for analyzing acoustic scattering as well. The impetus for this was given by the experimental observation by Barnard and McKinney [8] that a single underwater sound pulse which is incident on a scatterer will produce a series of echo returns. A number of experiments [9] have corroborated the basic idea of these results: that the physical mechanism for acoustic scattering consists of a superposition of continuously radiating circumferential waves. The creeping-wave theory furnishes this physical picture in a most natural way, via the Sommerfeld-Watson transformation, whereas in the classical "normal mode solution" (an infinite-series expansion in terms of separable eigenfunctions of the wave equation) this interpretation is hidden. (See Ref. 10.) The creeping-wave solution is much more rapidly convergent than the normal-mode solution. Acoustic scattering is particularly suited for the experimental study of creeping waves, (better than electromagnetic scattering), because of the slow propagation speeds of acoustical waves.

---

*The theoretical summary presented here first appeared in W. G. Neubauer, P. Uginčius, and H. Überall, Z. Naturforsch. **24a**, 691 (1969), written in memory of Arnold Sommerfeld's hundredth birthday, December 5, 1968.

35

The theory of creeping waves has been applied to various problems in acoustic scattering from rigid, soft, or elastic cylinders or cylindrical shells using both continuous and pulsed waves [11-17]. Here we consider the creeping-wave analysis of acoustic scattering by an infinite elastic cylinder immersed in a fluid. Two types of circumferential waves emerge: (1) highly attenuated "Franz-type" waves which are slower than c — the speed of free acoustic waves in the liquid; and (2) very slightly attenuated "Rayleigh-type" waves which are faster than c. The former are the original creeping waves first identified by Franz, and depend mainly on the geometry of the scatterer, whereas the latter owe their existence to the scatterer's elastic properties.

THEORY

Figure 1 shows the geometry of the problem: a plane wave exp $i(k_1 x - \omega t)$ is incident from the negative x-axis on an elastic cylinder with radius a, density ρ_2, and Lamé constants λ, μ. The surrounding medium is an infinite homogeneous fluid with density ρ_1, in which the speed of propagation for the acoustic wave is

$$c_1 = \omega / k_1; \tag{1}$$

whereas inside the cylinder the longitudinal (compressional) and transverse (shear) waves have the respective speeds

$$c_1 = [\lambda + 2\mu)/\rho_2]^{1/2}; \quad c_t = (\mu/\rho_2)^{1/2}. \tag{2}$$

expressed in terms of the properties of the material. The total acoustic pressure at the general observation point $P(r, \theta)$ is obtained in the usual way, by subjecting the general separable solution in terms of the cylinder eigenfunctions to the elastic boundary conditions at $r = a$. After suppressing the time dependence exp $(-i\omega t)$ the result is (see Refs. 11,17):

$$p = p_{\text{inc}} + p_{sc}, \tag{3a}$$

where the incident and scattered pressures are given respectively by

$$p_{\text{inc}} = \sum_{n=0}^{\infty} i^n (2 - \delta_{n0}) J_n (k_1 r) \cos (n\theta), \tag{3b}$$

$$p_{sc} = \sum_{n=0}^{\infty} i^n (2 - \delta_{n0}) \frac{B_n}{D_n} H_n^{(1)} (k_1 r) \cos (n\theta), \tag{3c}$$

where Bessel and Hankel functions are represented by J_n and $H_n^{(1)}$ respectively. The expansion coefficients in Eq. (3c) are found to be

$$B_n = \begin{vmatrix} \beta_1 & \alpha_{12} & \alpha_{13} \\ \beta_2 & \alpha_{22} & \alpha_{23} \\ 0 & \alpha_{32} & \alpha_{33} \end{vmatrix}; \quad D_n = \begin{vmatrix} \alpha_{11} & \alpha_{12} & \alpha_{13} \\ \alpha_{21} & \alpha_{22} & \alpha_{23} \\ 0 & \alpha_{32} & \alpha_{33} \end{vmatrix}. \tag{4}$$

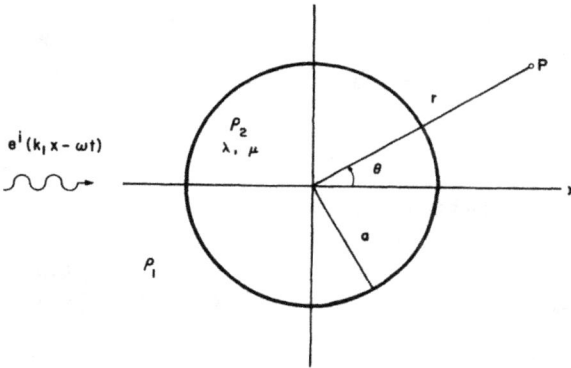

Fig. 1 — Scattering geometry.

With the introduction of the dimensionless parameters

$$x_i = ak_i = a\omega/c_i, \quad (i = 1, l, t) \tag{5}$$

the elements of these determinants are given by [20] (primes indicate differentiation with respect to the argument):

$$\beta_1 = (\rho_1/\rho_2)x_t^2 J_n(x_1),$$
$$\beta_2 = x_1 J_n'(x_1); \tag{6a}$$
$$\alpha_{11} = (\rho_1/\rho_2)x_t^2 H_n^{(1)}(x_1),$$
$$\alpha_{12} = -2x_1 J_n'(x_1) + (2n^2 - x_t^2)J_n(x_1), \tag{6b}$$
$$\alpha_{13} = 2n[J_n(x_t) - x_t J_n'(x_t)];$$
$$\alpha_{21} = -x_1 H_n^{(1)'}(x_1),$$
$$\alpha_{22} = -x_1 J_n'(x_1), \tag{6c}$$
$$\alpha_{23} = n_n J_n(x_t);$$
$$\alpha_{32} = 2n[x_1 J_n'(x_1) - J_n(x_1)],$$
$$\alpha_{33} = 2x_t J_n'(x_t) + (x_t^2 - 2n^2)J_n(x_t). \tag{6d}$$

A quantity of experimental interest is the differential scattering cross section defined by

$$d\sigma/d\theta = \lim_{r \to \infty} r|p_{sc}/p_{inc}|^2. \tag{7}$$

By using Hankel's asymptotic expansion for large argument

$$H_n^{(1)}(\rho) \cong (2/\pi\rho)^{1/2} \exp i\left[\rho - (\pi/2)\left(n + \frac{1}{2}\right)\right] \tag{8}$$

for the scattered pressure in Eq. (3c), we arrive at the

$$\frac{d\sigma}{d\theta} = \frac{2}{\pi k_1} \left| \sum_{n=0}^{\infty} (2 - \delta_{n0}) \frac{B_n}{D_n} \cos(n\theta) \right|^2. \tag{9}$$

In principle this equation could be summed to arrive at a value for cross section which is accurate to any desired degree. However, for large values of ka ($x_1 > 1$) Eq. (9) is limited by its very slow convergence. For example, Sommerfeld ([1], p. 282), in a similar problem finds that a normal-mode series like that of Eq. (9) would require more than 1000 terms before it would start to converge.

THE CREEPING-WAVE SOLUTION

The normal-mode solution can be reformulated via the Sommerfeld-Watson transformation. [1,21] This will not only improve the convergence of the solution but, more importantly, furnishes the physical picture of the circumferential creeping waves. The Sommerfeld-Watson transformation may be written in the form

$$\sum_{n=0}^{\infty} (2 - \delta_{n0}) f_n = iP \int_C \frac{d\nu}{\sin \pi\nu} e^{-\pi\nu} f(\nu), \tag{10}$$

where C is a contour which passes through the origin $\nu = 0$, surrounding the positive real axis in clockwise sense, and excludes all poles of the function $f(\nu)$. The Neumann factor $2 - \delta_{n0}$ requires that the principal value be taken for the integration through the origin. The next step is to transform the contour C (which encloses all the zeros of $\sin \pi\nu$ and no poles of $f(\nu)$) into a different contour which surrounds all poles of $f(\nu)$. To do this one must have an a priori knowledge of the complex zeros of the determinant $D_n \rightarrow D(\nu)$, because applying the transformation (10) to Eq. (3c) we see that the only poles of the function $f(\nu)$ are those zeros, all other cylinder functions which appear in B_n and $H_n^{(1)}$ being entire functions of their order in the complex ν-plane.

The determinant $D(\nu)$ is too complicated to yield any information about its roots by analytical methods. For special cases, however, it degenerates into simpler forms for which the asymptotic zeros are well known. Thus, for example, for rigid ($\mu \rightarrow \infty$) and soft ($\mu \rightarrow 0$) cylinders it is found [11] that they are the roots of $H_\nu^{(1)'}(x_1)$ and of $H_\nu^{(1)}(x_1)$ respectively. For large x_1 these roots are known from the work of Sommerfeld [1] and Franz [4]. There are infinitely many of them and they all lie in the first quadrant of the ν-plane on a line which when extrapolated intersects the real axis at $\nu \cong x_1$.

The zeros of the determinant $D(\nu)$, Eq. (4), can be found numerically (see Ref. 17). A procedure that is essentially the Newton-Raphson method, generalized to the complex plane, is used and converges quite rapidly if the initial estimate is fairly good. To make sure that no zeros are overlooked we also made use of the "Principle of the Argument"

$$2\pi (Z - P) = \Delta_C \text{ Arg } D(\nu), \qquad (11)$$

which states that the number of zeros Z minus the number of poles P (with their multiplicities taken into account) inside any closed contour C is equal to the net change (divided by 2π) of the argument of $D(\nu)$ in a complete traversal of C. With $P = 0$ Eq. (11) enables one to determine unambiguously the number of zeros in any given region of the complex plane. The results for an aluminum cylinder ($\lambda = 6.1 \times 10^{11}$ dyn/cm^2, $\mu = 2.5 \times 10^{11}$ dyn/cm^2, $\rho_2 = 2.7$ gm/cm^3) at $x_1 = k_1 a = 5$ are shown in Fig. 2. Two sets of zeros were found: (1) the set labelled F which lies entirely in in the first quadrant, and which differs numerically very little from the "rigid" Franz-type zeros — the roots of $H_\nu^{(1)'}(x_1) = 0$; and (2) the set labelled R which starts out with two zeros close to the positive real axis, and continues into the second quadrant approaching asymptotically the negative real axis. They seem to coalesce pairwise into the negative integers. This second set of Rayleigh-type zeros is absent from either rigid or soft cylinders, and their existence, therefore, must be attributed to the elastic properties of the scatterer. We also find that there may be zeros in the third quadrant; however, they were difficult to locate with the available numerical program. Fortunately, as will be seen below, only first-quadrant zeros close to the real axis can contribute in the theory. All these zeros are functions of x_1. The behavior of the Franz-type zeros vs x_1 is the same as that deduced from their asymptotic formulas (see Ref. 1 or Ref. 4). The line of Rayleigh-type zeros, however seems to be "pulled" into the first quadrant with increasing x_1, so that it appears that for $x_1 \to \infty$ there may be an infinite number of them in the first quadrant (see Fig. 5 in Ref. 15). For a cylindrical shell (15,17) the analogous function $D(\nu)$ is a six-by-six determinant. There we find the zeros are qualitatively the same as in Fig. 2 with the exception of an additional set in the fourth quadrant.

Figure 2 also shows the tranformation of the original contour C in the Sommerfeld-Watson integral (Eq. 10). This is done with the additional contours C_0 and C' closed up at ∞ in such a way that the new closed contour $C' + C_\infty + C_0 + C_\infty - C + C_\infty$ does not include any poles, so that the integral in Eq. (10) over this contour must vanish.

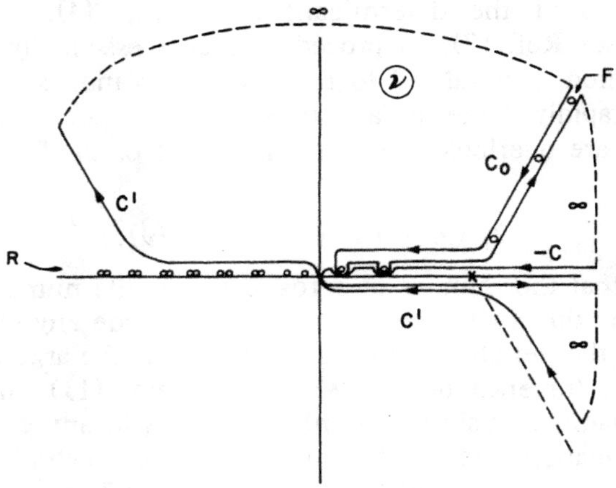

Fig. 2 — Contours for the Sommerfeld-Watson transformation. The cirles show the Franz (F) and Rayleigh-type (R) zeros of $D(\nu)$ for an aluminum cylinder with $k_1 a$ 5.

The peculiar shape of C' is dictated by the requirement (see the discussion after Eq. (22) that no second-quadrant poles should be included, and by the fact, which Franz [7] has shown, that the integral over C_∞ to the left of the dashed line in the fourth quadrant (which is the reflection of the F-line) would not converge. Vanishing of the integral over the portions C_∞ shown in Fig. 2 can be readily established. The transformation (10) applied on Eqs. (3) then yields for the total pressure

$$p = p_{\mathrm{I}} + p_{\mathrm{II}}; \tag{12a}$$

$$p_{\mathrm{I}} = i \int_{C_0} \frac{d\nu}{\sin \pi\nu}\, e^{-i\nu\pi/2} \cos(\nu\theta) H_\nu^{(1)}(k_1 r)\, \frac{B(\nu)}{D(\nu)}, \tag{12b}$$

$$p_{\mathrm{II}} = iP \int_{C'} \frac{d\nu}{\sin \pi\nu}\, e^{-i\nu\pi/2} \cos(\nu\theta)$$
$$\times\, \frac{J_\nu(k_1 r) D(\nu) + H_\nu^{(1)}(k_1 r) B(\nu)}{D(\nu)}. \tag{12c}$$

The "background" integral p_{II} of Eq. (12c) is absent for either soft or rigid cylinders. For an elastic (aluminum) cylinder it has been shown [13,14] that it is negligible (except maybe at critical angles) by comparing results evaluated without that integral with the exact normal-mode calculations of Faran. [22] We shall therefore neglect it. Note that the incident pressure which is proportional to $J_\nu(k_1 r)$ seems to have dropped out of Eq. 12b. We shall see below that this is not really so, but that it can be recovered by a saddle-point integration.

The integral p_1 could be evaluated by the residue method, but Franz [7] has shown that the resulting residue series would not converge in general, the reason being that p_1 also contains the geometrically reflected wave. That wave can be separated out of Eq. 12b by writing for $\cos(\nu\theta)$ the identity

$$\cos(\nu\theta) = e^{i\pi\nu} \cos \nu(\pi - \theta) - ie^{i\nu(\pi-\theta)} \sin(\pi\nu). \qquad (13)$$

The last term cancels the $\sin(\pi\nu)$ in the denominator of Eq. (12b). This results in an integral which has been shown [13] to represent the geometrically reflected wave. It is evaluated for large x_1 by the saddle-point method. Using Hankel's asymptotic expansion, Eq. (8), for $H_\nu^{(1)}$ in Eq. (12b), and Debye's asymptotic expansion

$$H_\nu^{1/2}(x) \cong (2/\pi x \sin x)^{1/2} e^{\pm i[x(\sin\alpha-\alpha\cos\alpha)-\pi/4]}, \qquad (14)$$

$$\nu \equiv x \cos x.$$

for the first-column elements of $B(\nu)$ and $D(\nu)$ we have shown [13,16] that the saddle point is located at $\alpha_s = \theta/2$,

$$\nu_s = x_1 \cos(\theta/2). \qquad (15)$$

Figure 2 reveals that at the "critical angles"

$$\theta_k^* = 2 \cos^{-1}(\text{Re } \nu_k/x_1), \qquad (16)$$

where ν_k is any first-quadrant Rayleigh pole, the saddle point (Eq. 15) will lie directly under a pole. This would mean that the resulting saddle-point evaluation for the geometrically reflected wave would be very inaccurate, because we would be unable to distort the contour C_0 in order to make it pass through the saddle point without coming too close to a pole of the integrand. (We have shown that the path of steepest descent must go through the saddle point at an angle of $3\pi/4$ with the real axis). To overcome this difficulty we break up the contour C_0 of Fig. 2 into the two separate contours C_1 and C_2, as shown in Fig. 3, before the separation of the geometric term by Eq. (13). We then evaluate the integral (Eq. (12b)) by the Residue Theorem in its "unseparated form" (leaving $\cos \nu\theta$ unaltered) over the contour C_1, and in the "separated form" (rewriting $\cos \nu\theta$ according to Eq. (13)) over the contour C_2. These two residue series will yield the creeping waves. The geometrically reflected wave can be evaluated by the saddle-point method over the contour C_s as outlined above. (Another saddle point is present which is shown schematically in Fig. 3 on the real axis to the right of x_1. Franz [5,7] has shown that this restores the contribution of the incident wave, but since we are only interested in the scattered wave, we shall ignore it here.)

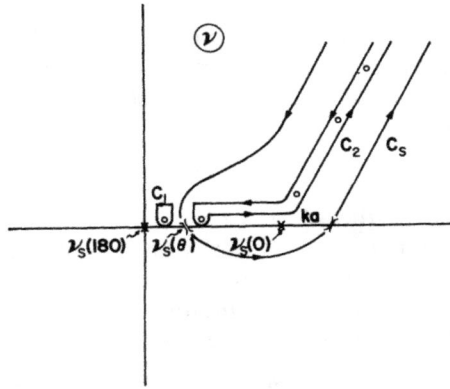

Fig. 3 — Contours for separation of the
geometrically reflected wave.

The results are:

$$p_1 = p_c + p_g; \tag{17a}$$

$$p_c = -2\pi \sum_{k=1}^{\infty} \frac{H_{\nu k}^{(1)}(k_1 r) B(\nu_k)}{\sin(\pi\nu_k)\dot{D}(\nu_k)} \begin{cases} \cos(\nu_k\theta)e^{-i\nu_k\pi/2} & \textcircled{1} \\ & ; \\ \cos\nu_k(\pi-\theta)e^{i\nu_k\pi/2} & \textcircled{2} \end{cases} \tag{17b}$$

$$p_g \cong -[\alpha\sin(\theta/2)/2r]R(\nu_s)e^{ik_1[r-2a\sin\theta/2]}. \tag{17c}$$

In the summation of Eq. (17b) $\dot{D}(\nu_k) = \partial D/\partial\nu$ evaluated at $\nu = \nu_k$; the forms ① and ② must be used when

$$\text{Re}(\nu_k) < x_1\cos(\theta/2): \text{① } Unseparated\ Form; \tag{18}$$

$$\text{Re}(\nu_k) > x_1\cos(\theta/2): \text{② } Separated\ Form.$$

The function $R(\nu)$ in Eq. (17c) was assumed to be constant in the saddle-point integration and is given explicitly by

$$R(\nu) = \frac{(\rho_1/\rho_2)x_t^2 D_1(\nu) + ix_1\sin\alpha D_2(\nu)}{(\rho_1/\rho_2)x_t^2 D_1(\nu) - ix_1\sin\alpha D_2(\nu)}, \tag{19}$$

where D_1, D_2 are the respective 2×2 minor determinants obtained by expanding $D(\nu)$ by its first column.

By the use of Eq. (8), Eq. (17b) can be written for $r \to \infty$ (the asymptotic approximation for $r \to \infty$ has already been made in Eq. (17c)) as

$$p_c \cong - \left(\frac{8\pi}{k_1 r}\right)^{1/2} e^{i(k_1 r - \pi/4)} \tag{20}$$

$$\times \sum_{k=1}^{\infty} \frac{1}{\sin \pi \nu_k} \frac{B(\nu_k)}{\dot{D}(\nu_k)} \begin{cases} \cos(\nu_k \theta) e^{-i\pi\nu_k} & \textcircled{1} \\ \cos \nu_k(\pi - \theta) & \textcircled{2} \end{cases}$$

At this point it is possible to demonstrate explicitly the circumferential character of the creeping waves. For the unseparated form $\textcircled{1}$ in Eq. (20) we rewrite the trigonometric terms

$$\frac{\cos(\nu_k \theta)}{\sin(\pi \nu k)} = - i \sum_{\lambda = \pm 1} \sum_{m=0}^{\infty} e^{i\nu_k(\lambda\theta + \pi + 2m\pi)}, \tag{21}$$

which is uniformly convergent for $\mathrm{Im}(\nu_k) > 0$. Inserting the time dependence $\exp(-i\omega t)$ we see that the residue series (20) is made up of terms having the form

$$e^{i(\pm\nu_k\theta - \omega t)} = e^{-\mathrm{Im}(\nu_k)(\pm\theta)} e^{i[\pm \mathrm{Re}(\nu_k)\theta - \omega t]}, \tag{22}$$

which are clearly circumferential waves traveling in the $\pm\theta$ directions, and are damped with an attenuation factor proportional to $\mathrm{Im}(\nu_k)$. Their phase velocities are given by

$$c_k^{ph} = \frac{\omega a}{\mathrm{Re}(\nu_k)} \frac{x_1}{\mathrm{Re}(\nu_k)} c_1. \tag{23}$$

For pulses we can also define the group velocities

$$c_k^{gr} = \frac{dx_1}{d\,\mathrm{Re}(\nu_k)} c_1. \tag{24}$$

The additional summation over m in Eq. (21) represents waves which have circumnavigated the cylinder m times. Equation (22) also shows that no second or fourth quadrant zeros ν_k are allowed since these would lead to physically unacceptable exponentially increasing waves. This was one of the main factors for determining the contours in Fig. 2.

The differential scattering cross section, Eq. (7), is now given by (approximately, because we have neglected the background integral Eq. (12c))

$$(d\sigma/d\theta) \cong \lim_{r \to \infty} r|p_g + p_c|^2, \tag{25}$$

which by using Eqs. (17c) and (20) can be put in the form

$$\frac{2}{a}\frac{d\sigma}{d\theta} \cong \left| [\sin(\theta/2)]^{1/2} R(\nu_s) + 4(\pi/k_1 a)^{1/2} e^{i[2k_1 a \sin(\theta/2) - \pi/4]} \right. \tag{26}$$

$$\left. \times \sum_{k=1}^{\infty} \frac{1}{\sin \pi \nu_k} \frac{B(\nu_k)}{\dot{D}(\nu_k)} \left\{ \begin{matrix} \cos(\nu_k \theta) e^{-i\pi\nu_k} & \text{①} \\ \cos \nu_k(\pi - \theta) & \text{②} \end{matrix} \right\} \right|^2 .$$

This is much more complicated than the relatively simple expression in Eq. (9). However, whereas the latter may need on the order of 1000 terms before starting to converge, we found that for $k_1 a \gtrsim 5$ we need sum only three or four terms (the first-quadrant Rayleigh-type poles plus at most two Franz-type poles) in Eq. (26) to obtain four-digit accuracy. A numerical evaluation of Eq. (26) for an aluminum cylinder is shown in Fig. 2. More extensive computations for aluminum shells can be found in Refs. 15 and 17.

Finally we consider the paths of the creeping waves around the cylinder, which determines the critical angles at which a creeping wave may be launched on the cylinder and radiate off to an observer after m circumnavigations. These critical angles have been derived intuitively in Keller's [23] "geometrical theory of diffraction" by noting that at these angles there exists a resonance effect: the incident wave velocity is equal to the component in the direction of incidence of the creeping-wave velocity. In Ref. 13 they have been established rigorously by following the path of a delta-function pulse and correlating its travel times with causality requirements. The picture which emerges from such an analysis is given in Fig. 4. Figure 4a shows a creeping wave (solid line) launched on the cylinder at the critical angle α (measured from the shadow boundary), then proceeding to the observer P, leaving the cylinder at the same critical angle α (measured from N — the normal to the observation direction). The rigorous theory [14] predicts these angles to be

$$\alpha_k^{ph} = \lim_{s \to \infty} \cos^{-1}(\nu_k/x_1)$$

$$\cong \cos^{-1}(c_1/c_k^{ph}) \quad \text{for continuous waves,} \tag{27a}$$

$$\alpha_k^{gr} = \lim_{s \to \infty} \cos^{-1}(d\nu_k/dx_1)$$

$$\cong \cos^{-1}(c_1/c_k^{gr}) \quad \text{for pulses,} \tag{27b}$$

where s is the integration variable in a Laplace transformation used to obtain a delta-function pulse from the plane wave. These angles in

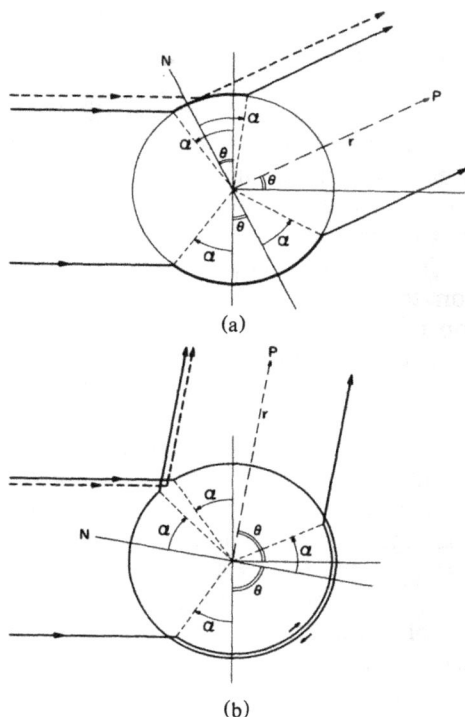

(a)

(b)

Fig. 4 — Geometry of the circumferential
waves: (a) $\theta < 2\alpha$ corresponding to the "unseparated form;" (b) $\theta > 2\alpha$ corresponding
to the "separated form." The geometrically
reflected wave is indicated by the dashed
line.

general are complex (except for the rigid or soft cylinder in which case
they become $\alpha_k = 0$, showing that the Franz waves are launched at the
shadow boundary). The approximation of taking their real parts (the
right-hand sides of Eq. (27)) gives meaningful physical angles if $\text{Im}(\nu_k)$
is small, which is the case for all Rayleigh-type zeros. For $\theta = 2\alpha$ the
distance which the upper creeping wave in Fig. 4a has to travel on the
cylinder shrinks to zero. It therefore has to make a full revolution
around the cylinder before proceeding to P, as shown in Fig. 4b. (This
means that m has increased by one in Eq. (21)). The condition $\theta = 2\alpha$
is precisely the same as that in Eq. (18) for the change-over from the
unseparated to the separated forms for the residue contributions. At
this point the saddle point (15) is directly under one of the Rayleigh-
type poles. The separated form of the solution (which is to be used for
$\theta > 2\alpha$) predicts exactly, from our causality arguments [14], the path
shown in Fig. 4b in agreement with physical intuition.

SCHLIEREN OBSERVATION OF FRANZ AND
RAYLEIGH RADIATION

Waves that contribute to the total acoustic field resulting from a wave incident on the curved circular cylindrical surface and normal to the cylinder axis can be isolated and identified by hydrophone experiments and also by schlieren visualization of the resultant radiation. The total field is a combination of many effects resulting from waves that are diffracted in the outer medium, travel circumferentially inside the cylinder or on the interface, and travel through the body of the cylinder. Isolation of specific waves and identification of exact causes for them has been largely achieved. A schlieren visualization of the entire radiated field resulting from an incident acoustic pulse is shown in Fig. 5. In this dark-field schlieren photograph the cylinder is seen in cross section as a dark disk and the acoustic pulse is seen as a white area. The schlieren system used to produce the photographs shown here is described in Chapter 22. In Fig. 5, a pulse of 7 MHz waves slightly longer than the cylinder diameter causes the complicated field that cannot be readily sorted out. Using a shorter pulse and insonifying only one side (quadrant) of the cylinder circumference causes considerable simplification of the radiated field and permits some isolation of the causative effects. Those are the conditions for Fig. 6 which shows photographs for a short acoustic incident pulse at two different times and the reradiated field resulting from that incident pulse.

Fig. 5 — The total field radiated from a cylinder when the entire cylinder is insonified with a pulse longer than the cylinder diameter.

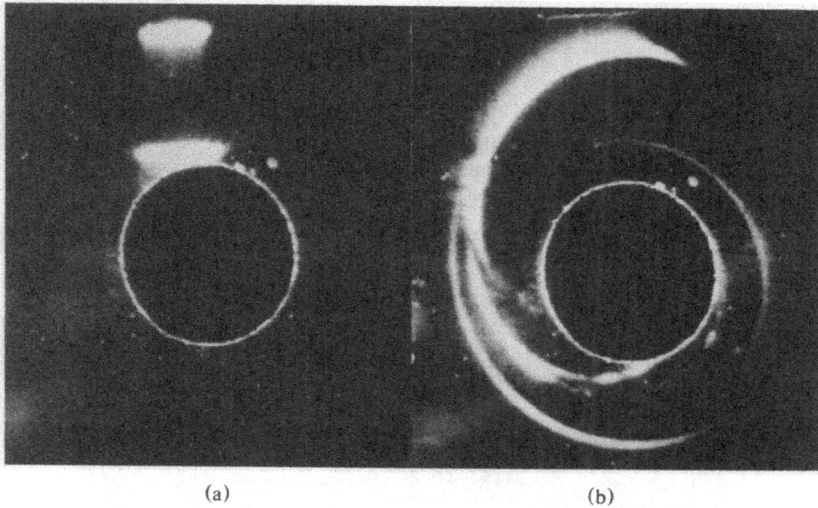

(a)   (b)

Fig. 6 — Schlieren photographs of a short (3) pulse of 6.6 MHz insonifying one quadrant of a solid aluminum cylinder in water ($ka = 353$). (a) The incident pulse shown at two different times in a multiple exposure before it strikes the cylinder. (b) The total radiated field of waves resulting from the incident pulse in (a).

Selectively insonifying only a small range of incidence angles can further simplify the radiated field. If the pulse falls incident only near the tangent to the cylinder the visualization of the Franz wave results in the shadow region. For cylinder sizes small enough, this diffracted wave can interfere with the directly reflected backscattering from the central region of the cylinder. This interference causes the characteristic oscillations in the differential scattering cross section or form function vs x_1 curves for rigid and elastic cylinders (see Chapter 4, Fig. 2). This behavior of the creeping or Franz wave will be discussed further in Chapter 4. In the schlieren photograph in Fig. 7, a pulse was incident only near the tangent. The creeping or Franz wave is seen to propagate into the shadow region on the opposite side from the source and attenuates rapidly from the intersection of the tangent to the point where it intersects the cylinder surface in the shadow. The striations in the wavefront to the left of the cylinder tangent have been explained by Marston and Kingsbury. [24] In the series of schlieren photographs in Fig. 8 the pulse was incident on the range of angles that included the Rayleigh angle and the tangent to the cylinder cross section. The resulting Franz wave radiation and the Rayleigh wave radiation are labeled in Fig. 8. The first dark gap in the wavefront, indicated by an arrow, is a result of the known dark strip caused by opposite phase of

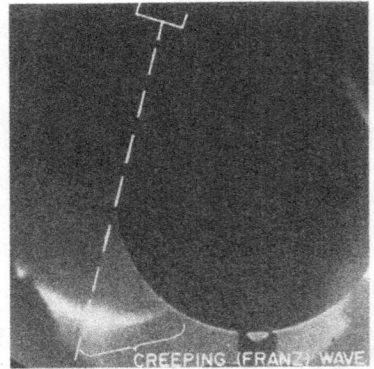

Fig. 7 — Schlieren photograph of the wavefront resulting from the Franz wave for an aluminum cylinder (diam = 10.16 cm) in water for ka of 1008.

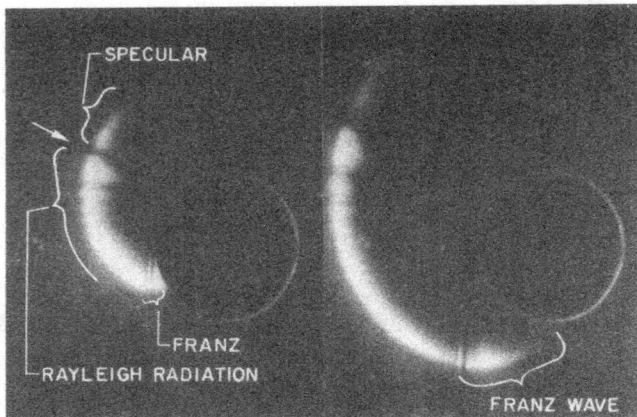

Fig. 8 — Schlieren photographs of the pulsed creeping wave propagated into the geometric shadow of an aluminum cylinder at a frequency of 5 MHz (ka = 324) showing two successive positions of the wave. The source is at the top radiating downward with its beam incident at the Rayleigh angle and on the tangent of the cross section at the far left of the cylinder.

the specular and Rayleigh radiation which will be discussed in detail for a flat surface in Chapter 18. One advantage of schlieren visualization is that it gives one an ability to examine the entire radiated field. However, shortcomings exist for this method of examining acoustic fields. It is largely qualitative rather than quantitative in amplitude, since schlieren is not a linear process over all of its dynamic range.

Accurate quantitative information about the radiated waves can be obtained from hydrophone experiments. A diagram of an experiment that permitted determination of creeping wave attenuation and circumferential speed is shown in Fig. 9a. The pulse in Fig. 9b is the one that was used, having a 1 MHz center frequency. The pulse was measured at six different positions for which the circumferential path on the cylinder, that the wave travels from source to receiver, was successively increased. A multiple exposure of the oscilloscope traces received at positions 2 through 6 is given in Fig. 10. The dashed line connects the peaks of the signals indicating the decay of the creeping wave.

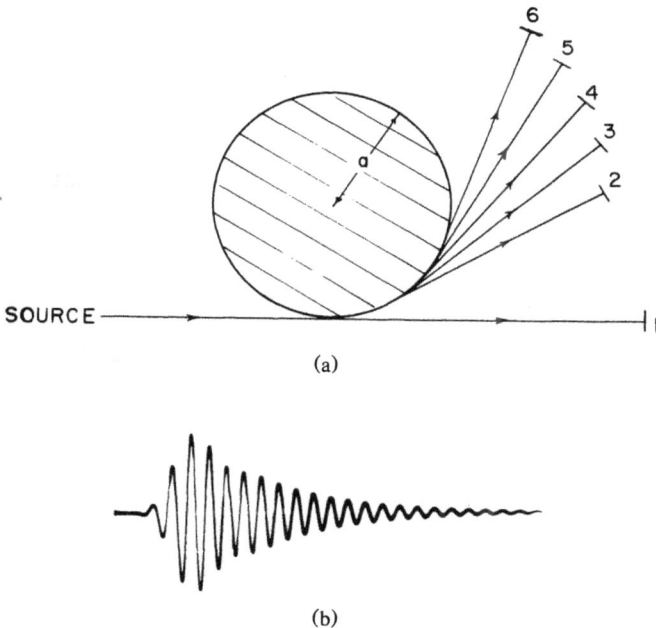

(a)

(b)

Fig. 9 — (a) Diagram of an experiment used to measure circumferential wave properties. (b) The pulse received at position (1) in (a) in the absence of the cylinder.

Fig. 10 — A multiple exposure of successive receptions of the "Franz-type" or "creeping" waves at receiver positions 2-6 in Fig. 9(a), showing the attenuation that occurs in the path along the cylinder.

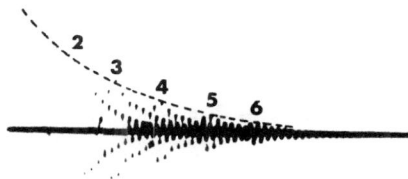

Accurate measurement of the time differences associated with the difference in path between the adjacent receiver positions allows a determination of circumferential wave speed. If peaks of the individual cycles of the pulse do not change their position within the pulse for the different path lengths on the cylinder, the speed that is measured represents both the phase and the group velocity value to within the accuracy of this experiment. A double exposure in Fig. 11 of the pulse at position 1 and at position 6 shows that to be the case for this experiment.

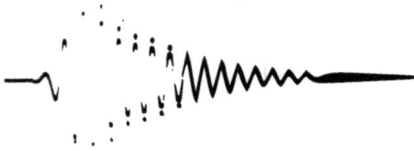

Fig. 11 — Double exposure of a received acoustic pulse in the absence of a cylinder and the same pulse with part of its transmission path along the circumference of the cylinder.

The creeping wave speed determined by this means along with theoretically computed values and other experimental values are plotted in Fig. 12. A similar plot for creeping wave attenuation is shown in Fig. 13.

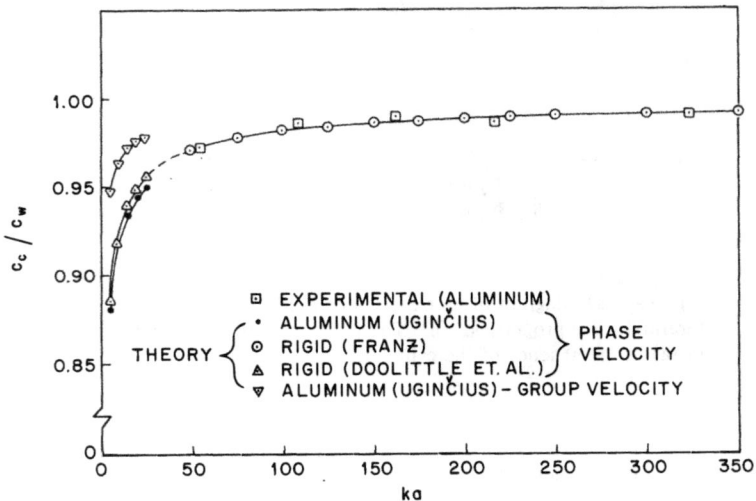

Fig. 12 — The comparison of experimental measurements and various theoretical calculations of the Franz-wave velocity is a function of size parameter (ka).

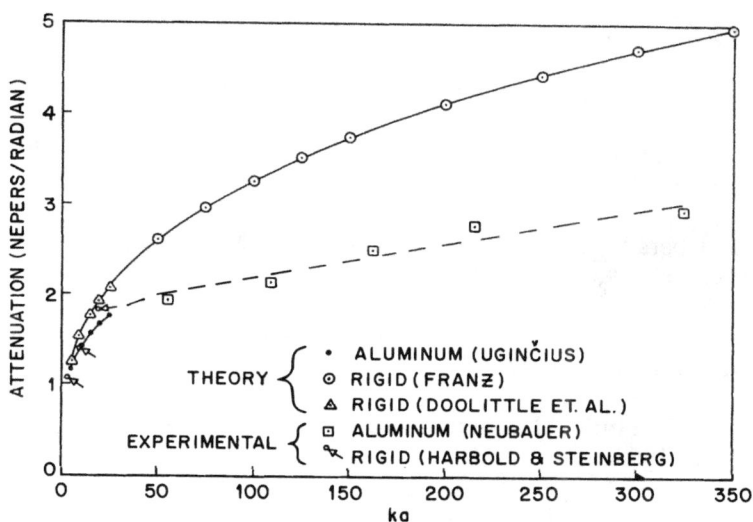

Fig. 13 — The comparison of experimental measurements and theoretical calculations of Franz-wave attenuation as a function of size parameter (ka).

REFERENCES

1. A. Sommerfeld, *Partial Differential Equations in Physics,* Academic Press, New York-London 1964. See particularly the appendix after Chapter 6.

2. B. van der Pol and H. Bremmer, Phil. Mag. **24**, 141, 825 (1937).

3. W. Franz and K. Deppermann, Ann. Phys. Leipzig **10**, 361 (1952).

4. W. Franz, Z. Naturforsch. **9a**, 705 (1954).

5. W. Franz and P. Beckmann, IRE Trans. Antennas and Propagation AP-**4**, 203 (1956).

6. P. Beckmann and W. Franz, Z. Naturforsch. **12a**, 257 (1957).

7. W. Franz, "Theorie der Beugung Elektromagnetischer Wellen," Springer-Verlag, Berlin 1957.

8. G. R. Barnard and C. M. McKinney, J. Acoust. Soc. Am. **33**, 226 (1961).

9. See the bibliographies in Refs. 11 and 13.

10. A. J. Rudgers, J. Acoust. Soc. Am. **45**, 900-910 (1969).

11. R. D. Doolittle and H. Überall, J. Acoust. Soc. Am. **39**, 272 (1966).

12. H. Überall, R. D. Doolittle, and J. V. McNicholas, J. Acoust. Soc. Am. **39**, 564 (1966).

13. R. D. Doolittle, J. V. McNicholas, H. Überall, and P. Uginčius, J. Acoust. Soc. Am. **42**, 522 (L) (1967).

14. R. D. Doolittle, H. Überall, and P. Ugincius, J. Acoust. Soc. Am. **43**, 1 (1968).

15. P. Uginčius and H. Überall, J. Acoust. Soc. Am. **43**, 1025 (1968).

16. J. V. McNicholas, H. Überall, and K. Choate, J. Acoust Soc. Am. **44**, 752 (1968).

17. P. Uginčius, Ph.D. thesis, Phys. Dept., The Catholic Univ. of America, 1968; and U. S. Naval Weapons Lab. Tech. Report No. 2128, Feb. 1968, Dahlgren, VA 22448.

18. W. G. Neubauer, J. Acoust. Soc. Am. **44**, 298 (L) (1968).

19. W. G. Neubauer, Ph.D thesis, Physics Dept., The Catholic Univ. of America, 1968; Experimental Examination, By Means of Pulses, of Circumferential Waves on Aluminum Cylinders, Naval Research Laboratory Report 6791, November 15, 1968; and J. Acoust. Soc. Am. **45**, 1134-1144 (1969).

20. These differ from the corresponding elements in Refs. 10 and 13 by the overall factor of μ, thereby making them dimensionless. See also Refs. 14 or 16 where the correctly nondimensionalized elements are given for the more general 6X6 determinants for a cylindrical shell.

21. G. N. Watson, Proc. Roy. Soc. London **A59**, 83, 546 (1919).

22. J. J. Faran, J. Acoust. Soc. Am. **23**, 405 (1951).

23. B. R. Levy and J. B. Keller, Commun. Pure Appl. Math. **12**, 159 (1959).

24. P. L. Marston and D. L. Kingsbury, J. Acoust. Soc. Am. **70**, 1488-1495 (1981).

Chapter 4

CIRCUMFERENTIAL WAVE INTERPRETATIONS IN CYLINDER REFLECTIONS*

I. INTRODUCTION

Theoretical solutions to the problem of the scattering of sound by rigid, immovable cylinders, nonrigid cylinders in a fluid medium, and small cylindrical obstacles in a solid medium were formulated by Rayleigh [1]. The solutions he presented described goemetries in which the diameters of the cylinders were small compared to the acoustic wavelength in the surrounding medium, although he outlined a more general method for finding the solution for larger diameter cylinders in terms of cylindrical harmonics. This method, called the harmonic series or the Rayleigh series method of solving acoustic scattering problems is in theory applicable to targets whose shape conforms to any of the eleven separable coordinate systems. In practice it has been extensively applied only to scattering from spherical and infinite cylindrical geometries, since cylindrical and spherical harmonics are readily available. Solutions to the problems of the scattering from rigid cylinders and rigid spheres which have radii up to the order of a wavelength ($ka \simeq 6$) were given by Morse, [2]; here $ka = 2\pi a/\lambda$, a is the radius of the scatterer, and λ is the acoustic wavelength in water. Exact solutions to the scattering of a plane sound wave by homogeneous, isotropic cylinders and spheres capable of supporting both shear and compressional waves (elastic scatterers) were first given by Faran, [3] who obtained expressions in terms of normal mode series. Faran presented comparisons of computed bistatic patterns and experimental measurements at $ka = 5$. Extensions of the normal mode calculations to higher ka [4,5] and experimental measurements to determine the degree to which the normal mode theory and experiment agreed, over a broad ka range, were first made on solid elastic spheres and spherical shells [6-9]. Hickling [4] was the first to make extensive use of a digital computer to evaluate the normal mode series expressions, although

*These results first appeared as NRL Report 8216 by Louis R. Dragonette (1978).

his computations were hampered by the slow convergence of the harmonic series solution, which led to computation difficulties with the computers available at that time. Hickling gave computed curves that describe the steady-state-backscattered pressure vs ka, which results when the target is a solid-elastic sphere in water. The computations in his work ranged generally from $0 < ka \leqslant 30$, and he extended the formulation to include the scattering of incident spherical as well as incident plane waves. He also included both near-field and far-field formulations. The results presented by Hickling are given in terms of the form function, f_∞. This dimensionless quantity is obtained by normalizing the reflected pressure with respect to the radius (a) of the target and the range (r) of the field point from the center of the target. Hickling also computed acoustic reflections from elastic spherical shells [5]. Empirical results on solid metal spheres in water were given by Hampton and McKinney [6], who demonstrated that the reflection from metal spheres immersed in water could not be described by purely geometric theory, and by Diercks [7], who demonstrated qualitative agreement between the computations of Hickling and measurements made in a lake. Precise quantitative comparisons between normal mode theory and experiment were first carried out by Neubauer et al, [8] who performed a series of precise steady-state measurements on solid metal spheres in a controlled acoustic tank facility. These measurements demonstrated quantitative agreement, between computations, based on the normal mode series, and experiment, to within the known accuracy of the shear wave speeds of the materials used in the sphere fabrications. This work [8] covered the ka range $0 < ka \leqslant 30$. Dragonette et al. [9] demonstrated empirically that quantitative steady-state results could be obtained from measurements made with short broadband-incident-acoustic pulses. Comparisons between normal-mode theory and experiment for elastic cylinders in water are more recent, [10] but again demonstrated excellent agreement between the theory based on the infinite elastic cylinder and near-real-time experiments performed with finite length cylinders in a laboratory tank. The preceding theoretical and empirical papers [1-10] established that the normal-mode series formulation of the acoustic reflection from elastic-metal targets quantitatively describes measured results up to at least $ka = 30$, without the necessity of material absorption being included in the theory.

Empirical observations by Barnard and McKinney [11] demonstrated periodic, multiple echo returns when solid and hollow brass cylinders ($ka \simeq 40$) were illuminated by short acoustic pulses. Subsequent empirical work and analysis by Diercks et al. [12], Horton et al. [13], and others [14,15] proposed the existence of two types of circumferential waves that were compared to flexural and longitudinal modes on infinite plates. A similarity was recognized between the circumferential behavior of the waves they observed and the waves discussed by Franz [16] in his work on the diffraction of electromagnetic waves by conducting cylinders and spheres. The original analogy between the acoustic waves observed on cylinders and the purely geometrically diffracted circumferential waves considered by Franz broke down, because the speeds of the observed acoustic circumferential waves were from 33% to 300% higher than the speed of sound in the medium surrounding the targets; whereas, analogy with the "creeping waves" of Franz would have predicted a speed slower than that in the surrounding medium.

Überall and collaborators at Catholic University also noted the similarity between the circumferential behavior of empirically observed acoustic waves and waves studied in electromagnetic theory. A modified Sommerfeld-Watson [17] transformation had been used in the study of the propagation of radio waves around the earth [18] and in the studies of the diffraction of electromagnetic waves by cylinders [16]. The Sommerfeld-Watson transformation offered certain advantages, namely, the opportunity to isolate the individual mechanisms responsible for the empirically observed circumferential waves and rapid convergence of the solution. This latter advantage was particularly significant at the time since the normal mode series was considered to be practical only at low ka, because of its slow convergence and the expense of computation. (Advances in computer technology make present high ka Rayleigh series computations both possible and economical [19].) Überall et al. applied the Sommerfeld-Watson transformation to cylinders with rigid and soft boundary conditions and predicted the existence of true ,Franz type, or purely geometrically diffracted, circumferential waves [20]. Application of the Sommerfeld-Watson method to solid elastic cylinders [21] revealed two groups of poles corresponding to two types of circumferential waves discussed in Chapter 3. The Franz-type or geometrically diffracted waves again appeared, and, in addition, poles related to elastic circumferential waves, called R or Rayleigh-type waves, whose speed and properties depend primarily on the elastic constants of the target, were found. Grace and Goodman [22] also presented theoretical evidence for the existence of R-type waves.

Experimental detection of the acoustic Franz-type or purely geometrically diffracted wave was accomplished by Neubauer [23] and by Harbold and Steinberg [24]. The first experiments designed to demonstrate elastic R-type circumferential waves, that is, those related to R-type poles, were performed by Bunney et al. [25] and by Neubauer [26]. Both of these researchers [25,26] used short incident pulses and narrow beam sources to observe the scattering from solid aluminum cylinders. Their results demonstrated the existence of a train of periodic echoes with a circumferential speed close to the shear wave speed in aluminum. Neubauer's [26] work included schlieren visualization of wavefronts resulting from the circumferentially traveling waves. These experiments gave mainly high ka results (ka values between 50 and 500). Originally the mechanism responsible for the periodic pulse trains observed [25,26] was, in fact, considered to be multiple circumnavigations of the cylinder by the Rayleigh wave. Later work by Neubauer and Dragonette [27] showed that multiple internal reflections of shear waves could produce the observed effect, and this multiple reflection analysis was supported by the theoretical work of Brill and Überall [28], who demonstrated the circumferential behavior of the radiation from multiply internally reflected waves. Theoretical [29] and measured attenuation [30] of the Rayleigh wave on submerged flat surfaces also gave attenuations too large to support the conclusion that Rayleigh waves were the source of the multiple returns observed at high ka in Refs. 25 and 26. Dragonette [31] first predicted and observed the ka range at which a Rayleigh circumferential wave can be significant.

The normal mode solutions give a straightforward method of obtaining the scattered acoustic pressure vs frequency, limited only by the expense involved in summing a slowly convergent series. Experimental results have been obtained which agree with the computation to a high degree of accuracy. The major disadvantage of this approach is that individual physical phenomena, such as surface waves, which make up the solution are not immediately obvious. The Sommerfeld-Watson transformation of the normal mode series has the advantage that it isolates individual circumferential waves and the disadvantage that the poles must be found and interpreted and their significance judged.

GENERAL THEORETICAL FOUNDATION

The Rayleigh series expression for the scattered acoustic pressure, $p_s(\theta)$, which results when a plane wave, $p_0 e^{ikx}$, illuminates an infinite elastic cylinder, in the geometry described by Fig. 1, is given in many publications [3,10,21,32,33]. The following form is found in Refs. 32 and 33:

$$p_s(\theta) = -p_0 \sum_{n=0}^{\infty} \epsilon_n (i)^n \left[\frac{J_n(Z) \, L_n - Z \, J_n'(Z)}{H_n(Z) \, L_n - Z \, H_n'(Z)} \right] H_n(kr) \cos n\theta. \quad (1)$$

The time dependence $e^{-i\omega t}$ is suppressed. In Eq. 1, ϵ_n is the Neumann factor ($\epsilon_n = 2$, $n = 0$; $\epsilon_n = 1$, $n > 0$), J_n is a Bessel function, H_n is a

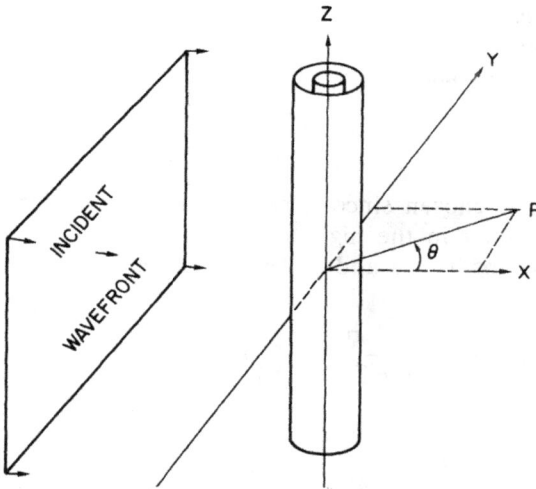

Fig. 1 — The geometry used in the description of the scattering of a plane wave by an infinitely long cylinder.

Hankel function of the first kind, $Z \equiv ka$, and the L_n are the quotients of two 2-by-2 matrices:

$$L_n = \frac{\rho}{\rho_s} \frac{D_n^{(1)}[Z]}{D_n^{(2)}[Z]} = \frac{\rho}{\rho_s} \frac{\begin{vmatrix} a_{11} & a_{13} \\ a_{21} & a_{23} \end{vmatrix}}{\begin{vmatrix} a_{11} & a_{13} \\ a_{31} & a_{33} \end{vmatrix}} \qquad (1a)$$

where the matrix elements a_{ij} are given in Ref. 32. In the far field where $r \gg a$, $H_n(kr)$ may be written in its asymptotic form

$$H_n(kr) = \left(\frac{2}{\pi kr} \right)^{1/2} e^{ikr - in\pi/2 - i\pi/4} \qquad (2)$$

and, defining

$$\left[\frac{J_n(Z) L_n - Z J_n'(Z)}{H_n(Z) L_n - Z H_n'(Z)} \right] = G_n(Z), \qquad (3)$$

the far-field pressure scattered by an infinite cylinder illuminated by a plane incident wave may be written

$$p_s(\theta) = -p_0 e^{ikr} \left(\frac{2}{\pi kr} \right)^{1/2} e^{i\pi/4} \sum_{n=0}^{\infty} \epsilon_n G_n(Z) \cos(n\theta). \qquad (4)$$

For backscattering, $\theta = \pi$ and

$$p_s(\pi) = -p_0 e^{ikr} \left[\frac{2}{\pi kr} \right]^{1/2} e^{i\pi/4} \sum_{n=0}^{\infty} \epsilon_n (-1)^n G_n(Z). \qquad (5)$$

A quantity called the far field form function, f_∞, is defined to give a nondimensional representation of the scattered pressure. In keeping with the definition used extensively in the literature [4,5,7-10,32,33]

$$f_\infty(\theta) = \left[\frac{2r}{a}\right]^{1/2} \frac{p_s(\theta)}{p_0}. \tag{6}$$

This definition is chosen since it results in $|f_\infty| = 1$ for the case of a purely rigid cylinder in the high frequency limit. From Eqs. 4 and 5 the expressions for $f_\infty(\theta)$ and $f_\infty(\pi)$ for an elastic cylinder are given by

$$f_\infty(\theta) = \frac{-2}{(i\pi Z)^{1/2}} \sum_{n=0}^{\infty} \epsilon_n G_n(Z) \cos(n\theta) \tag{7a}$$

and

$$f_\infty(\pi) = \frac{-2}{(i\pi Z)^{1/2}} \sum_{n=0}^{\infty} \epsilon_n(-1)^n G_n(Z). \tag{7b}$$

Using Eq. 7b, the individual normal modes or partial waves which make up the backscattered form function are defined as

$$f_n(\pi) = \frac{-2}{(i\pi Z)^{1/2}} \epsilon_n(-1)^n G_n(Z) \tag{7c}$$

where

$$f_\infty(\pi) \equiv \sum_{n=0}^{\infty} f_n(\pi). \tag{7d}$$

Computed plots of f_∞ vs ka, obtained from Eq. 7b are called reflection function plots, and such curves give a dimensionless representation of the scattered steady state pressure vs frequency. This representation can describe the scattering at any combination of radius and frequency within the ka limits of the calculation. Equations 7a and 7b give steady state values of f_∞, so that a continuous wave or very long pulse experiment can be used to obtain a direct comparison between experimental and calculated results [8]. Such an experimental method is tedious and excessively time consuming, as each experiment at each single frequency gives one point on the reflection function curve. To overcome this practical deficiency, methods to obtain the steady state quantity, f_∞, from short broadband incident pulses were developed [9,10].

If the incident sound wave in the geometry described by Fig. 1 is not steady state but a pulse, $p_i(\tau)$, with a Fourier transform $g_i(ka)$ given by

$$g_i(ka) = \int_{-\infty}^{\infty} p_i(\tau) e^{ika\tau} d\tau \tag{8}$$

then from Eqs. 6 and 8, the backscattered pressure has a Fourier transform $g_s(ka, \pi)$ given by

$$g_s(ka, \pi) = \left(\frac{a}{2r}\right)^{1/2} f_\infty(ka, \pi) \, g_i(ka) \qquad (9)$$

and $|f_\infty(ka, \pi)|$ can be obtained from

$$|f_\infty(ka, \pi)| = \left(\frac{2r}{a}\right)^{1/2} \frac{|g_s(ka, \pi)|}{|g_i(ka)|}. \qquad (10)$$

The quantity τ is a dimensionless time parameter

$$\tau = \frac{ct - r}{a} \qquad (11)$$

which is normalized to be zero when the incident pulse is coincident with the position of the center of the cylinder. The wave speed in the ambient fluid is c and t is the time. Equation 10 is the basis by which a steady state quantity $|f_\infty(ka, \pi)|$ can be obtained over a broad frequency range by a single short pulse experiment. The incident and reflected pulses are digitized, their transforms computed, and the division indicated in Eq. 10 carried out [9]. With present minicomputer technology this entire procedure can be accomplished in a near real time framework [10]. The ka range over which f_∞ is obtained depends of course on the bandwidth of the incident pulse. Theoretical computations of the scattered echoes which result when a short incident pulse with a known spectrum, $|g_i(ka)|$, is used to insonify a target with a known $f_\infty(\theta, ka)$ can be obtained by using Eq. 9. This computational procedure allows the isolation of the individual mechanisms which contribute to the steady state scattered pressure. These pulse calculations are of significant value for many reasons, the most important of which is, that the theoretically formulated incident pulses used, can be made shorter than any which can be reasonably achieved in the laboratory. This allows isolation of closely spaced echoes which cannot be accomplished at a reasonable cost in the laboratory. In addition theoretical computations can simulate experimental measurements over a large number of frequencies, target materials, target sizes, and target shell thickness that would be impossible to duplicate economically in a laboratory. In the theoretical procedure an incident pulse $p_i(\tau)$ with a known spectrum, $|g_i(ka)|$, is used to insonify a target whose form function can be computed. Computation of the form function and the procedure indicated in Eq. 9 are accomplished by the computer, and the scattered echo, $p_s(\tau)$, is described by

$$p_s(\tau) = 1/2\pi \int_{-\infty}^{\infty} g_s(ka) \, e^{-ika\tau} \, dka. \qquad (12)$$

Solutions to Eq. 12 are obtained by using fast Fourier transform techniques in the computer.

NORMAL MODES OF VIBRATION OF A CYLINDER AND CREEPING WAVES

The acoustic reflection from a rigid cylinder has been well understood since the prediction [20] and empirical observation [23,24] of the Franz wave, and only a few ideas relating to the direct use of the form function curves to derive Franz wave properties can be added. In the case of a rigid cylinder, $G_n(Z)$ as defined in Eq. 3 reduces to

$$G_n^{(R)} (Z) = \frac{J_n' (Z)}{H_n' (Z)}$$ (13)

and the form function for a rigid cylinder is given by

$$f_\infty^{(R)} (Z) = \frac{-2}{(i \pi ka)^{1/2}} \sum_{n=0}^{\infty} \epsilon_n (-1)^n G_n^{(R)} (Z).$$ (14)

A plot of $f_\infty(\pi)$ vs ka computed from Eq. 14 is given in Fig. 2. Since by definition the boundary conditions imposed to obtain the curve in Fig. 2 preclude penetration into the cylinder, the backscattered reflection function curve can include contributions only from specular reflection and diffraction.

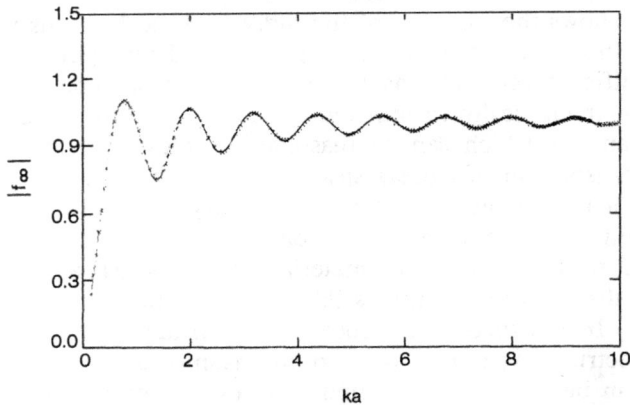

Fig. 2 — The form function for a rigid cylinder.

In the creeping wave solution Eq. 14 is transformed from an infinite series of n terms into a series of "creeping surface waves" by the Sommerfeld-Watson transformation [18,20]. The creeping waves arise as the residue of poles in the complex ν plane determined from the equation

$$H_\nu' (Z) = 0 \tag{15}$$

with solutions

$$\nu_l(\omega) = Z + (Z/6)^{1/3} e^{i\pi/3} q_l - \left(\frac{6}{Z}\right)^{1/3} e^{-i\pi/3} \left[\frac{1}{10q_l} + \frac{1}{180} + \frac{q_l^2}{180}\right] \tag{16}$$

where the q_l are the zeroes of the first derivates of the Airy function as defined by Franz [18]. The index $l = 1, 2, 3...$, and l increases in the direction of increasing real and imaginary parts of ν_l. The attenuation, α_l^F, of the lth Franz wave in Np/rad is given by

$$\alpha_l^F(\text{Np/rad}) = \text{Im } \nu_l \tag{17a}$$

and the phase speed, c_p^F, is given by [20]

$$c_p^F/c = \frac{Z}{R_e \nu_l} \tag{17b}$$

where c is the speed of sound in water. Calculations based on Eqs. 16 and 17a demonstrate that only the $l = 1$, or first Franz wave is of significant magnitude, and Figs. 3 and 4 give computations of the attenuation and phase velocity of this first Franz wave as a function of ka as computed from Eqs. 17b and 17a.

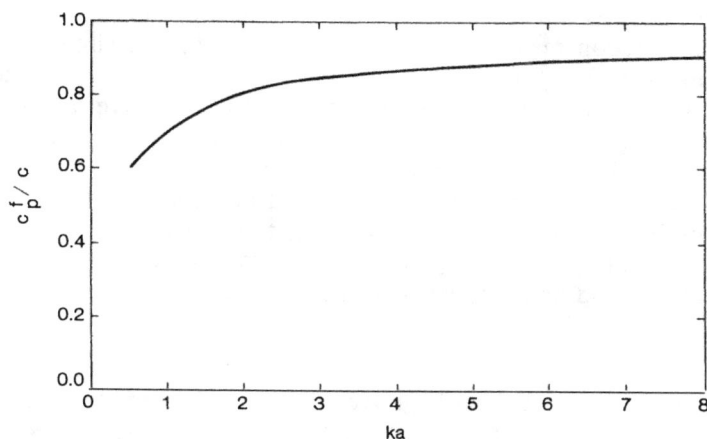

Fig. 3 — The Franz wave phase velocity vs ka for a rigid cylinder.

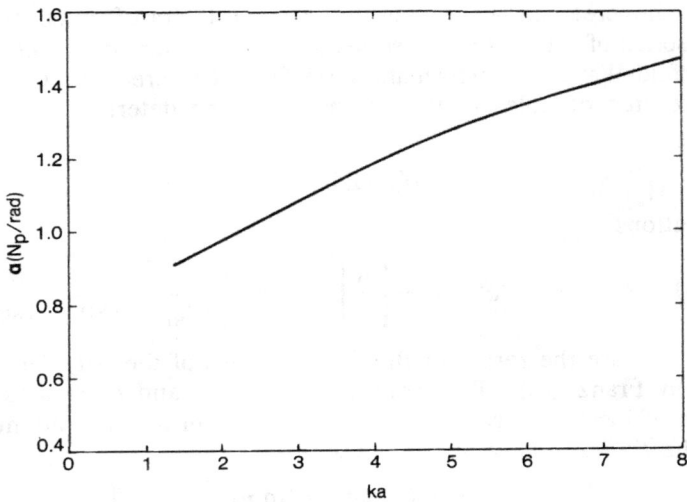

Fig. 4 — The Franz wave attenuation vs *ka* for a rigid cylinder.

In the case of the rigid cylinder the connection between the steady state form function f_∞^R, given in Fig. 2 and the Franz wave with properties described by Figs. 3 and 4 is not difficult to determine. Sound does not penetrate a rigid cylinder, thus the form function must be made up entirely of specular reflection plus a pure geometrically diffracted contribution. The backscattered return must then be as described by Fig. 5. A specular reflection begins at point A of Fig. 5, and two Franz waves begin at points B and C and take the paths shown. This well understood result can, however, be taken further. The reflection function given in Fig. 2 is a steady state function, and the knowledge that this reflection function results from sound waves taking the circumferential paths shown in Fig. 5 leads to the following analysis. Because the two diffracted waves BC and CB take the same path and travel at the same speed, they are always in phase with each other in the backscattered direction. Computations based on Eqs. 16 and 17a demonstrate that for $ka > 1$ only these first Franz returns need be considered; that is, the contribution from succeeding circumnavigations of the cylinder are too small in amplitude to be significant. The difference in the time of arrival, at the field point P, between the specular and diffracted waves is

$$\Delta t = \frac{2a}{c} + \frac{\pi a}{c_p^F}. \tag{18}$$

The difference in path lengths traveled by the backscattered specular and Franz wave contributions at the field point P is expressed as

$$\Delta d = 2a + \frac{\pi a\, c}{c_p^F}. \tag{19}$$

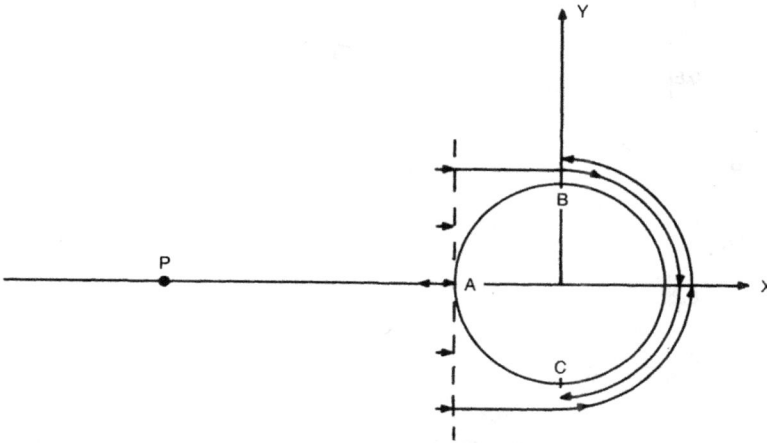

Fig. 5 — Source of backscattered echoes from a rigid cylinder.

If it is assumed that the peaks in the reflection function curve (Fig. 2) occur when the specular and Franz contributions add in phase at P, then peaks occur when

$$\Delta d = a (2 + \pi c / c_p^F) = n\lambda = \frac{n \, 2\pi}{k} \qquad (20)$$

which leads to

$$c / c_p^F = \frac{2n}{(ka)_{peak}} - \frac{2}{\pi} . \qquad (21)$$

The notation $(ka)_{peak}$ is used to indicate the (ka) values at which a peak in $|f_\infty^R|$ as seen in Fig. 2 occurs. A similar expression can be derived by assuming that the nulls in the reflection function curve of Fig. 2 occur when $\Delta d = \frac{(2n-1)}{2} \lambda$. This expression is

$$c / c_p^F = \frac{(2n-1)}{(ka)_{null}} - \frac{2}{\pi} . \qquad (22)$$

A Franz wave speed can than be calculated directly from the reflection function curve by determination of $(ka)_{peak}$ and $(ka)_{null}$. The normalized phase velocity c_p^F / c (which is the reciprocal of the left-hand side of Eqs. 21 and 22) is computed from these equations and compared to the direct "creeping wave" theory computation, from Eq. 17b, in Fig. 6. Agreement between the two methods is excellent, demonstrating the possibility of obtaining Franz wave velocity directly from the normal mode form function curves. Thus Franz wave velocities can be

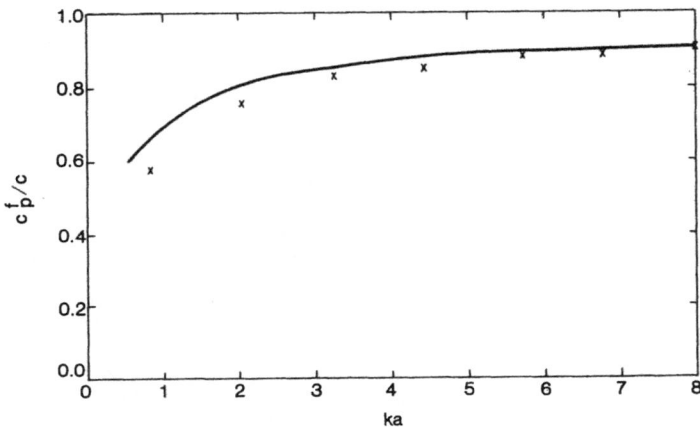

Fig. 6 — A comparison between the computed Franz wave wave speed for a rigid cylinder (—) and values estimated from the form function curve (×).

obtained for bodies for which no creeping wave analysis exists. Examples are given in Figs. 7 and 8, where computed form functions and derived Franz wave velocity for a rigid sphere and an aluminum oxide cylinder are given. The rigid sphere shows a much more rapid rise in Franz wave velocity with increasing frequency, than is observed for the rigid cylinder. The velocity for the aluminum oxide cylinder shows only minor deviations from the rigid cylinder curve.

The attenuation of the Franz wave as a function of ka can also be investigated directly from the normal mode calculation. Here the reduction in amplitude of the successive oscillations is assumed due to the increase in attenuation of the Franz wave as a function of ka. The reduction in magnitude of the oscillations in Fig. 2 with increasing ka should then give a measure of the Franz wave attenuation vs ka. A comparison of attenuation values obtained from creeping wave theory for a rigid cylinder and values obtained from the form function curve is given in Fig. 9a. In Fig. 9b the attenuation is given for an aluminum oxide cylinder for which no direct creeping wave data are available. The rigid cylinder curve is included in Fig. 9b, for comparison.

Empirical observations of Franz waves on a rigid cylinder were obtained in an air acoustic range. A solid 3.18-cm-diameter cylinder was used as the target, and the frequency of the incident pulse was 34.4 kHz. In air the impedance mismatch between the aluminum and surrounding air medium is so great that rigid boundary condition assumptions are successfully achieved.

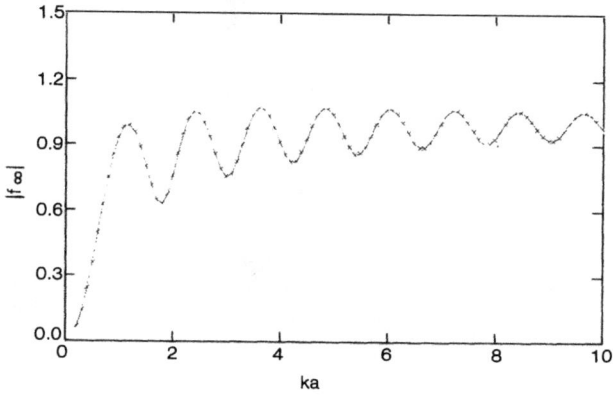

Fig. 7(a) — The form function vs ka for a rigid sphere.

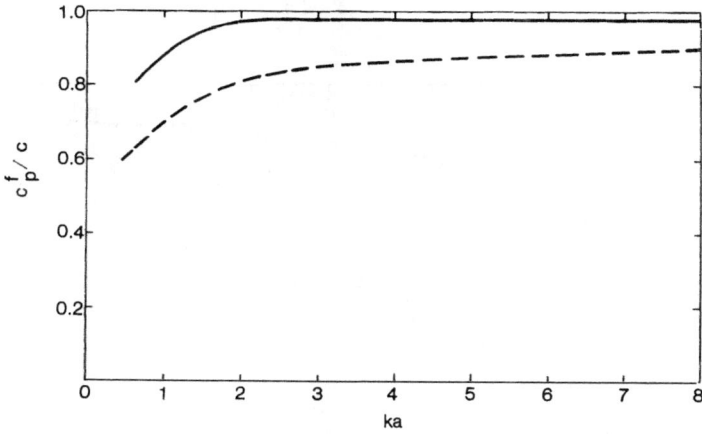

Fig. 7(b) — Franz wave velocity estimated from (7a) for a rigid sphere
(——) and Franz wave velocity for a rigid cylinder (- - -).

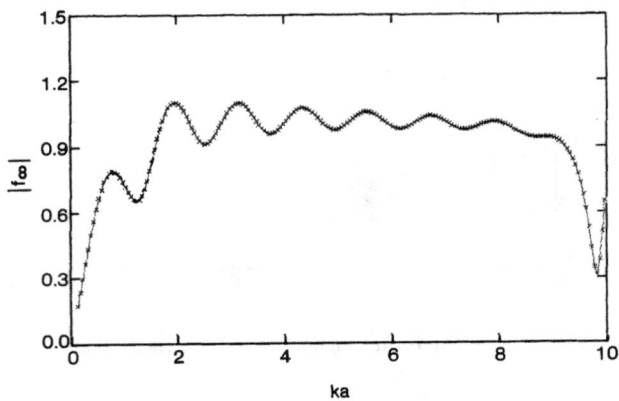

Fig. 8(a) — The form function for an aluminum oxide cylinder.

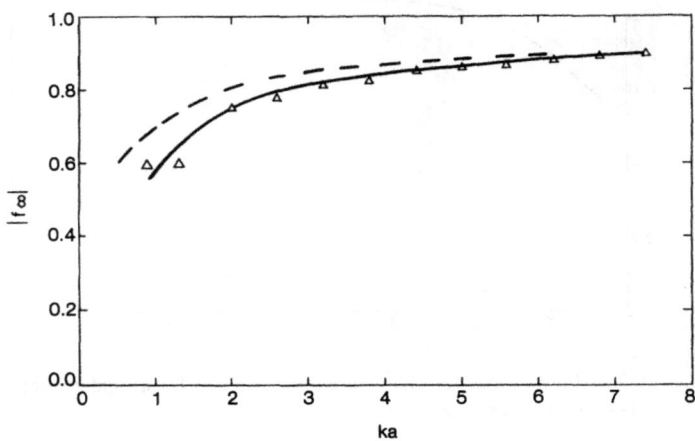

Fig. 8(b) — Franz wave velocity estimated from the form function (——)
and Franz wave velocity for a rigid cylinder (- - -).

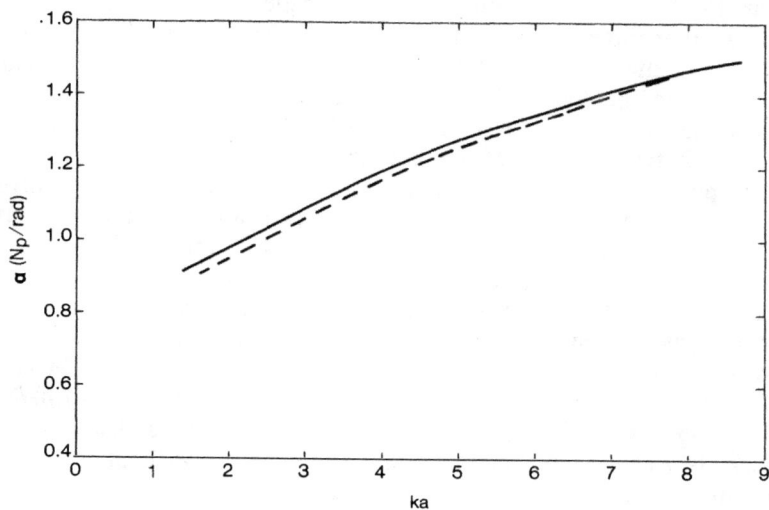

Fig. 9(a) — Franz wave attenuation for a rigid cylinder obtained directly (——)
and estimates obtained from the form function curve (- - -).

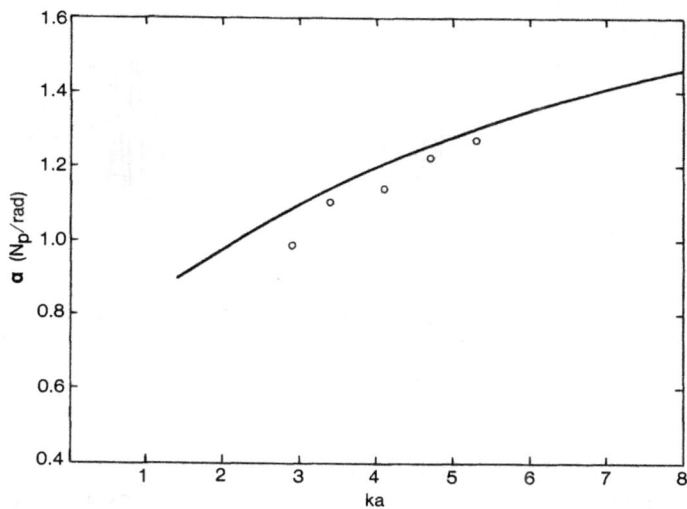

Fig. 9(b) — Franz attenuation estimates for an aluminum oxide cylinder (o) compared to the rigid cylinder (——).

Experimental observations of Franz waves on a rigid cylinder are given in Fig. 10. The measurements are made at aspect angles of 45° and 75° so that the changes, both in relative amplitude and time separation, between the specular reflection and the Franz wave can be clearly observed. The incident pulse is seen in Fig. 10(a) and the scattered echoes at 45° and 75° are seen in Figs. 10(b) and 10(c). The time separations between the specular and Franz waves are measured from the large positive going peak in each echo. The measured time differences are 81 μs at 45° and 125 μs at 75°. The ratio of the Franz wave to specular amplitude is p_F/p_{spec} = 0.56 at 45° and = 0.13 at 75°. Plots of theoretical computations, based on Eqs. 9 and 12, for the echoes from a rigid cylinder are shown in Fig. 11. The theoretical computations agree closely with experiments in the relative magnitudes and positions of the Franz waves. The values obtained from Fig. 11 are a time separation of 78 μs at 45° and 125 μs at 75°, and a ratio p_F/p_{sec} of 0.57 at 45° and 0.14 at 75°. The center dimensionless frequency of the pulse is $k_o a$ = 10. Echo computations agree with experimental measurements, and can be used to isolate mechanisms or to supplement measurements when f_∞ is known or can be computed.

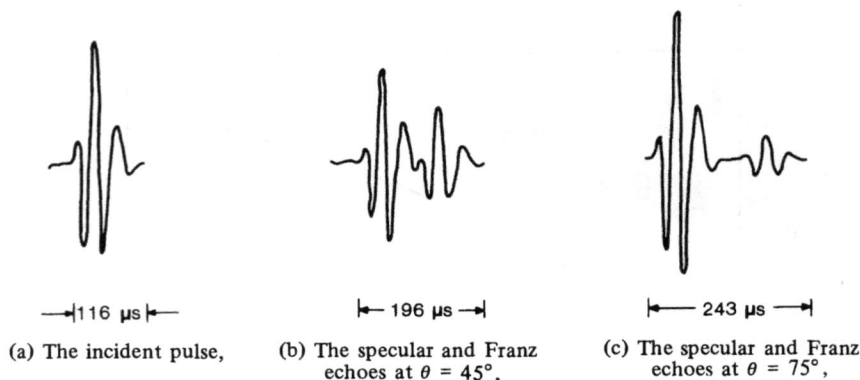

⟶⊣116 μs⊢⟵	⊢⟵ 196 μs ⟶⊣	⊢⟵ 243 μs ⟶⊣
(a) The incident pulse,	(b) The specular and Franz echoes at θ = 45°,	(c) The specular and Franz echoes at θ = 75°,

Fig. 10 – The experimental observation of the Franz wave radiation from a rigid cylinder.

For elastic cylinders, resonances in the normal mode solution can be identified with specific R-type circumferential waves predicted by creeping wave theory. The ka range over which the Rayleigh wave makes a significant backscattering contribution has been the subject of much conjecture [25-28,34] in the literature. There is a limited, low ka, region of importance for the Rayleigh circumferential wave in contrast to previous hypotheses [25,26,4]. The first experimental observation of backscattered circumferential radiation from the true Rayleigh

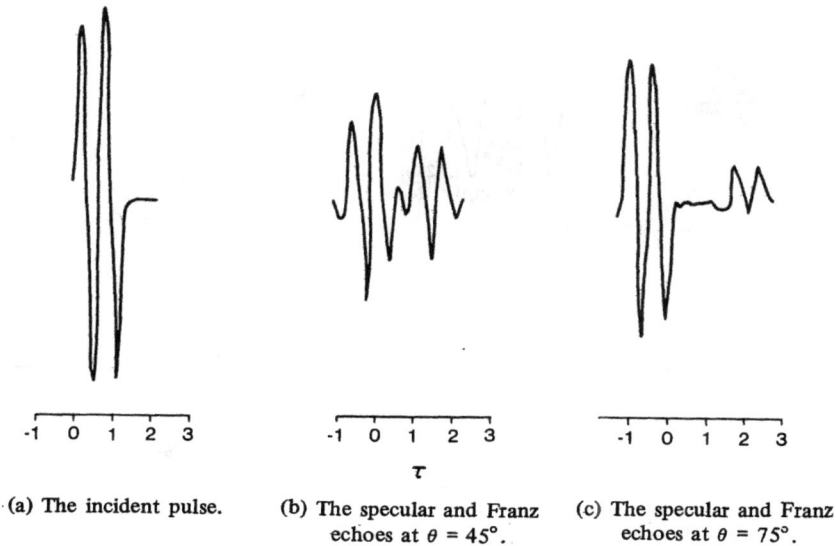

-1 0 1 2 3	-1 0 1 2 3	-1 0 1 2 3

τ

(a) The incident pulse.   (b) The specular and Franz   (c) The specular and Franz
                              echoes at $\theta = 45°$.   echoes at $\theta = 75°$.

Fig. 11 — Computation of the scattering of a two cycle
incident pulse by a rigid cylinder.

wave was described by Dragonette [31] and was accomplished in the low ka region where it is predicted. Computation and analysis of the effects of the normal mode resonances on the backscattered $|f_\infty|$ demonstrate further that the form function is made up of a rigid background on which narrow resonances are superimposed.

The backscattered form function for an elastic aluminum cylinder in water is calculated from Eq. 7b and given in Fig. 12 over the range from $0.2 \leqslant ka \leqslant 20$. The curve is calculated in ka steps of $\Delta ka = 0.01$. A comparison between this theoretical computation and an experiment using the short pulse experimental technique is given in Fig. 13. The theoretical curve is computed in intervals of $\Delta ka = 0.05$ which is compatible with the Δka resolution that can be achieved experimentally. Agreement between theory and experiment is within 2%. The form function curve seen in Fig. 12, shows that over the range $0.2 \leqslant ka \leqslant 4.7$ the aluminum reflection curve is very similar to that of the rigid cylinder (Fig. 2). This region of similarity is followed by marked irregular oscillations and these oscillations continue as (see Chapter 13) $ka \rightarrow \infty$, if no absorption is included in the theory. Damping of the oscillations as ka increases will occur when absorption becomes significant [35], but for metals such as aluminum, experimental results indicate [8-10] that absorption need not be included over the ranges of ka that will be considered here.

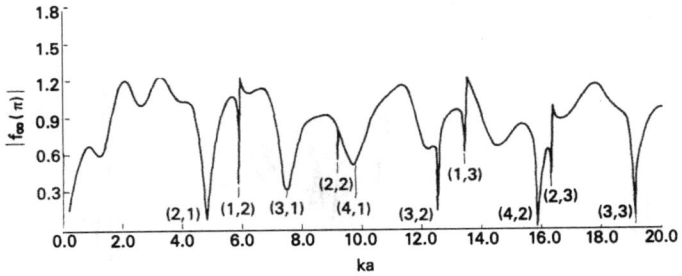

Fig. 12 — The form function for an aluminum cylinder in water.

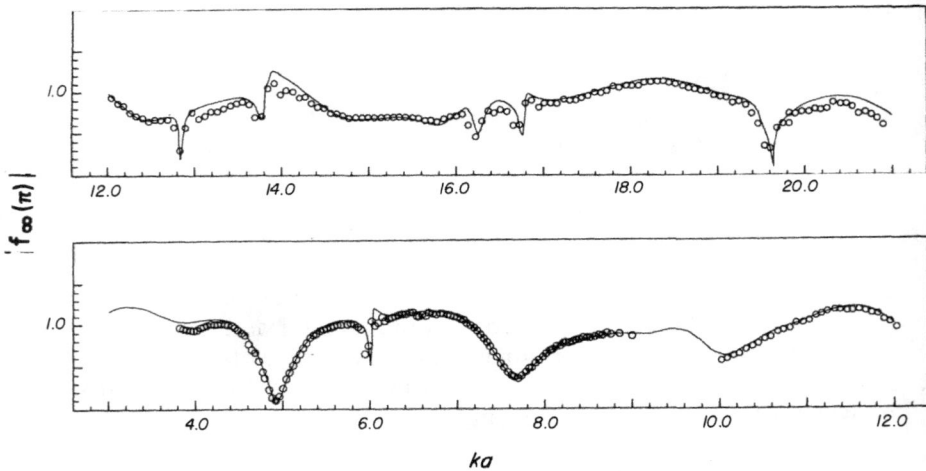

Fig. 13 — Comparison of theory (——) and experimental observations (the points)
for an aluminum cylinder in water.

The irregular oscillations in the aluminum form function, which begin at $ka \simeq 4.5$, are related to excitation of the individual normal modes, f_n, as defined in Eq. 7c. Resonances occur at ka values at which

$$D_n^{(2)} [ka] \equiv D^{(2)} [n, \ ka] = 0 \qquad (23)$$

with $D_n^{(2)} [n, \ ka]$ the matrix introduced in Eq. 1a. Solutions to Eq. 23 give the ka position of the free modes of vibration (resonances) of the cylinder. The correspondence between the ka values at which irregularities in the form function occur and the ka values at which resonances occur is indicated in Fig. 12, where the resonances are identified by the subscript $(n, \ l)$. Here n is the mode number and l is the eigenfrequency, for example $(n, \ 1)$ means the fundamental resonance of the

nth normal mode, $(n, 2)$ the first harmonic, etc. The "creeping wave" solution to the problem of the scattering of an incident plane wave by a solid aluminum cylinder was carried out by Doolittle et al. [21], who transformed Eq. 7b using the Sommerfeld-Watson technique. Doolittle found poles in the complex ν plane at the positions

$$D^{(2)} [\nu, ka] = 0. \tag{24}$$

He computed a series of R type, or elasticity related, poles and gave a table of the positions of the first six R-type poles in the complex ν plane as a function of ka. Each pole gives rise to a circumferentially traveling wave [21]. The relationship between Eqs. 23 and 24 form the basis for the correspondence between normal mode resonances and the individual circumferential waves predicted by the creeping wave theory. To obtain this relationship, it is necessary to consider the properties of the normal modes individually. Equation 7c describes the nth normal mode. The $n = 0$ term is the breathing mode, $n = 1$ the dipole term, $n = 2$ the quadrupole, etc. The individual motions can be represented by a pair of standing waves $e^{i \pm (n\theta - \omega t)}$ traveling in the opposite directions with phase velocities

$$c_n(ka) = \frac{kac}{n} \tag{25}$$

and group velocities

$$c_n^g(ka) = c \frac{d(ka)}{dn}. \tag{26}$$

At a resonance of the nth mode exactly n wavelengths fit over the circumference of the body, and the lth eigenfrequency of the nth mode, $(ka)_{n,l}$ is the lth solution to Eq. (23). A comparison of Eqs. (23) and (24) leads to the connection between the creeping wave solution. The complex quantity

$$\nu = \nu_l(ka), \tag{27}$$

which is related to R type circumferential waves with phase velocities

$$c_l(ka) = \left[\frac{kac}{\text{Re } \nu_l} \right] \tag{28}$$

and group velocities

$$c_l^g(ka) = \frac{c}{d\text{Re } \nu_l / dka}. \tag{29}$$

If now Re $\nu_l ka = n$, Eqs. 23 and 24 become identical in form and the modal velocities (Eqs. 25, 26) are identical to the creeping wave velocities (Eqs. 28, 29). Thus, when Re $\nu_l = n$, the lth Rayleigh type circumferential wave coincides with the wave speed, $c_n(ka)$, of the nth modal vibration. This hypothesis is demonstrated below. Table 1 gives

Table 1 — Modal Eigenvalues
$(ka)_{nl}$ and Mode Speeds c_n/c for an
Aluminum Cylinder in Water

n \ l	(ka_{nl})			c_n/c		
	1	2	3	1	2	3
0	—	—	9.43	—	—	∞
1	—	5.87	13.44	—	5.87	13.44
2	4.78	9.17	16.31	2.39	4.59	8.16
3	7.38	12.53	19.12	2.46	4.18	6.37
4	9.65	15.84	—	2.41	3.96	—
5	11.78	19.02	—	2.36	3.80	—

the modal eigenfrequencies obtained from Eq. 23, that are identified in Fig. 12, and the corresponding modal phase velocities are computed from Eq. 25. The breathing modes $(0, l)$ are not strongly excited, as evidenced from Fig. 12, but are included in Table 1. The $(1, 1)$ mode is generated in the region where the Franz wave or rigid reflection predominates and is also not observed in the curve of Fig. 12. The ka values at which Re ν_l (ka) = n, are extrapolated from the work of Doolittle et al (Table 2 of Ref. 21), and comparisons are made between these extrapolated values and the normal mode resonances identified in Fig. 12. This comparison is shown in Table 2, which also gives the computed values of the creeping wave phase velocity [21]. Tables 1 and 2 demonstrate that the ka values at which resonances occur correspond to ka values at which Re ν_l (ka) = n. They demonstrate further the equality of the modal and circumferential wave velocities c_n (ka) and $c_l(ka)$. The relationship is thus established between the (n, l) normal modes and the elastic or R-type poles found by the "creeping wave" theory. For $n = 2$ (i.e., the $(2,1)$ mode), the circumference of the cylinder is 2 wavelengths of the R_1 type wave, at the $(3, 1)$ resonance, the circumference of the cylinder is exactly 3 wavelengths, etc. The R_2 circumferential wave is similarly related to the $(n, 2)$ normal mode resonances, and so on, with the (n, l) normal mode resonances

Table 2 — The correspondence between the normal mode
resonances and the ka values at which $\mathrm{Re}\nu_l = n$.
The target is and elastic aluminum cylinder in water.

Normal mode resonances			(ka) values at which $\mathrm{Re}\nu_1 = n$ from the Sommerfeld-Watson formulation of Ref. 21		Phase velocities c_l/c of the R_L circumferential wave when $\mathrm{Re}\nu_l = n$				
N	l	$(ka)_{nl}$	L	$(ka)_{nl}$	R_1	R_2	R_3	R_4	R_5
0	3	09.43	3	09.40	—	—	∞	—	—
	4	10.46	4	10.40	—	—	—	∞	—
	5	17.14	5	17.08	—	—	—	—	∞
1	2	05.87	2	05.80	5.85	—	—	—	—
	3	13.44	3	13.39	—	13.24	—	—	—
	4	16.02	4	15.90	—	—	16.04	—	—
2	1	04.78	1	04.85	2.37	—	—	—	—
	2	09.17	2	09.10	—	04.58	—	—	—
	3	16.31	3	16.28	—	—	08.16	—	—
3	1	07.38	1	07.30	2.45	—	—	—	—
	2	12.53	2	12.47	—	04.15	—	—	—
	3	19.12	3	19.05	—	—	06.36	—	—
4	1	09.65	1	09.70	2.40	—	—	—	—
	2	15.84	2	15.80	—	03.94	—	—	—
5	1	11.78	1	11.80	2.35	—	—	—	—
	2	19.02	2	18.90	—	03.79	—	—	—

related to the lth order Rayleigh or R_l Rayleigh type circumferential
wave.

The R_1 circumferential wave is related to the leaky Rayleigh wave
[36] on a flat surface (see Chapter 18). This is the surface wave which,
as ka increases, approaches the phase velocity of the Rayleigh wave on
a flat infinite half space. The higher order R-type waves are called
"whispering gallery" waves and become lateral waves in the limit as ka
$\rightarrow \infty$ [35]. Since the R_1 or Rayleigh wave is related to the $(n, 1)$ resonances in the normal mode solution, the influence of the circumferentially traveling Rayleigh wave on the backscattering from an aluminum
cylinder can be inferred from the relative influence of the $(n, 1)$ modes
on the form function seen in the curve of Fig. 12. The $(2, 1)$, $(3, 1)$,
and $(4, 1)$ resonances are observed to have a marked effect on the form
function in the ka range from $4 \leqslant ka \leqslant 10$. For $n > 4$ the effects of
the $(n, 1)$ resonances on $|f_\infty|$ are small, and in fact for $ka > 20$ no $(n,
1)$ modes were observed to influence $|f_\infty|$. The result of the calculations
plotted in Fig. 12 and calculations of $|f_\infty|$ carried out between $20 \leqslant ka
\leqslant 40$ (not shown) strongly indicate that the Rayleigh wave will not

contribute significantly to the observed backscattering from an aluminim cylinder at ka values above $ka = 20$. This is a significant point, since the possibility of Rayleigh wave generation at high (ka) has been a matter of dispute in the literature [25-28,33]. This point will be explored in more detail. Figure 12 does show, however, that the $(2,1)$, $(3,1)$, and $(4,1)$ resonances are major features of the form function curve for $ka < 10$. Nulls in the form function at the ka positions of these three resonances should be related to the interference between specular reflection and the circumferentially traveling Rayleigh wave. Specular reflection and Rayleigh wave radiation are known to be 180° out of phase with each other in the flat surface case [30,36,37]. Verification of the above explanation of the nulls at the $(2,1)$, $(3,1)$, and $(4,1)$ resonance values should be possible both by computing the low ka echo response of an aluminum cylinder to an incident short acoustic pulse and by experimentally determining the echo response of an aluminum cylinder at low ka. Both measurement and computation were done, and this experimental observation of the backscattered reradiation from a circumferentialy traveling Rayleigh wave was the first observation of this phenomenon [31].

The major difficulty in achieving an experimental observation of the Rayleigh wave at low ka is in obtaining a practical and possible combination of cylinder radius, frequency, and pulse length that allows the Rayleigh wave to be separated from the specular reflection. The best available combinations were an aluminum cylinder of radius $a = 0.635$ cm measured with a short pulse centered at frequencies $f_0 = 386$ kHz and $f_0 = 500$ kHz. Commercially available lead zirconate titinate transducers, with active elements 1.905 cm square were driven with a rectangular pulse, and the achieved pulse length was 5 cycles.

Figure 14 shows the experimentally observed backscattering from the 0.625-cm-radius-aluminum cylinder at $f_0 = 500$ kHz, or $k_0 a = 13.5$. The backscattered echo consists of a specular return followed by a Rayleigh circumferential wave which is 180° out of phase with the specular return. Even at this relatively low ka value the second traversal of the cylinder by the Rayleigh wave is already almost entirely in the noise 25 dB below specular reflection. The experimental backscattering result at $f_0 = 386$ kHz, $(k_0 a) = 10.4$ is given in Fig. 15. Here there is a slight overlap of the final cycle of the specular reflection and the first cycle of the Rayleigh wave, but the 180° phase shift can still be observed. At this ka, direct measurement can be obtained of the group velocity and attenuation, as the second transversal of the Rayleigh wave is now visible. The path difference between the first and second manifestations of the Rayleigh wave is $\Delta d = 2\pi a$. The measured group

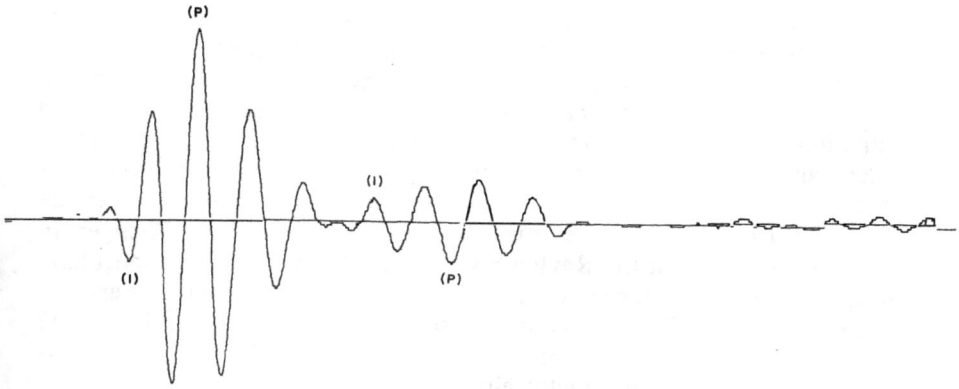

Fig. 14 — Experimental observation of the Rayleigh circumferential wave on an aluminum cylinder at $k_o a = 13.5$.

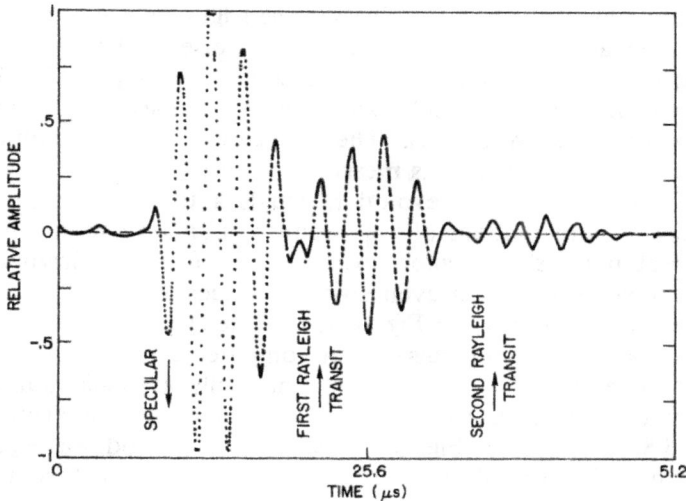

Fig. 15 — Experimental observation of the Rayleigh circmferential wave on an aluminum cylinder at $k_o a = 10.4$.

velocity is $c_R^g/c = 1.9$, in comparison to the estimated value of $c_R^g/c = 2.0$ obtained from Ref. 21. It is expected that the group velocity would be higher [39,40] at these ka values than the infinite half space Rayleigh wave velocity, which is $c_R^g/c = 1.8$ [30]. Equality between the group velocities of the flat surface and circumferential Rayleigh wave does not occur until the ka value of the cylinder reaches, at least $ka \simeq 30$, when the cylinder circumference is greater than 10 wavelengths of the Rayleigh wave [39].

The fact that the circumferential speed of the R_1 wave is a function of frequency, or ka, means that changes in the pulse shape should occur between the specular reflection and the R_1 circumferential wave. The speed of sound in water is not a dispersive quantity, i.e. not a function of frequency, and the specular reflection has the same pulse shape as the incident wave. The Rayleigh velocity is a function of frequency over the range $0 \leqslant ka \simeq 30$, and thus is not constant over the range of frequencies represented by the incident pulse. Changes in pulse shape between the Rayleigh wave and the incident and specularly reflected waves are expected. Figures 14 and 15 are digital representations of the experimental recovered signal sampled at intervals of 0.02 μs, or at about 100 points per cycle of the received pulse. No two digital representations are exactly alike, but with so many points having been taken, the only differences that are noticeable occur in the flattening effects at some of the peaks and valleys. In amplitude measurements these effects are negated by averaging many measurements. The phase shift of 180° was determined by comparing the pulse cycles labeled (1) and (P) in Fig. 14. Despite the slight change in pulse shape as discussed above, the beginning of the pulse labeled (1) and the characteristically large amplitude at the center of the pulse labeled (P) are present in both the specular and Rayleigh echoes and are 180° out of phase for these two echoes. The attenuation of the circumferential Rayleigh wave at $ka = 10.4$ is measured from Fig. 15 to be $\alpha_R = 1.69$ Np/revolution where one revolution equals a travel path of one circumference. Computations of the response of an aluminum cylinder to an incident pulse should allow examination of the Rayleigh circumferential wave properties at even lower ka values than $k_o a = 10.4$ (the experimental conditions for Fig. 15). A one-cycle pulse can be programmed as the incident pulse for a computer even though it is not readily attainable in a laboratory with normally available transducers. The computation of the reflection of a single-cycle pulse centered at $k_o a = 8.8$ is shown in Fig. 16. Again specular and Rayleigh wave echoes are 180° out of phase. The group velocity obtained from Fig. 16 is $c_R^g/c = 2.4$, which compares to $c_R^g/c = 2.3$ estimated from Ref. 21.

Estimates of the expected Rayleigh wave attenuation due to radiation into the water can be made from the flat surface formula given by Dransfeld [29]:

$$\alpha_R = \frac{\rho c}{\rho_s \, c_R \, \lambda_R}. \qquad (30)$$

The nonsubscripted variables refer to water, and the subscript R refers to the Rayleigh wave. Just as the limiting velocity is not achieved until $ka > 30$ for aluminum, it is not expected that computations made from

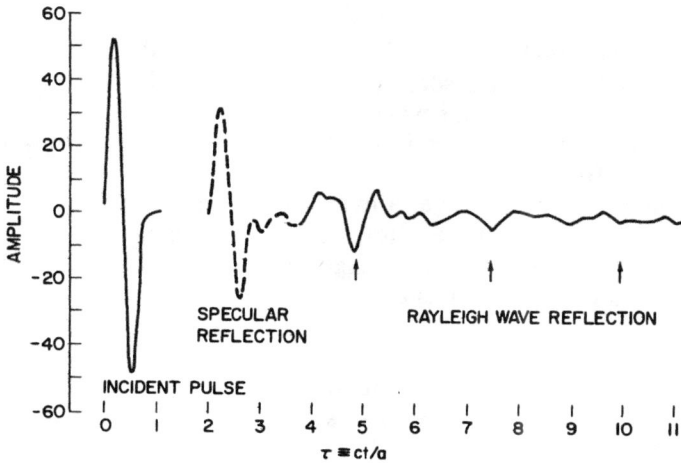

Fig. 16 — Computation of the echoes scattered by an aluminum cylinder at $k_o a = 8.8$.

Eq. 30 be exact for low ka. For aluminum the flat surface values for the variables in Eq. 30 are

$$\rho/\rho_s = 0.37; \; c/c_R = 0.55, \text{ and } \lambda_R = 1.8\lambda$$

yielding

$$\alpha_R = \frac{0.113}{\lambda} = \frac{0.113k}{2\pi} \; cm^{-1} \tag{31a}$$

or

$$(2\pi a) \, \alpha_R \equiv \alpha_R' = (0.113) \, (ka) \tag{31b}$$

where α_R' is dimensionless.

Equation 31b gives $\alpha_R' = 1.0$ at $ka = 8.8$ in comparison with the value $\alpha_R = 1.0$ obtained from Fig. 16. At $ka = 10.4$, Eq. 31b gives $\alpha_R = 1.2$ compared with $\alpha_R' = 1.7$ obtained from the measurement in Fig. 15. The results indicate that a reasonable estimate of α_R can be made using the flat surface attenuation formula, even at low ka. The empirical observations given in Figs. 14 and 15 are the first observations of the true Rayleigh circumferential wave. Previous observations of circumferential waves on solid elastic cylinders have been made at ka values in the range $40 \leqslant ka \leqslant 400$ [25,26]. This high ka range was chosen because of the ease of pulse separation of any circumferential effects and, also, because Rayleigh-like circumferential wave properties would have more closely approximated the flat surface Rayleigh wave at high ka. The circumferential waves that were observed on aluminum were identified as Rayleigh [25] or "Rayleigh type" [26] waves. Further analysis of high ka circumferential results lead to an explanation of the effects seen in Refs. 25 and 26 in terms of multiple internal reflections of shear waves in the cylinder [27]. This view was further supported by theoretical calculations of the circumferentially radiated wavefronts which result from internal reflections [28]. The subject has remained however a matter of some conjecture, [25-28,34], but the results given here show that the Rayleigh wave has large enough amplitude to contribute significantly to the backscattering by the cylinder only at ka values below 20.

The excitation of the (2,1) resonance, which corresponds to the ka value at which the cylinder circumference is two wavelengths of the Rayleigh circumferential wave, marks the highest ka at which the Franz wave contribution can be isolated. A comparison of the form function curves for the rigid cylinder (Fig. 2), the aluminum oxide cylinder (Fig. 8a), and the aluminum cylinder (Fig. 12) shows that, in all these cases, there exists a region where the behavior of the form function is purely rigid, i.e., dominated by the interference of specular reflection and the Franz wave. For an aluminum cylinder this behavior exists up to $ka \simeq 4.5$, where the (2,1) resonance minima begin. For aluminum oxide the specular plus Franz wave behavior persists up to $ka \simeq 9.9$, and the resonance null at $ka = 9.9$ in Fig. 8a is the (2,1) mode for aluminum oxide. Similar curves were computed for copper, brass, and tungsten carbide, and in all cases the generation of the (2,1) mode marks the end of the purely rigid behavior. The ka value at which the purely rigid behavior will end for a cylinder of a given material can be inferred by using aluminum as a reference. The (2,1) mode will be excited at

$$ Z_{2,1}(\text{material}) = Z_{2,1}(Al) \cdot \frac{c_R(\text{material})}{c_R(\text{aluminum})}. \qquad (31c) $$

The Rayleigh wave speeds used in Eq. 31c are flat surface numbers, and the equation assumes that the effect of curvature is the same for all materials, i.e., that the flat surface limit is reached when the circumference is greater than 10 Rayleigh wavelengths. Using the simple formula given in Eq. 31c, the ka position of the (2,1) resonance was predicted to within 1% for the materials discussed above.

SEPARATION INTO RIGID BACKGROUND AND RESONANCE PARTS

Junger and Feit [40] qualitatively considered the resonance features of the acoustic scattering by elastic bodies. They showed that resonances should appear where the sum of the mechanical and radiation impedance goes to zero. The acoustic scattering from a submerged aluminum cylinder can be described in terms of a rigid background term, with a resonance contribution superimposed on that background. The observed phenomenon can be formalized by using the methods of nuclear scattering theory, so that mathematically explicit forms for the resonances and background, as well as expressions for the resonance widths, are obtained. It is necessary to establish the nature of the background before the formalism is developed because a parallel formalism could have been developed, using a soft or an intermediate background, which would have had no physical significance for the problem of the solid metal cylinder. The conclusions and formalism developed for the scattering from an aluminum cylinder apply to the scattering from any submerged solid cylinder whose density is greater then that of the surrounding fluid and whose shear and compressional wave speeds are greater than the speed of sound in the fluid.

It has been shown in the previous section that the irregular characteristics of the form function for solid elastic cylinders are related to the normal modes of free vibration of the body, and these resonances often occur over a narrow frequency range as was seen in Fig. 12. It will be demonstrated in Figs. 17 through 20 that the resonances are superimposed on a background of reflection resulting from rigid boundary conditions, so that the elastic body can be regarded as a rigid body except in the frequency interval over which the resonances occur. In Fig. 17, the individual partial waves, $|f_n|$, from $n = 0$ through $n = 5$ are plotted vs ka for an aluminum cylinder. The f_n are described by Eq. 7c. The curves in Fig. 17 show that the amplitude of the individual partial waves $|f_n|$ have distinctive behavior in regions where the resonances occur. The eigenvalues l, are labeled along the curve in Fig. 17. The individual partial waves for both the infinitely rigid and the infinitely soft cylinder have no resonance irregularities, as seen in Fig. 18(a) and 18(b) respectively. The demonstration that the individual partial waves for a metal elastic cylinder consist of resonances superimposed on a rigid background is seen in Fig. 19. Here the $|f_2|$ and $|f_3|$ individual partial waves for rigid, soft, and elastic boundary conditions are plotted. It is clearly observed in Fig. 19 that the rigid and elastic curves are the same except in the region where resonances occur. The resonances for $l \geqslant 2$ are narrow resonances; the $l = 1$ eigenvalue which corresponds to the R_1 or Rayleigh surface wave is a broader resonance. A more dramatic example of the relationship between the elastic and the rigid solutions for a solid cylinder is seen in Fig. 20. Here the quantity $|f_2(\pi)^{\text{elastic}} - f_2(\pi)^{\text{rigid}}|$ is plotted vs ka, and the $(2,1)$, $(2,2)$, and $(2,3)$ resonances are clearly isolated. As was noted,

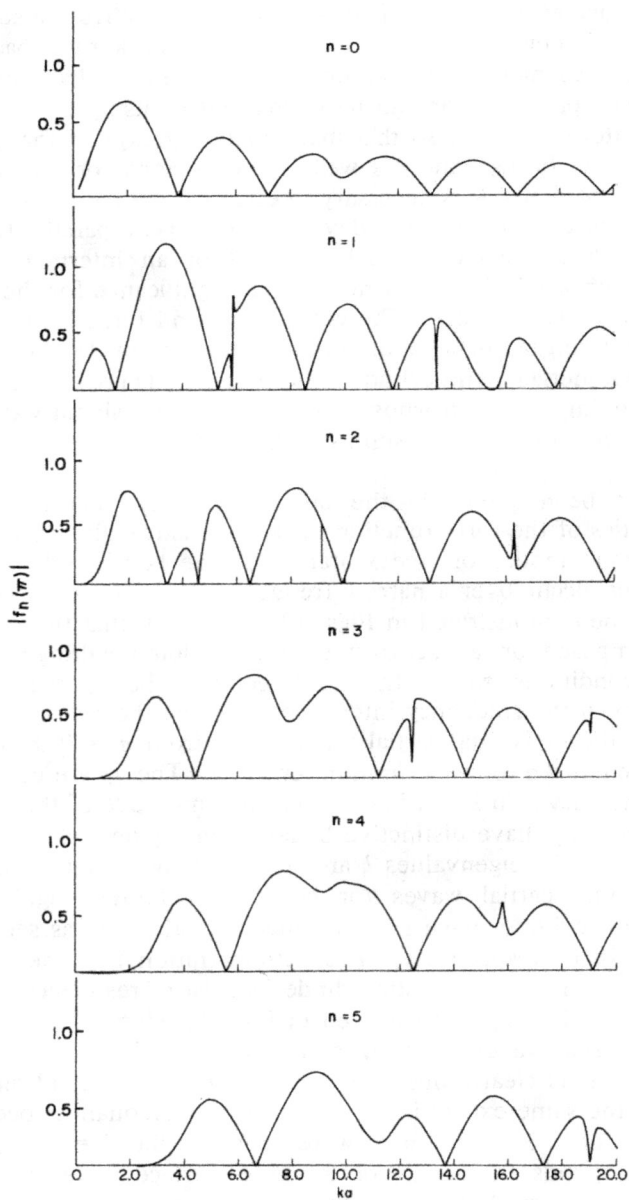

Fig. 17 — The individual partial wave amplitudes from
$n = 0$ to $n = 5$ for the aluminum cylinder.

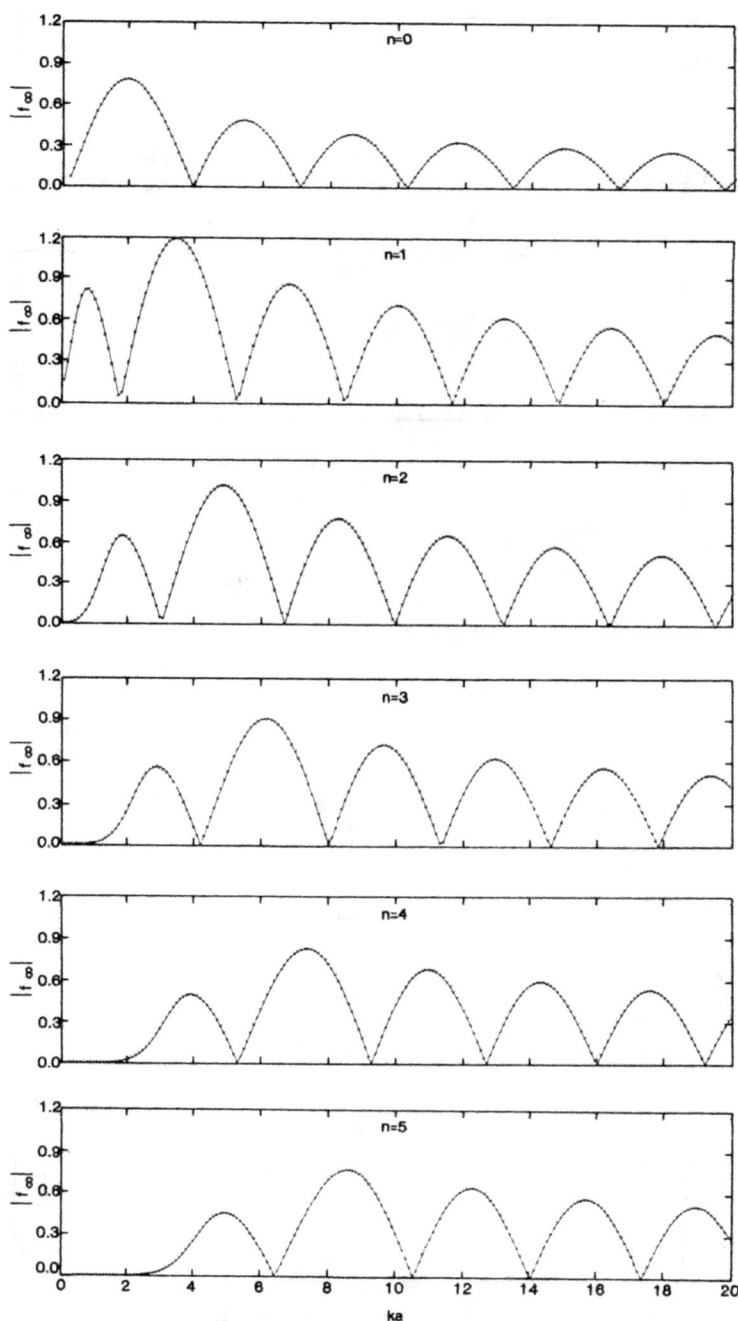

Fig. 18(a) — The individual partial wave amplitudes from
$n = 0$ to $n = 5$ for a rigid cylinder.

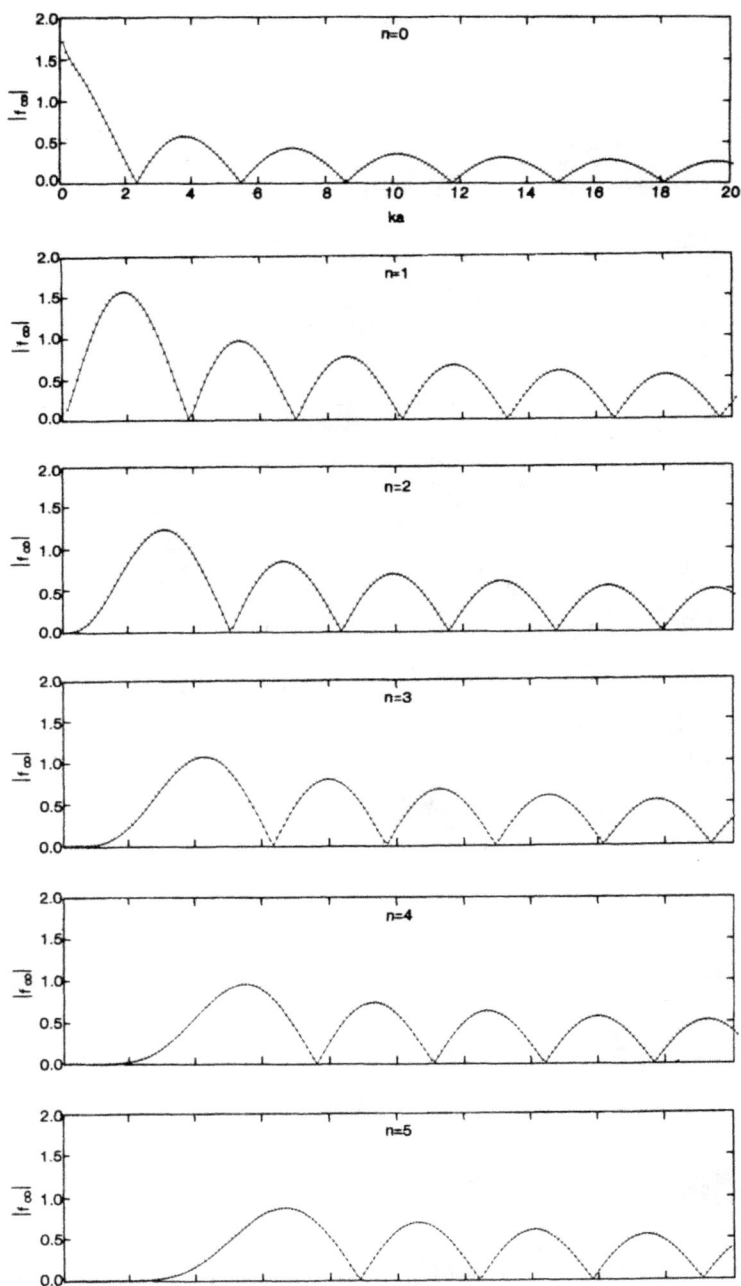

Fig. 18(b) — The individual partial wave amplitudes from
$n = 0$ to $n = 5$ for a soft cylinder.

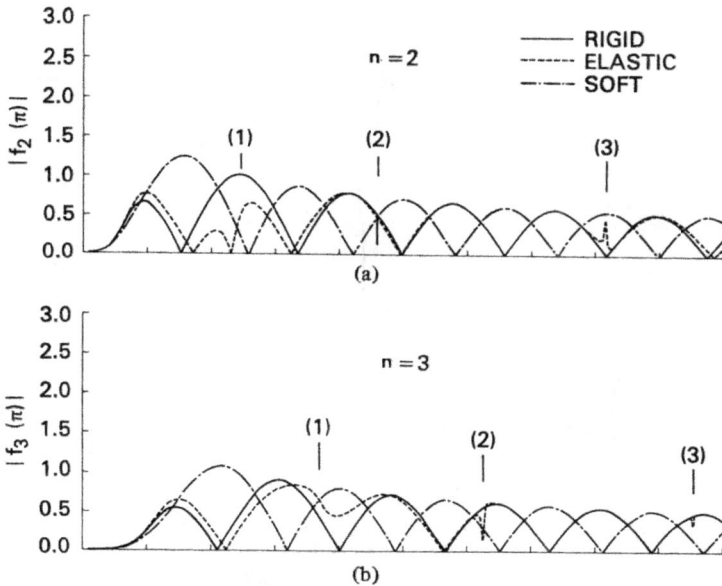

Fig. 19 — A comparison of the rigid (———), soft (— · —) and elastic (- - -) partial wave amplitudes for (a) $n = 2$ and (b) $n = 3$.

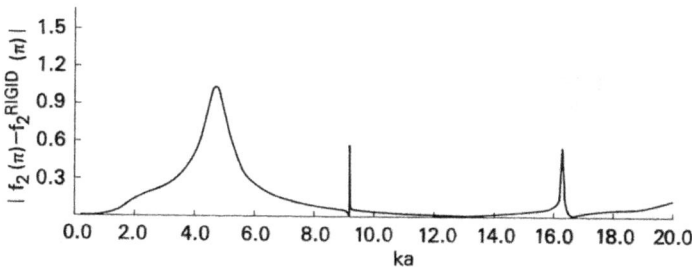

Fig. 20 — A plot of the difference between the elastic and rigid partial wave amplitudes for $n = 2$, over the range $0.2 \leqslant ka \leqslant 20$.

the $(2,1)$ resonance and in fact the $(n, 1)$ resonances in general are broader than the narrow resonances which occur for higher order eigenfrequencies, i.e. $l > 1$. A method of computing resonance widths will be described later. The results seen in Figs. 17 through 20 are, of course, not restricted to the backscattered direction.

Fig. 21a shows a bistatic form function $|f_\infty(\theta)|$ curve for a rigid cylinder at $ka = 12.53$, which is the ka value at which the $(3,2)$ resonance occurs in the aluminum case. In Fig. 21b this bistatic form function for a rigid cylinder is compared with the bistatic form function

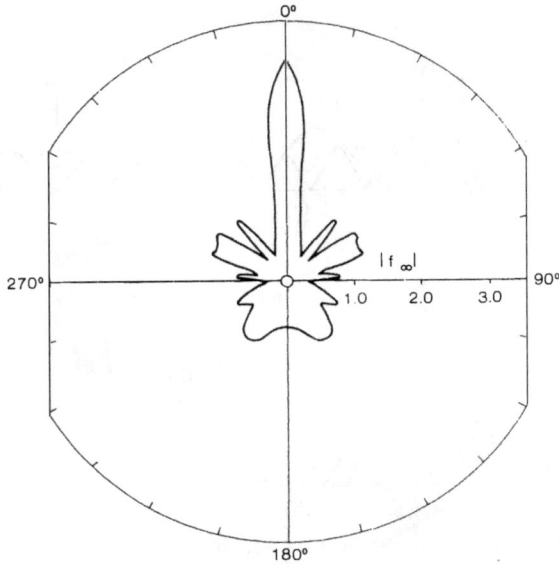

Fig. 21(a) — A polar plot of $|f_\infty|$ vs θ for a
rigid cylinder at $ka = 12.53$.

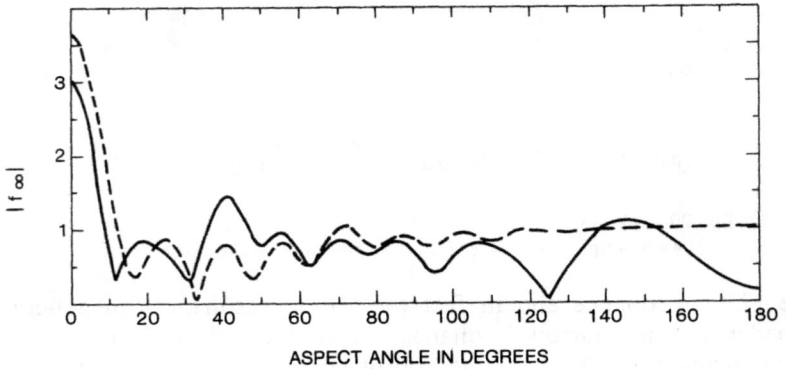

Fig. 21(b) — A comparison of $|f_\infty(\theta)|$ for rigid and
aluminum (—) cylinders at $ka = 12.53$.

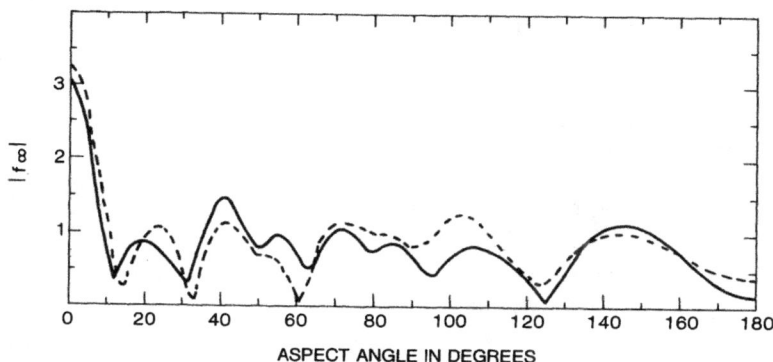

Fig. 21(c) — The same comparison as (b) except that the $n = 3$ term in the rigid series is replaced by the $n = 3$ elastic term.

for an aluminum cylinder. Here the results are plotted on a linear rather than polar plot, and due to the symmetry apparent in Fig. 21a only the range $0° \leqslant \theta \leqslant 180°$ is plotted. The rigid and elastic solutions plotted in Fig. 21b were obtained with 23 terms ($n = 0$ through 22 from Eq. 7a). If the single $n = 3$ term from the elastic solution is substituted for the $n = 3$ term in the rigid solution, the result seen in Fig. 21c is obtained. A similar procedure was carried out by Vogt and Neubauer [41] for a sphere in a monostatic geometry. In Fig. 21c the exact bistatic solution for the aluminum cylinder at $ka = 12.53$ is compared with the hybrid solution formed by taking 22 terms from the solution for the rigid cylinder ($n = 0$ through 2 and $n = 4$ through 22) and adding the $n = 3$ term from the solution for the elastic cylinder. The form function resulting from the modified solution for the rigid cylinder vs aspect curve and the same plot for the elastic form function are seen in Fig. 21c as very similar. The differences between Figs. 21b and 21c are most noticeable in the backscattered half space, $90° < \theta < 270°$ (recall that by symmetry the results seen in Fig. 21c from $90° < \theta < 180°$ are exactly the same as the results between $180° < \theta < 270°$). Figure 21 again indicates the probability that the scattering from an aluminum cylinder can be treated as resulting from a rigid background term with resonances superimposed. It indicates further that this behavior is not limited to the single monostatic angle ($\theta = 180°$) but can be utilized at any bistatic aspect angle.

These results indicate that the solution to the scattering from solid elastic bodies as expressed in Eqs. 4, 5, and 7 should be separable into two terms, the rigid background term and the resonance term.

Consider the scattering function

$$S_n \equiv \exp(2i\,\delta_n). \tag{32}$$

Now the solution describing the scattering by an elastic cylinder may be written in the form familiar to nuclear reaction theory [43] as

$$p_s = 1/2 \sum_{n=0}^{\infty} \epsilon_n\, i^n (S_n - 1)\, H_n(kr)\, \cos n\theta. \tag{33}$$

The δ_n of Eq. (32) are called scattering phase shifts, and a comparison of Eqs. 1, 3, and 33 shows that

$$S_n - 1 = -2G_n(ka). \tag{34}$$

For the case of a rigid cylinder, defining ξ_n as the scattering phase shift for a rigid cylinder and defining

$$S_n^{(R)} \equiv \exp(2i\,\xi_n) \tag{35a}$$

as the scattering function for a rigid cylinder using Eqs. 34 and 13, leads to

$$S_n^{(R)} - 1 = -2G_n^{(R)}(ka) \tag{35b}$$

and thus

$$S_n^{(R)} = \frac{H_n^{(2)'}(Z)}{H_n'(Z)}. \tag{35c}$$

If the rigid portion of the scattering function as given in Eq. 35c is factored out of the expression for the elastic scattering function Eq. (34), we have using Eqs. 3 and 35c

$$S_n = S_n^{(R)} \left\{ \frac{L_n^{-1} - z_2^{-1}}{L_n^{-1} - z_1^{-1}} \right\}. \tag{36}$$

The L_n's were previously defined in Eq. (1a), and the z's are defined by

$$z_1^{-1} = (Z)\, \frac{H_n(Z)}{H_n'(Z)} \tag{37a}$$

and

$$z_2^{-1} = (Z)\, \frac{H_n^{(2)}(Z)}{H_n^{(2)'}(Z)} \tag{37b}$$

the primes in Eqs. 37 and 35c represent derivatives with respect to the argument, and as defined previously H and $H^{(2)}$ are Hankel function of the first and second kind respectively. The quantities z_i^{-1} of Eq. 37 can be separated into real and imaginary parts:

$$z_{1,2}^{-1} = \Delta_n \pm is_n \tag{38}$$

with

$$\Delta_n = 1/(Z) \frac{J_n(Z) J_n'(Z) + Y_n(Z) Y_n'(Z)}{[J_n'(Z)]^2 + [Y_n'(Z)]^2} \tag{39}$$

and

$$s_n = \frac{-2}{\pi Z^2} \left\{ \frac{1}{[J_n'(Z)]^2 + [Y_n'(Z)]^2} \right\}. \tag{40}$$

Equation 36 may be rewritten using Eq. 38 as

$$S_n = S_n^{(R)} \left\{ \frac{L_n^{-1} - \Delta_n + is_n}{L_n^{-1} - \Delta_n - is_n} \right\}. \tag{41}$$

The linear approximation method of nuclear resonance theory is used in which the resonance frequencies Z_n are defined by the condition

$$L_n^{-1}(Z_n) = \Delta_n. \tag{42a}$$

The quantity $(L_n^{-1} - \Delta_n)$ is assumed to be linearly varying with frequency so that it can be expanded in a Taylor series in Z in the vicinity of any one of the resonance frequencies:

$$L_n^{-1} \simeq \Delta_n + \beta_n(Z - Z_n). \tag{42b}$$

A resonance width is defined by

$$\Gamma_n = \frac{-2s_n}{\beta_n} \tag{43}$$

and the scattering-function S_n, may be rewritten in resonance form as

$$S_n \equiv e^{2i\delta_n} = S_n^{(R)} \left\{ \frac{Z - Z_n - i\Gamma_n/2}{Z - Z_n + i\Gamma_n/2} \right\}. \tag{44}$$

From Eq. 44 the S_n are seen to have resonance poles at the complex frequencies $Z = Z_{pole}$ given by

$$Z_{pole} = Z_n - i\Gamma_n/2 \tag{45}$$

and a resonance zero, $Z = Z_{zero}$, at

$$Z_{zero} = Z_n + i\Gamma_n/2. \tag{46}$$

The resonance width Γ_n defined in Eq. 43 is a positive quantity. Thus Z_{pole} is located in the lower half of the complex Z-plane a distance $(1/2)\Gamma_n$ from the real axis, and Z_{zero} is located in the upper half plane at the same distance above the axis.

The quantity $S_n - 1$ which appears in Eq. 33 can be written, recalling the definition in Eq. 32, as

$$S_n - 1 \equiv 2i \, e^{i\delta_n} \sin \delta_n. \tag{46a}$$

Using Eq. 36 and the expressions for $z_{1,2}^{-1}$ given in Eq. 38 $(S_n - 1)$ can be rewritten in the form

$$S_n - 1 = 2i \, e^{i\xi_n} \left\{ \frac{S_n}{L_n^{-1} - \Delta_n - is_n} + e^{i\xi_n} \sin \xi_n \right\} \tag{46b}$$

or, using Eqs. 44 and 46a, $S_n - 1$ may be written in the resonance form

$$\frac{S_n - 1}{2i} = e^{2i\xi_n} \left\{ \frac{1/2 \, \Gamma_n}{Z_n - Z - 1/2 \, \Gamma_n} + e^{-i\xi_n} \sin \xi_n \right\}. \tag{47}$$

The individual partial waves, $f_n(\theta)$, of Eq. 7c can thus be written, using Eq. 47, as

$$f_n(\theta) = \frac{2i \, \xi_n}{(i\pi \, ka)^{1/2}} \, e^{2i\xi_n} \left[\frac{1/2 \Gamma_n}{Z_n - Z - 1/2 \, i\Gamma_n} + e^{-i\xi_n} \sin \xi_n \right] \cos(n\theta). \tag{48}$$

The first term of Eq. 48 represents the resonance contribution, and the second term represents the rigid boundary contributions. The representation of $f_n(\theta)$ given in Eq. 48 shows that the complex eigenfrequencies of the scatterers are the locations of the resonance poles in the complex frequency plane, whose real parts determine the resonance frequencies in the scattering amplitudes.

A consideration of the field within the elastic cylinder can also be made in light of the above results. The displacement $\mathbf{u}$ within the cylinder is represented by a scalar potential Ψ and a vector potential $\mathbf{A}$, and is written

$$\mathbf{u} = -\nabla \Psi + \nabla \times \mathbf{A} \tag{49}$$

with solutions [20]

$$\Psi = \sum_{n=0}^{\infty} \epsilon_n \, i^n \, C_n \, J_n(k_L \, r) \cos n\theta \tag{50a}$$

and

$$A_Z = \sum_{n=0}^{\infty} \epsilon_n \, i^n \, B_n \, J_n(k_T \, r) \sin n\theta. \tag{50b}$$

The subscripts L and T are longitudinal and shear respectively. The coefficients C_n and B_n are given by

$$C_n = \frac{2i}{\pi p \, \omega^2} \; \frac{1}{ZH_n'(Z) \, D_n^{(1)}} \; \frac{a_{33}}{z^{-1} - L_n^{-1}} \qquad (51a)$$

and

$$B_n = \frac{2i}{\pi p \, \omega^2} \; \frac{1}{ZH_n'(Z) \, D_n^{(1)}} \; \frac{a_{32}}{z^{-1} - L_n^{-1}}. \qquad (51b)$$

The expansion of L_n^{-1} in Eq. 42 leads in to the resonance expressions for Ψ and A_z, which are

$$\Psi = \frac{2}{i\pi \, p \, \omega^2} \sum \frac{\epsilon_n \, i^n}{\beta_n} \; \frac{a_{33}}{ZH_n'(Z) \, D_n^{(1)}} \; \frac{J_n(k_L r) \cos n\phi}{Z - Z_n + 1/2 \, i\Gamma_n} \qquad (52a)$$

and

$$A_z = \frac{2}{i\pi \, p\omega^2} \sum_{n=0}^{\infty} \frac{\epsilon_n \, i^n}{\beta_n} \; \frac{a_{32}}{ZH_n'(Z) \, D_n^{(1)}} \; \frac{J_n(k_T r) \sin n\phi}{Z - Z_n + 1/2 \, i\Gamma_n}. \qquad (52b)$$

Equations 52a and 52b show that the internal solutions are of a pure resonance form only. This is as expected, since by definition a rigid body is impenetrable; thus no rigid background term is expected for the internal solution.

CIRCUMFERENTIAL WAVES ON CYLINDRICAL SHELLS

Numerous empirical observations of circumferential radiation from cylindrical shells exist in the literature [11-15,25-27], but these have left many serious voids in the understanding of circumferential waves as well as erroneous information concerning the properties of the waves. A connection can be made between circumferential waves and the exact Rayleigh series solution. It can be demonstrated that the Lamb wave dispersion curves on plates predict the range of possible excitation of circumferential waves and that the velocity of circum-ferential waves may be obtained directly from the form function vs ka curves. Lamb dispersion curves also give immediate knowledge of the ka region over which a particular circumferential mode is significant. Calculations of the backscattered echoes from shells can be used to obtain curves relating the amplitude of the specular and circumferential contributions as a function of ka, and, contrary to earlier literature, the circumferential waves will be shown to be of most importance in the low ka region generally avoided in that work. A target classification scheme has been proposed in the literature which relied on the assump-tion that a hollow shell acts as a soft body, in that its specular reflection is 180° out of phase with an incident wave [44]. By consideration of the interference between specular and circumferential radiation, the actual *ka* range over which such a hypothesis is valid is determined.

Barnard and McKinney [11] were the first to observe backscattered acoustic echoes with circumferential properties. This observation was a significant contribution done in the absence of guiding theory. They attempted to link the observed acoustic phenomenon with the geometric diffraction phenomenon observed by Franz and Doypermann [18] in the electromagnetic domain; however, the analogy broke down, since the acoustic phenomenon was a predominately elastic effect, the geometry serving in the capacity of a waveguide. Horton, et al. [13] made the initial attempt to relate circumferential waves on cylindrical shells to the elastic properties of flat plates. They compared their observation of a circumferential wave on an aluminum shell to a theoretically computed flexural plate mode and found a 10% difference between measured and predicted velocities. A similar comparison was attempted with a brass shell [13], but the circumferential wave could not be excited. The *ka* range considered was $21 < ka < 38$, with $b/a = 0.96$, where b and a are the inner and outer shell radius respectively. The circumferential wave observed by Horton et al. will be related in this work to the first antisymmetric Lamb mode, whose properties and *ka* region of possible excitation are discussed later. Diercks, et al. [12] made empirical observations of circumferential waves on both brass and aluminum shells near $ka \simeq 50$, with $b/a = 0.96$. They established the existence of circumferential waves with two different group velocities. The faster velocity wave was called a longitudinal mode, and the lower velocity wave was called the flexural mode. This paper by Diercks et al. [12] was significant in that it was the first to clearly state that more than one circumferential mode existed. Their conclusions, concerning which mode is dominant and whether both modes can be simultaneously excited, were thought to be general but are limited strictly to the conditions of the observations. It has been demonstrated [31] that the so-called longitudinal mode is actually many different modes, and therein lies much of the confusion about the frequency range of excitation and/or dominance of a particular mode. Goldsberry [14] demonstrated that the circumferential waves observed previously [11-13] would reflect from slits cut in the shell. He called the wave with lower group velocity a low frequency wave, and the faster wave a high frequency wave. Again these generalizations do not survive beyond his experimental conditions. The ranges he considered were $32 < ka < 38$ (flexural, slow, low frequency wave) and $70 < ka < 76$ (compressional, fast, high frequency wave). The b/a was 0.96. Überall et al.

predicted circumferential waves of different types on rigid [20] and elastic cylinders [21] and shells [45]. These predicted wave types were Franz-type waves [20] with properties similar to the electromagnetic case of Franz [16], and R-type or Rayleigh-type [21,45] waves, which depend on the elastic properties of the target. Neubauer [23] empirically isolated the Franz-type wave on a solid elastic cylinder in water, and Harbold and Steinberg [24] isolated the wave on a rigid cylinder in air.

Bunney, et al. [25] used narrow beam sources to illuminate cylindrical shells over narrow ranges of incidence angles and "directional" receivers to observe circumferential-wave radiation. The ka range they considered was between $50 < ka < 320$ with $b/a = 0.95$. Many observations of a low velocity circumferential wave were compared to the antisymmetric Lamb mode, and the single observation of a higher velocity mode was related to the symmetric plate mode.

Neubauer and Dragonette [27] and Dragonette [46] empirically demonstrated that the velocity of the observed circumferential waves on cylindrical shells [27] and the velocity of Lamb waves on plates [46] could be predicted by considering guided wave propagation within cylindrical shells and plates. Dragonette [46] also established that Lamb modes were most easily excited in the frequency thickness regions where the phase velocity reached a constant plateau. This result is significant in a consideration of the so-called fast circumferential wave on cylindrical shells and was the basis for correcting some erroneous conclusions in the literature.

Shirley and Diercks [47] compared measured and predicted values of the steady state response of spherical shells over the range $25 < ka < 65$ with $b/a = 0.95$. Differences of the order of 10 dB, or 300% in pressure amplitude, were found between theory and prediction, but similarities in shape between the theoretical and empirical curves were observed.

Horton and Mechler [15] attempted to measure phase velocity of circumferential waves on aluminum cylindrical shells by setting up a long pulse or steady state interference pattern between the successive circumferential pulses. The significance of their paper was that it offered a possible approach to phase velocity determination which, as will be discussed later, is a difficult parameter to obtain when waves are excited on a curved surface.

Figure 22 shows the geometry of the cylindrical shell problem. It is similar to Fig. 1 except that the target now has a finite thickness h

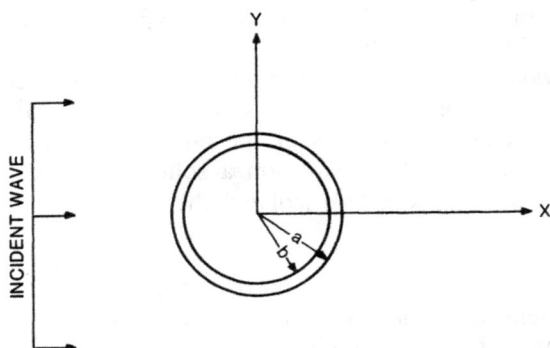

Fig. 22 — The geometry used in the solution to the
reflection from a cylindrical shell.

given by $h = a - b$, where b is the inner radius and a is the outer
radius. The shell is air filled. An experimental observation of circumferential waves using a pressure sensor is seen in Fig. 23. Here the target is a stainless steel cylindrical shell with $b/a = 0.96$. The radius of
the shell is 1.27 cm, and the center frequency of the pulses seen in Fig.
23 is 1.5 MHz, leading to $k_o a = 80$. The specular reflection is not
shown, since it is 40 dB greater than the largest echo seen in Fig. 23,
and was gated out of the return so that the echoes seen in the figure
could be amplified to the maximum extent for display. The
source/receiver is located 15 diameters from the target. The backscattered echo seen in Fig. 23 was digitized at the rate of 13 points per
cycle and stored on magnetic tape. The first, third, and all successive
echoes in Fig. 23 result from a circumferential wave which circumnavi-
gates the cylinder, with little attenuation, continually radiating into the
water as it travels. The second echo in Fig. 23 is the result of a second
type of circumferential wave so highly attenuated at the ka and/or kh
value of this experiment that only its first traversal around the cylinder
is observed before it attenuates into the noise. (Observation of Fig. 23
alone would not allow identification of the second echo as a circumferential wave. This identification was based on many experimental
measurements, some of which will be given below.) Measurements of
the circumferential velocity of the persistent series of equally spaced
echoes (1-7) in Fig. 23 straightforward. The circumferential group
velocity c_g^* is obtained from Fig. 23 by

$$c_g^* = \frac{2\pi a}{\Delta t}$$

where Δt is the time between echoes and $2\pi a$ is the circumference of
the shell. The measured value from Fig. 23 is $c_g^* = 5.48 \times 10^5$ cm/s or

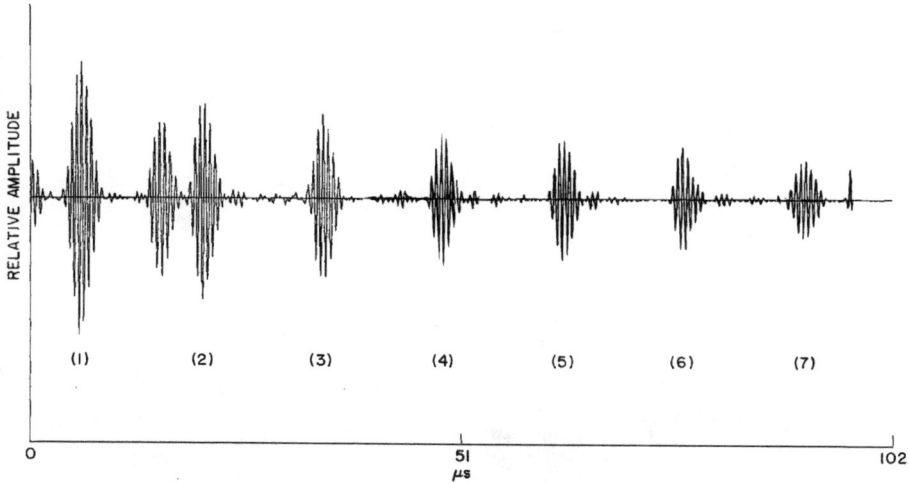

Fig. 23 — Experimental observation of circumferential waves on a
stainless steel shell at $k_o a = 80$.

$c_g^*/c = 3.7$. This value for c_g^* identifies this wave as that which had
been called the "fast", "high frequency", or "compressional" wave [13-
15]. As will be discussed, these names can be misleading or in error.
Attenuation measurements from Fig. 23 are also straightforward. The
successive amplitudes from Fig. 23 are plotted on semilog paper in Fig.
24, yielding an attenuation of

$$\alpha_z^* = 0.14 \text{ Np/revolution.}$$

The use of the digitizing procedure and display makes possible observation of the individual cycles of the successive echoes in Fig. 23. The
empirical observations in Figs. 25 through 28 are photographs of scope
traces. Time scales needed to show many successive echoes do not
allow observation of the individual cycles within the echo.

Figures 25, 26, 27, and 28 show further results of reflection measurements on steel shells, and the experimental conditions and results
are summarized in Table 3, which also includes the results described
for Fig. 23. The hydrophone measurements seen in Figs. 23 and 25
through 28 show the acoustic reflection in the backscattered direction, θ
$= \pi$. Observation of the entire scattered field, $0 \leqslant \theta \leqslant 2\pi$ can be
obtained simultaneously by schlieren visualization. Figure 29 shows a
schlieren visualization of the scattered field of a stainless steel cylindrical shell of radius 0.9525 cm with $b/a = 0.95$, and $k_o a = 202$. The
incident pulse is seen in Fig. 29a, and the time sequence of photographs shows the scattered field at later times. The specular reflection

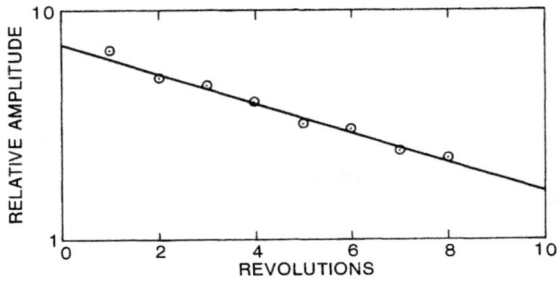

Fig. 24 — The amplitude of the circumferential
waves seen in Fig. 23 plotted on semilog paper.

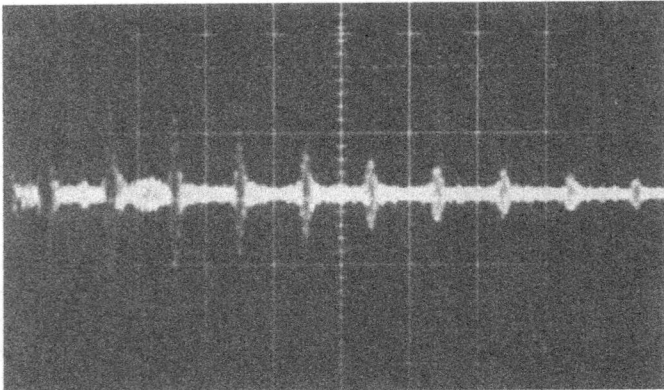

Fig. 25 — Experimental observation of a "fast" circumferential wave
at $k_o a = 11$, on a stainless steel shell. The time scale is
20μs/division; the amplitude scale is 5 mv/division.

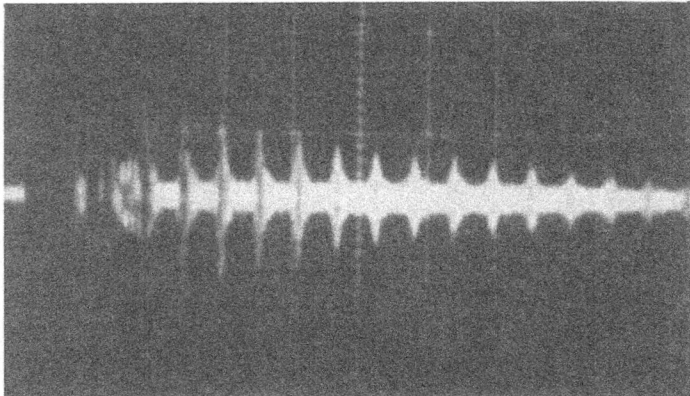

Fig. 26 — Experimental observation of a "fast" circumferential wave
at $k_o a = 69$, on a stainless steel shell. The time scale is
20μs/division; the amplitude scale is 2 mv/division.

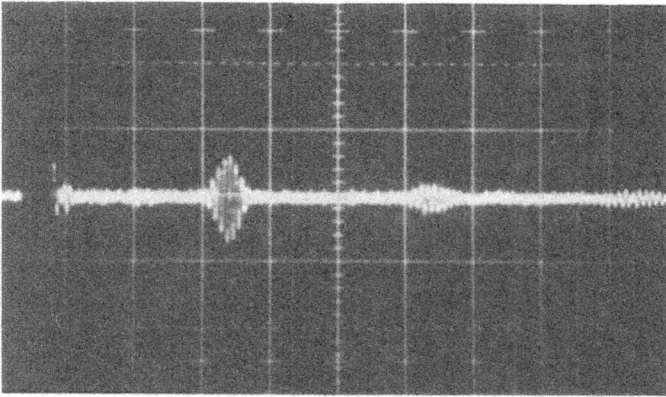

Fig. 27 — Experimental observation of a "slow" circumferential wave at $k_o a = 17$, on a stainless steel shell. The time scale is 20 μs/division; the amplitude scale is 200 mv/division.

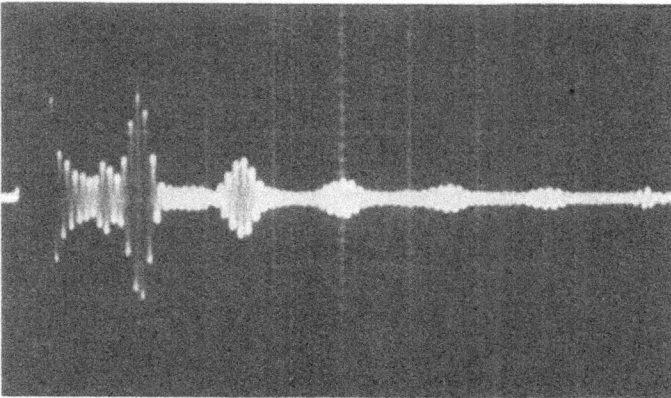

Fig. 28 — Experimental observation of a "slow" circumferential wave at $k_o a = 43$, on a stainless steel shell. The time scale is 20 μs/division; the amplitude scale is 2 mv/division.

Table 3 — Summary of Cylindrical Shell Observations

Figure	Radius (cm)	$k_o a$	b/a	c_g/c	α^* Np/revolution
23	1.27	80.8	.96	3.7	0.142
25	1.59	113.0	.97	3.5	0.209
26	0.9525	69.0	.95	3.6	0.126
27	0.9525	17.0	.95	1.35	1.000
28	1.905	43.0	.97	1.35	0.600

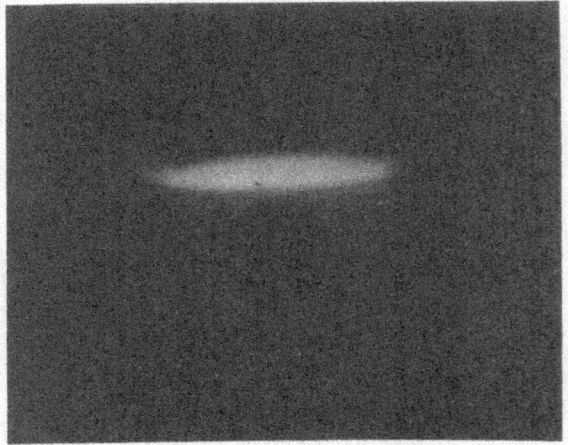

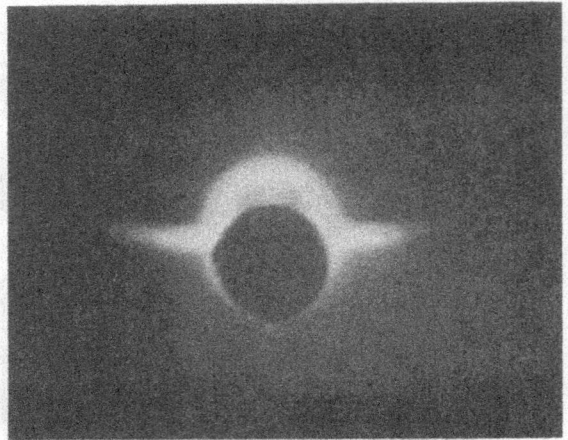

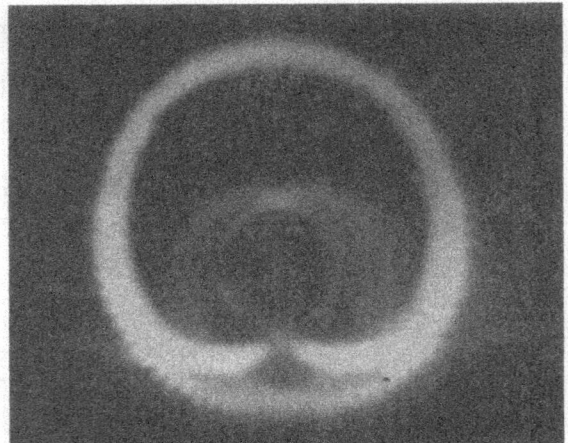

Fig. 29 — Schlieren visualization of a "fast" circumferential wave at $k_o a = 202$ on a stainless steel shell.

and the beginning of the radiation from a circumferential wave are seen in Fig. 29b, and Fig. 29c shows the reradiation from the first complete traversal of the circumferential wave into the backscattered direction. At the bottom of Fig. 29c the diffraction around the shell can also be observed. The group velocity of the wave seen in Fig. 29 is $c_g^*/c = 3.6$. In Fig. 29 the incident pulse insonified the entire cylinder, so that the same effect was generated on both the upper left and upper right quadrants of the cylinder.

The properties of the circumferential waves seen in Figs. 23 and 25 through 29 are similar to circumferential wave properties previously reported in the literature [11-15,25-27]. Specifically, circumferential waves with group velocities $c_g^*/c \simeq 3.6$ (Figs. 23, 25, and 26) and $c_g^*/c \simeq 1.3$ (Figs. 27 and 28) are observed, with the faster group velocity observed at the higher frequency (or higher ka) and the slower group velocity wave at the lower frequency. This simple interpretation of the results is, however, misleading and demonstrates some of the practical difficulties in a predominantly empirical approach to this problem, where broad generalizations are made, based on limited measurements. The experimental measurements here, and reported previously, are generally made at high ka values. At very low frequencies, or ka values below $ka = 20$, it is a practical impossibility to achieve short enough pulse lengths to separate circumferential waves with transducers generally available; hence, high ka measurements are made as a matter of necessity. This low-ka limitation on the isolation of separate echoes is especially true of high speed circumferential waves. In addition, most empirical measurements are made on shells with $b/a \leqslant 0.96$, since thinner shells are more difficult to fabricate and maintain. Finally, as a practical matter it is not possible to measure enough combinations of shells and frequencies to do a complete empirical study. This latter statement is true not only because of the low ka separation difficulty mentioned above but also because in the case of a curved shell there are two frequency variables. For a flat plate, frequency times thickness, fh may be considered a single variable. The radiation from a given plate may be examined as a function of fh, simply by varying the frequency of the incident pulse or continuous wave. A similar experiment on a curved shell is not as unambiguous. As the frequency of the incoming wave is varied, the fh or kh of the shell changes accordingly, but in addition the ka value changes. As will be seen, there are certain effects which are strong functions of ka, and others which depend almost entirely on kh. These had not previously been differentiated successfully in the literature, and this could not have been reasonably accomplished empirically. Because previous work has been limited to the high *ka* region, where the flat surface limit is approached, the *kh* variable has generally been considered most significant in all empirical observations of isolated circumferential waves. In fact the low *ka* region which has been avoided is the only *ka* region where a high velocity circumferential wave plays a significant role in the acoustic scattering by a cylindrical shell, as first shown by Dragonette [31].

Analysis of circumferential waves on cylindrical shells can best be accomplished by determining the relationship between the circumferential waves and exact steady state theory. In addition, Lamb theory for plates can be used to predict the possible ranges of excitation of circumferential waves on shells.

The geometry of the cylindrical shell problem was given in Fig. 22. The formulation of the exact normal mode solution to the scattering of sound by an elastic cylindrical shell exists in the literature and can be presented in a form similar to that of Eq. 1, which described the scattering from solid elastic cylinders [33]:

$$p_s(\theta) = -p_o \sum \epsilon_n (i)^n \left\{ \frac{J_n(Z) \, Q_n - Z \, J_n'(Z)}{H_n(Z) \, Q_n - Z \, H_n'(Z)} \right\} H_n(kr) \cos n\theta. \quad (53)$$

This expression differs from Eq. (1) only in the replacement of L_n, which involved the division of two 2-by-2 matrices, by Q_n, which involves the division of two, 4-by-4 matrices. The larger matrix results from the extra boundary condition on the surface $r = b$, and the expression for Q_n is

$$Q_n = \frac{\rho}{\rho_s} \frac{\begin{vmatrix} a_{21} & a_{22} & a_{23} & a_{24} \\ a_{31} & a_{32} & a_{33} & a_{34} \\ a_{41} & a_{42} & a_{43} & a_{44} \\ a_{61} & a_{62} & a_{63} & a_{64} \end{vmatrix}}{\begin{vmatrix} a_{11} & a_{12} & a_{13} & a_{14} \\ a_{31} & a_{32} & a_{33} & a_{34} \\ a_{41} & a_{42} & a_{43} & a_{44} \\ a_{61} & a_{62} & a_{63} & a_{64} \end{vmatrix}} \qquad (54)$$

with the matrix elements a_{ij} given in Chapter 5 [33]. Computations are given in Fig. 30 of the form function vs *ka* for stainless steel shells with $b/a = 0.99$ (Fig. 30a) and $b/a = 0.98$ (Fig. 30b). The curves cover the *ka* range $0.2 \leqslant ka \leqslant 50$. A similar set of curves for aluminum are given in Fig. 31, and the two figures demonstrate the similarity

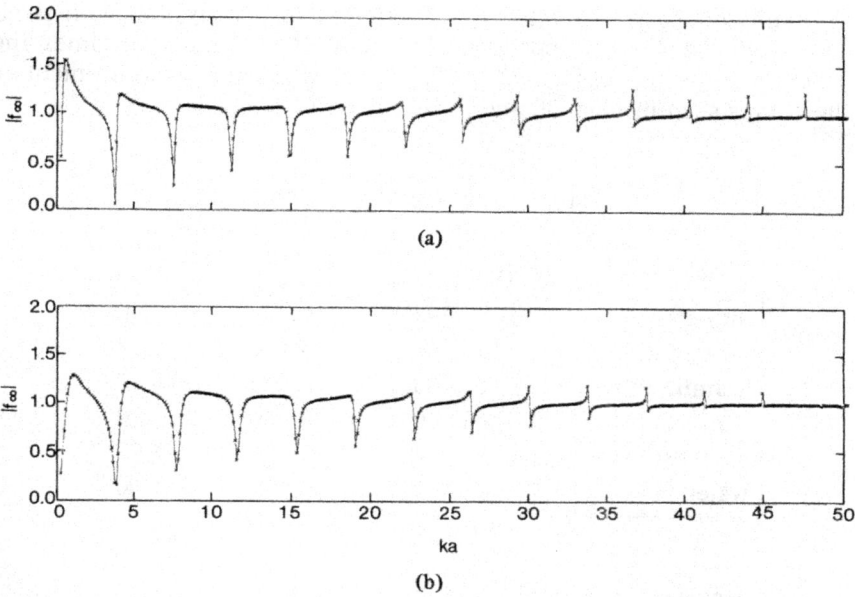

(a)

(b)

Fig. 30 — The form function for stainless steel shells over the range $0.2 \leqslant ka \leqslant 50$ for shell thickness of (a) 0.99 and (b) 0.98.

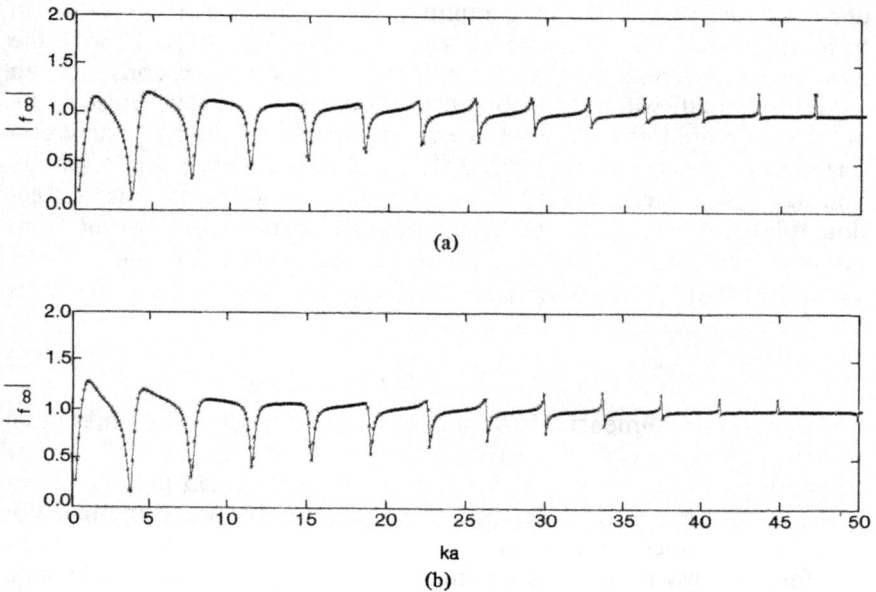

(a)

(b)

Fig. 31 — The form function for aluminum shells over the range $0.2 \leqslant ka \leqslant 50$ for shell thicknesses of (a) 0.99 and (b) 0.98.

of results obtained on metals quite different in density, but with shear and longitudinal wave speeds approximately twice and four times the water wave speed, respectively. The elastic constants used in obtaining the curves seen in Figs. 30 and 31 are given in Table 4.

Table 4 — Table of Constants

Material	c_L $(10^5$ cm/s$)$	c_T $(10^5$ cm/s$)$	ρ (g/cc)
air	00.343	0.000	0.00
aluminum	06.370	3.120	2.17
aluminum oxide	10.700	6.300	3.92
iron	05.950	3.240	7.70
stainless steel	05.5940	3.106	7.90
water	01.493	0.000	1.00

The curves in Figs. 30 and 31 are made up of the steady state interference of specular reflection and a single circumferential mode which makes many circumnavigations of the cylinder before attenuating into the noise. The above explanation of Figs. 30 and 31 can be demonstrated analytically by computing the response of the shell to an incident pulse. The computation is made using Eqs. 9 and 12 with the procedures described earlier. Figure 32 shows the response of an aluminum shell, with $b/a = 0.99$, to an incident acoustic pulse. The pulse is centered at a dimensionless frequency $k_o a = 10$. The backscattered echoes are seen in Fig. 32(a), and the incident wave is seen in Fig. 32b. The backscattered return is made up of the specular reflection followed by a series of equally spaced echoes which result from multiple circumnavigations of a circumferential wave. The ratio of the circumferential wave group velocity c_g^* to the water wave speed c is given by

$$c_g^*/c = \frac{2\pi}{\Delta\tau} \tag{55}$$

where $\Delta\tau$ is the dimensionless time between successive circumferential echoes. The result obtained from Eqs. 32 and 55 is $c_g^*/c = 3.7$. The deviations from $|f_\infty| = 1$ in Figs. 30 and 31 occur when the circumference, $2\pi a$, is an integral number of circumferential wavelengths. For ka values at which $2\pi a = n\lambda^*$, where λ^* is the wavelength of the circumferential wave and n is an integer, the long pulse or steady state interference of the circumferential waves gives a maximum contribution, since all add in phase with one another. At these ka values peaks

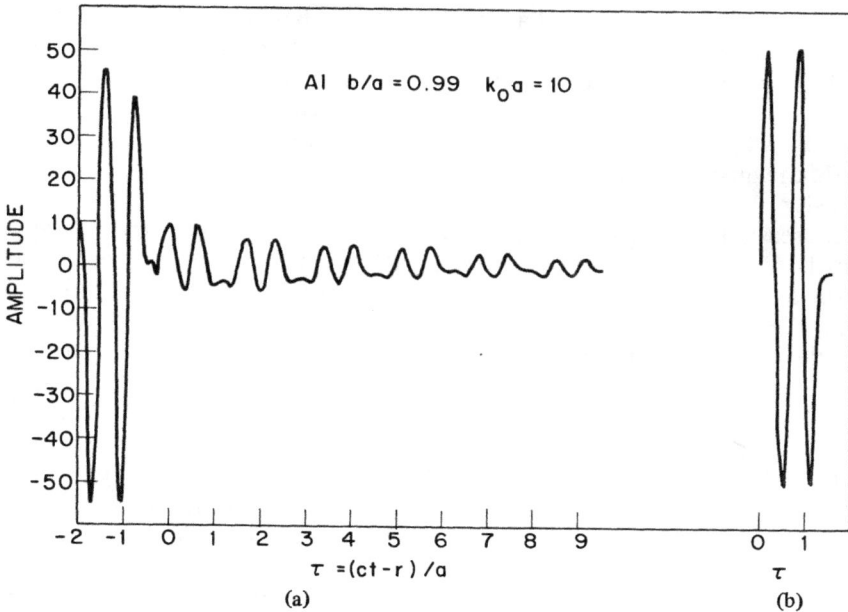

Fig. 32 — Computations of the echoes scattered when a short incident pulse impinges on an aluminum shell with $b/a = 0.98$. The pulse is centered at $k_o a = 10$. The scattered echoes are seen in (a) the incident wave in (b).

will occur in the form function if the specular reflection and circumferential waves are in phase and nulls will occur if they are out of phase. The ka difference, Δka, between the successive fluctuations in the $|f_\infty|$ vs ka curve are directly related to the circumferential wave phase velocity c_p^* by

$$c_p^*/c = \Delta ka \qquad (56)$$

which for Figs. 30 and 31 gives $c_p/c = 3.7$. The nearly constant spacing of the fluctuations $\Delta ka \simeq 3.7$ indicates a constant or slowly varying phase velocity, so that the approximation $c_g^* \simeq c_p^*$ is valid over the ka range $0 < ka < 50$ seen in Figs. 30 and 31. The phase velocity of the circumferential wave on an aluminum cylindrical shell with $b/a = 0.99$, is obtained from Fig. 31a and Eq. 56 as $c_p/c = 3.7$, which is identical to the group velocity obtained from Fig. 32 and Eq. 55.

The previous paragraphs demonstrated the significance of the Δka spacing between the deviations in the form function for cylindrical shells. The direction of the deviations from $|f_\infty| = 1$ in the form function curves, as seen in Figs. 30 and 31, is also significant. The hypothesis of Tucker and Barnickle [44] was based on the assumption that a hollow air-filled shell will act as a soft body in that its specular reflection will be 180° out of phase with an incident wave. This would then distinguish hollow body echoes from echoes scattered by solid bodies, which would act rigidly, that is, give a specular return in phase with the incident wave. In fact, the return from solid bodies whose density is greater than the density of water and whose shear and compressional wave speeds are greater than the speed of sound in water can be described in terms of a rigid background term plus a resonance term over all $ka > 0$. Thus, for solid bodies with these elastic properties, the rigid background portion of the hypothesis of Tucker and Barnickle would be correct. With regard to the "soft" scattering by a hollow shell, however, the hypothesis breaks down, as can be determined form the work described in the previous paragraph. Figures 30 and 31 demonstrate that the ka range over which cylindrical shells will act as a soft body is a function of frequency. As frequency is increased, the thickness h of a given shell becomes greater with respect to a wavelength, and whether a shell acts as a "soft" body (specular reflection 180° out of phase with the incident wave) or a "rigid" body (specular reflection in phase with the incident wave) depends both on the frequency and shell thickness. For example, as was discussed the fluctuations in $|f_\infty|$ for the shell described in Fig. 31b occur at intervals $\Delta ka \simeq 3.7$. The deviations from $|f_\infty| = 1$ are, however, not uniform in direction either in Fig. 31b or in any of the other form function curves shown in Figs. 30 and 31. Three separate background regions exist.

The shell acts as a soft body over the ka range where the fluctuations in $|f_\infty|$, at $\Delta ka = 3.7$, are in the negative direction. Here the specular and incident wave are 180° out of phase. Recall that the incident wave and the circumferential wave are in phase (as seen in Fig. 32), and further pulse calculations such as that in Fig. 32 show that they remain in phase over the ka range from $0.2 \leqslant ka \leqslant 50$. For an aluminum shell with $b/a = 0.98$ the ka region over which the shell acts as a "soft" body (specular reflection and the incident wave 180° out of phase) is seen from Fig. 31b to be $0 \leqslant ka \leqslant 23$.

As ka increases, the shell passes through a transition region during which a single fluctuation has both positive and negative aspects. This occurs over the range $23 \leqslant ka \leqslant 37$ for the 0.98 Aluminum shell. Finally, for $ka > 40$ the deviations from $|f_\infty| = 1$ are positive and the shell is a rigid reflector with respect to its specular reflection. The extent in ka of the three background regions will vary with thickness and with material. Later in this section the advent of higher order modes will be discussed, but it can generally be said that if the product

of frequency and thickness is large enough to allow more than one circumferential mode to be excited, the shell has already reached the rigid
background region. Development of the formalism for the resonance
scattering from hollow shells with intermediate background is the subject of work by Murphy et al. [48].

The generation of a single circumferential wave under the conditions present in the computation of the results seen in Figs. 30 through
32 are consistent with the dispersion curves for Lamb waves on plates.
Figure 33 shows dispersion curves for the first four symmetric and
antisymmetric Lamb [49] waves for aluminum plates. The symmetric
Lamb modes satisfy the frequency equation [46,49,50]

$$\frac{\tanh\{(\pi fh/V)\,[(c_T^2-V^2)/c_T^2]^{1/2}\}}{\tanh\{(\pi fh/V)\,[(c_L^2-V^2)/c_L^2]\}^{1/2}}$$

$$= \frac{4\{[(c_L^2-V^2)/c_L^2]^{1/2}\cdot[c_T^2-V^2)/c_T^2]^{1/2}\}}{[(2c_T^2-V^2)/c_T^2]^2} \qquad (57)$$

and the antisymmetric mode satisfies the equation

$$\frac{\tanh\{(\pi fh/V)\,[(c_T^2-V^2)/c_T^2]^{1/2}\}}{\tanh\{(\pi fh/V)\,[(c_2^2-V^2)/c_2^2]^{1/2}\}}$$

$$= \frac{[(2c_T^2-V^2)/c_T^2]^2}{4\{[(c_L^2-V^2)/c_L^2]^{1/2}\cdot[(c_T^2-V^2)/c_T^2]^{1/2}\}} . \qquad (58)$$

In Eqs. 57 and 58, V is the Lamb phase velocity. The group velocity of
the Lamb wave, V_g, is related to the phase velocity by

$$V_g = V\left[1 - \frac{1}{1-(fh)\,dV/d(fh)}\right]. \qquad (59)$$

It has been demonstrated that Eqs. 57 through 59 describing Lamb
waves on plates in vacuo are not strongly modified when the plate is
immersed in water [46,50,51] and that Lamb waves can be generated
by illuminating a plate in water by an incident pulse [46,50]. Radiation
of the Lamb wave into the water can be observed either with a hydrophone [50] or by schlieren visualization [46].

Grigsby and Tajchman [52] gave dimensionless curves for the
phase and group velocities of Lamb waves on a plate whose ratio of
longitudinal to shear speeds is 1.8. Their curves are seen in Fig. 34.
Special attention is directed to the group velocity curves in Fig. 34b,
where all the modes show a flat peak in the group velocity at $V_{gn} \simeq$
1.8. In Fig. 34 the ordinate V_n is Lamb phase velocity divided by shear
wave speed i.e. $V_n = V/c_T$, and the abscissa $(fh)_n$ is fh/c_T. In Fig.
34b the ordinate $V_{gn} = V_g/c_T$.

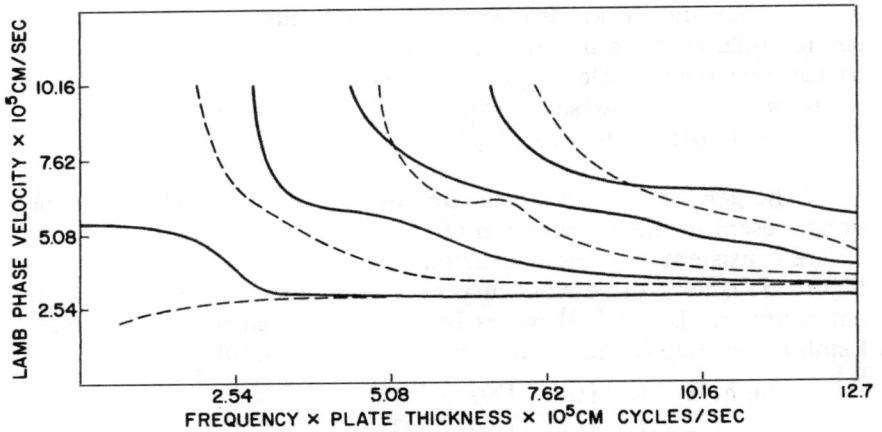

Fig. 33 — The Lamb dispersion curves for the first four symmetric (———) and antisymmetric (- - -) Lamb waves on aluminum plates.

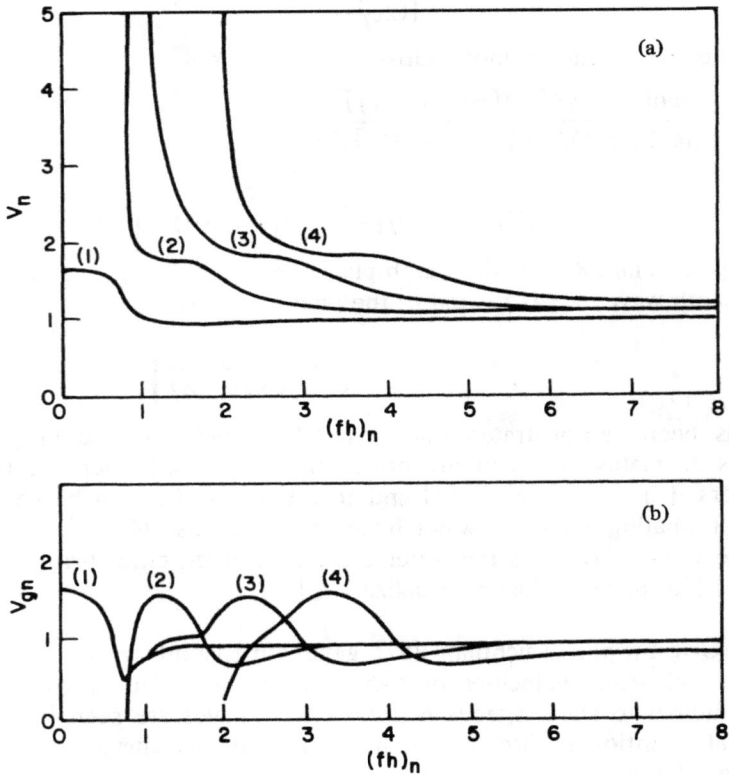

Fig. 34 — Dimensionless curves of (a) Lamb phase velocity and (b) Lamb velocity curves for materials with $c_L/c_T = 1.8$.

Dragonette [46] demonstrated that strong generation of a Lamb mode takes place in the region where the phase velocity curves reach a flat plateau (at approximately $V_n = 1.8$ in Fig. 34a). This plateau region corresponds to the frequency thickness region where the group velocity curve for a particular mode reaches a flat maximum (at approximately $V_{gn} = 1.8$ in Fig. 34b). Dragonette [46] demonstrated further that this strong generation of a Lamb mode, in the fh region where phase velocity is approximately equal to the group velocity, persisted as the plate was curved.

In Fig. 35 the Lamb phase velocity curves describing the first symmetric and first antisymmetric mode curves for an aluminum plate are isolated. The ordinate is given in terms of the Lamb phase velocity V, and also in terms of the angle of incidence θ_i at which a Lamb wave, with that phase velocity, can be generated by a plane wave incident from water onto the plate surface. This angle θ_i satisfies the equation

$$\sin \theta_i = c/V, \tag{60}$$

and a Lamb mode cannot be generated by an acoustic wave incident from water to the plate unless $V \geq c$. The frequency thickness variable, fh, which is the abscissa of the Lamb curves seen in Fig. 35 may be written in terms of ka for a specific cylindrical shell by a simple algebraic manipulation:

$$kh = \frac{2\pi fh}{c} = ka\,[1 - (b/a)] \tag{61a}$$

$$fh = \frac{(ka)\,c\,(1 - b/a)}{2\pi}. \tag{61b}$$

Using Eq. 61b the abscissa, fh, for the flat plate case may be transformed from fh into ka for shells with various b/a values. Figure 35 shows the abscissa written in equivalent ka values for an aluminum shell with $b/a = 0.98$ (see Fig. 31b). The results seen in Fig. 35 predict that the first symmetric mode can be generated at any ka value, and in fact since the flat plateau in group velocity occurs at the low end of the frequency thickness or ka scale, this wave should be strongly generated at low ka. The arrow in Fig. 35 points to the place where $V = c$ for the first antisymmetric mode. For the shell with $b/a = 0.98$, this curve predicts that the antisymmetric mode cannot be excited at ka values below $ka \simeq 50$. The phase and group velocity of the symmetric mode is predicted to be $V/c = 3.7$ by the curves in Fig. 35, in excellent agreement with the circumferential wave observed for the cylindrical shell (Fig. 32). The circumferential wave related to the first symmetric Lamb mode has died out by $ka \simeq 50$, as seen in Fig. 31b, and, as seen above, the results in Fig. 35 predict that the onset of a circumferential wave related to the first antisymmetric mode cannot occur at a

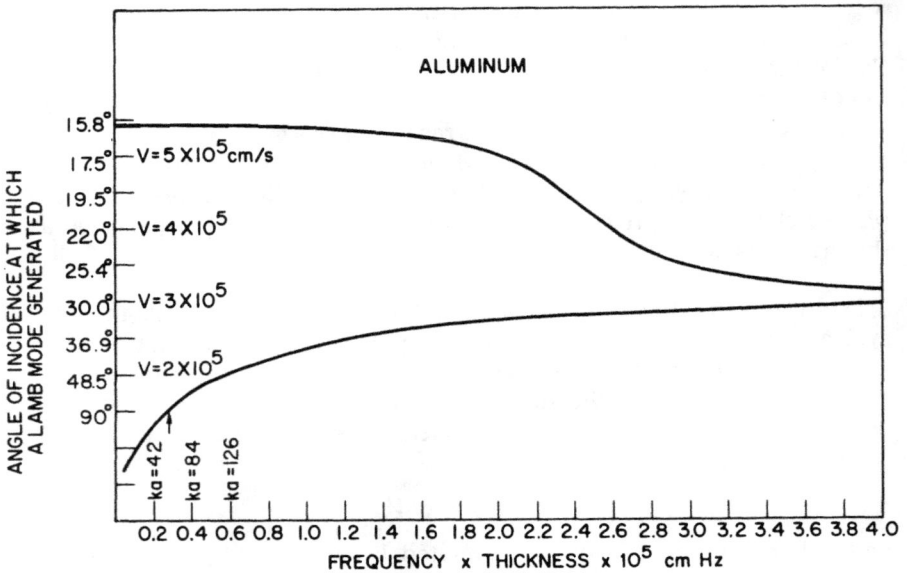

Fig. 35 — The angle of incidence at which the first symmetric and
antisymmetric Lamb modes may be generated.

value lower than $ka = 50$. Such a wave would have a lower group
velocity; i.e., oscillations in $|f_\infty|$ vs ka would occur at closer intervals
than those observed in Fig. 31b. Recall that $c_p^*/c = \Delta ka$.

A plot of $|f_\infty|$ vs ka for an aluminum shell with $b/a = 0.98$ is
given over the range $50 \leqslant ka \leqslant 90$ in Fig. 36. This is then a continua-
tion of the curve given in Fig. 31b, and it shows the onset of a circumferential wave with the properties related to the first antisymmetric
mode. The oscillation in the form function curve predict a circumferential wave with $c_p^*/c = \Delta ka = 1.3$.

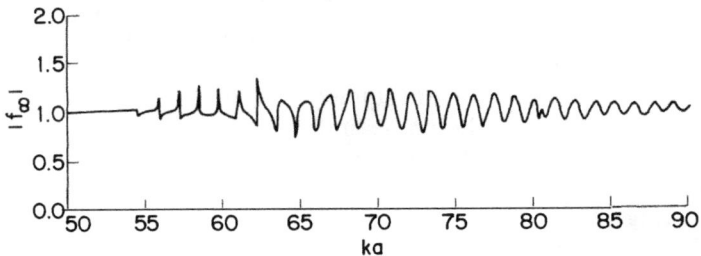

Fig. 36 — The form function vs ka over the range $50 \leqslant ka \leqslant 90$
for an aluminum shell with $b/a = 0.98$.

In general, then, a circumferential wave related to the first symmetric mode should always be generated for a shell of any thickness. Its influence is restricted to the low ka region over which the phase velocity has a flat plateau. The location of this region is a function of thickness, as will be described below. A circumferential wave related to the first antisymmetric mode can only be generated at ka values higher than the coincidence frequency $ka = (ka)_c$, where $V = c$. Thus the thicker the shell, the lower the ka value at which this mode can be generated.

These conclusions should be reflected in the form function curves for aluminum cylindrical shells of various thicknesses. Differences from $|f_\infty| = 1$ should occur at intervals $\Delta ka \simeq 3.7$ for all thin shells, and these differences should die out more quickly with ka as thickness increases, since the plateau region in Fig. 35 corresponds to a smaller ka range for thicker shells. Differences at $\Delta ka \simeq 1.3$ should begin to occur at lower ka values as thickness is increased. Fig. 37 shows the form function curves for aluminum cylindrical shells with $b/a = 0.99$, 0.98, 0.96, 0.94, 0.92, 0.90, and 0.85. The above conclusions are demonstrated by the curves in Fig. 37.

As thickness increases, the antisymmetric mode is seen to occur at lower ka values. The low velocity circumferential wave observed in Figs. 27 and 28, and by others [12-15,25-27], is related to the first antisymmetric Lamb mode for a plate. The ka range over which it is generated depends on the thickness of the shell. The first symmetric mode is strongly generated on aluminum plates in the thickness region where it has a phase and group velocity ratio of $V/c = 3.7$. This is carried over to the shell case, where a circumferential wave having the properties of the first symmetric mode are observed at low ka on all thicknesses of shells from $0.85 \leqslant b/a \leqslant 0.99$. Closer spaced oscillations in $|f_\infty|$ vs ka are also seen to occur at relatively lower ka values as shell thickness is increased. These oscillations are related to a circumferential wave with $c_p^*/c \simeq 1.3$. If one returns now to the Grigsby-Tajchman [52] group velocity curves seen in Fig. 34b, the reason for the association of low frequency, with low velocity and high frequency with high velocity in the literature becomes apparent. All the higher order Lamb modes in Fig. 34b are most strongly generated when their group velocity is $V_{gn} \simeq 1.8$. This is also the group velocity of the first symmetric mode at its region of strong excitation. Moreover, the same is true of all higher order antisymmetric modes (see Fig. 33). Thus the first symmetric mode and all higher order symmetric and antisymmetric modes cannot be distinguished from one another by measurement of group velocity alone, and these waves collectively have been identified

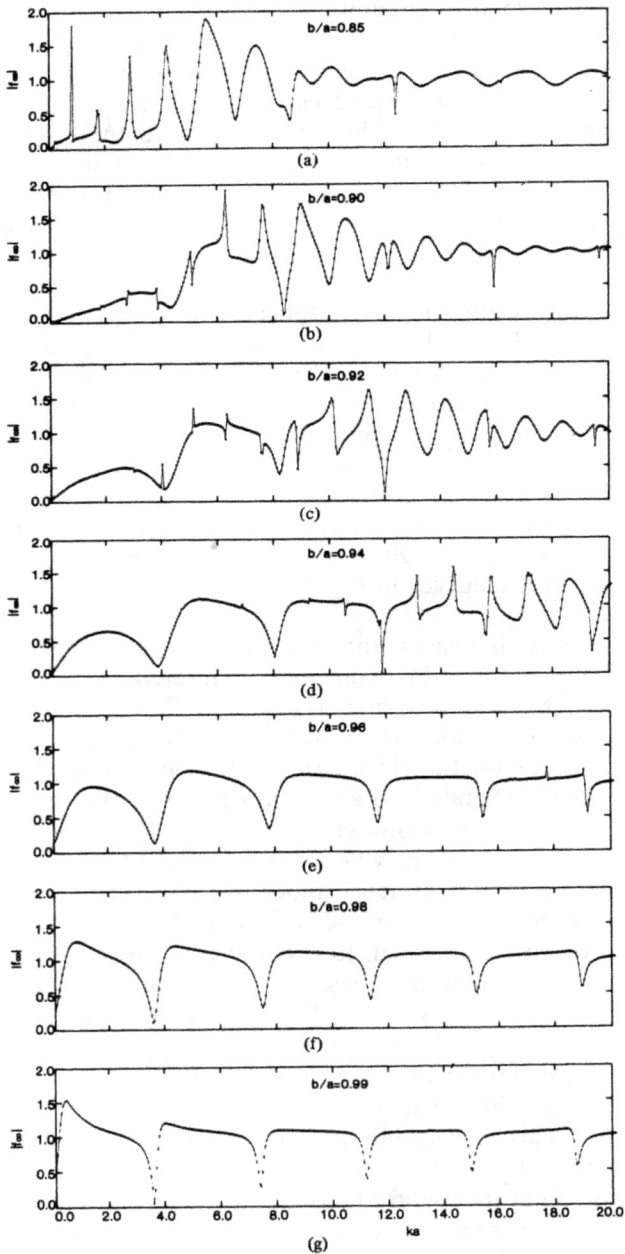

Fig. 37 — The form function vs *ka* over the range $0.2 \leqslant ka$ $\leqslant 20$ for aluminum shells with b/a values of (a) 0.85, (b) 0.90 (c) 0.92 (d) 0.94 (c) 0.96 (f) 0.98 (g) 0.99.

as the fast circumferential wave. The particular mode generated depends on the shell thickness and frequency, but, as discussed earlier, practical considerations generally preclude isolation of the first symmetric mode. All of the measurements of a low velocity wave are related to the first antisymmetric mode, which for thin shells is generated at a higher frequency than the first symmetric mode but at a lower frequency than any of the higher order modes. In the empirical observations of circumferential waves, the wave related to the first antisymmetric mode is obtained at a frequency which depends on the thickness (see Fig. 37).

Because of the slow speed of this wave, there is a greater time difference between successive traversals of the circumferential wave (a factor of 3 as compared to the faster waves); hence, this mode when present can be isolated at lower ka values than a mode traveling with a velocity 3 times higher. Thus in past pulse hydrophone measurements [12-15,27] low velocity corresponded to low frequency in the experimental observations. As frequency was increased it became possible to isolate higher velocity modes, all of which were strongly generated with the same group velocity; hence, high velocity corresponded experimentally to high frequency.

The circumferential wave related to the first symmetric mode is the only one of the "fast" circumferential waves whose amplitude approaches the amplitude of the specular reflection, and, while it is not practical to isolate it experimentally, its contribution to the steady state pressure or form function at low ka is apparent. As demonstrated in Fig. 33, it is possible to isolate the first symmetric mode by computation of the response of a shell to a short incident pulse. The results of computations similar to those carried out to produce Fig. 33 are given in Figs. 38 and 39 for various thicknesses shell at various center frequencies. The purpose of these calculations is to demonstrate that the relative amplitude of the circumferential and specular contributions is a function of ka. In Figs. 38 the responses of three aluminum shells with thickness $b/a = 0.99$, 0.96, and 0.90 are presented. The center dimensionless frequency of the calculation is $k_o a = 10$. A circumferential wave related to the first symmetric mode is seen in Fig. 38a ($b/a = 0.99$) and Fig. 38b ($b/a = 0.96$). The measured attenuation in these two cases is 0.50 Np/revolution (Fig. 38b) and 0.43 Np/revolution (Fig. 38a). In Fig. 38c the first antisymmetric mode is generated simultaneously with the symmetric mode. This figure demonstrates that the antisymmetric mode is in fact generated at lower frequency as thickness is increased, as predicted in the discussion of Fig. 37. It also shows that at the same $k_o a$ the attenuation of the symmetric mode increases with

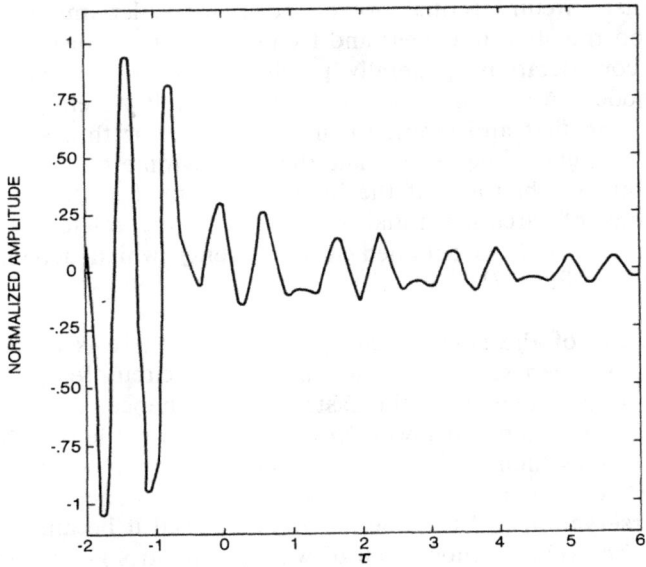

Fig. 38(a) — Computation of the echoes scattered by an aluminum shell at $k_o a = 10$; the shell thickness is $b/a = 0.99$.

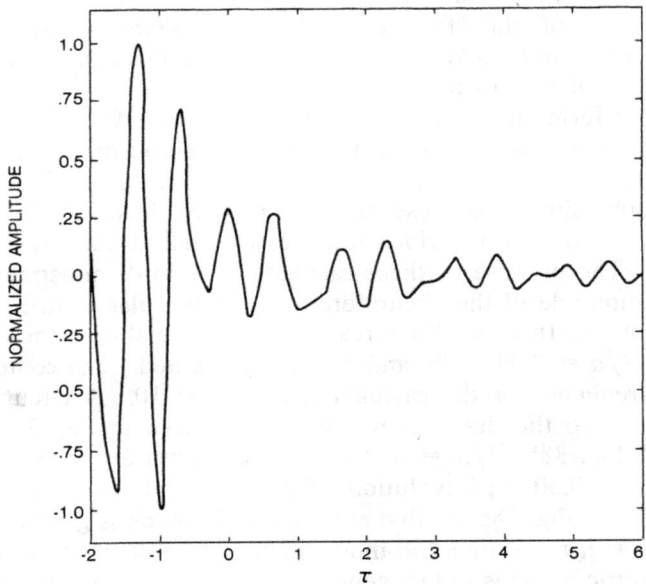

Fig. 38(b) — Computation of the echoes scattered by an aluminum shell at $k_o a = 10$; the shell thickness is $b/a = 0.96$.

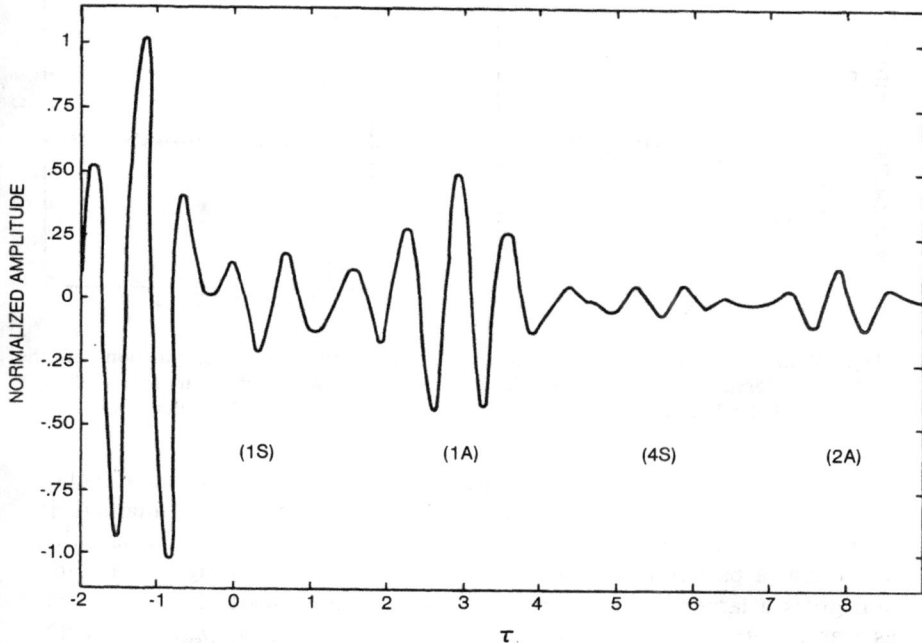

Fig. 38(c) — Computation of the echoes scattered by an aluminum shell at $k_o a = 10$; the shell thickness is $b/a = 0.90$.

increasing thickness, and finally it demonstrates that the large oscillations in $|f_\infty(\pi)|$ which appear with increasing thickness in Fig. 37 are due to the larger magnitude with which the antisymmetric mode is generated. The attenuation of the antisymmetric mode in Fig. 38c is 1.0 Np/revolution, which is much larger then that of the symmetric mode, but the magnitude of the first-antisymmetric-mode-echo is more than 6 dB larger than that of the first-symmetric-mode-echo.

The responses of an iron shell with $b/a = 0.99$ at $k_o a$ values of 11 and 20 are seen in Figs. 39(a) and 39(b) respectively. Here the attenuation is 0.46 Np/revolution at $k_o a = 11$ and 0.23 Np/revolution at $k_o a = 20$. Thus the attenuation decreases with higher frequency, an observation similar to that of Horton and Mechler [15], who observed this phenomenon at $ka \simeq 30$ for the wave identified here as the antisymmetric wave.

As may be observed in Fig. 38c, a determination of the attenuation of a particular circumferential wave is not necessarily a measure of its relative importance to the total scattered field. If the ratio of the

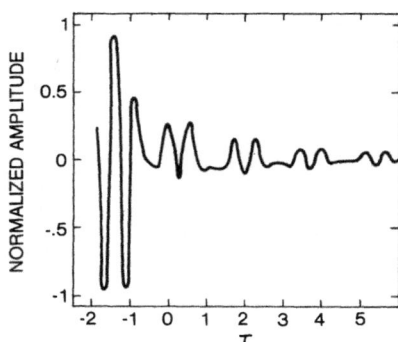

Fig. 39(a) — Computation of the echoes scattered by an iron shell with $b/a = 0.99$ at a $k_o a$ value of 11.

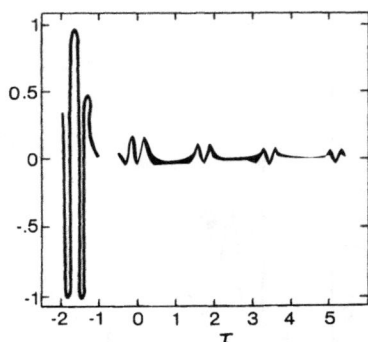

Fig. 39(b) — Computation of the echoes scattered by an iron shell with $b/a = 0.99$ a ta $k_o a$ value of 20.

amplitude of the first circumferential echo, p_1, to the specular echo, p_{spec}, is taken from Figs. 39(a) and 39(b), the result obtained is $p_1/p_{spec} = 0.2$ at $k_o a = 11$ and $p_1/p_{spec} = 0.1$ at $k_o a = 20$. These ratios show that a particular circumferential wave is more strongly generated and gives a larger contribution to the steady state scattering at low ka, as was observed in Figs. 30, 31, and 37. The ratio p_1/p_{spec} is directly related to the relative contribution or importance of the wave to the steady state solution. Figure 40 shows a plot of p_1/p_{spec} for various thicknesses of stainless steel shells as a function of ka. The points were obtained from pulse computations such as those in Figs. 38-39. In general for shells with $b/a \leqslant 0.99$ the circumferential wave related to the first symmetric mode has a backscattered amplitude more than 20 dB down from specular for $ka > 20$.

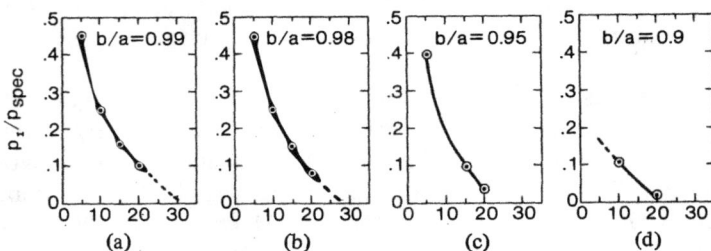

Fig. 40 — The ratio (p_1/p_{spec}) vs ka for stainless steel shells of thicknesses (a) 0.99 (b) 0.98 (c) 0.95 (d) 0.90.

For higher order modes generated above $ka = 20$, the ratio p_1/p_{spec} is more than 20 dB down from specular. This is consistent with the observations made here of a "fast" circumferential wave (Figs. 23, 25, 26), where p_1 is greater than 30 dB below specular, and with all reported observations in the literature [12-15,26].

The circumferential wave related to the fundamental antisymmetric mode also has its largest influence on $|f_\infty|$ at low ka. It can, however, be generated only at low ka for thick shells. The general observation in the literature that the "slow" circumferential wave is more strongly generated in thicker shells, is simply because of the results discussed in connection with Fig. 37, namely that as shells become thicker, it is possible to excite the first antisymmetric mode at lower ka.

The results considered here were for thin shells. Figures 33 and 34 show that as fh becomes larger, all the Lamb modes tend toward a final velocity $V \simeq 2.0c$. Since the phase velocity curves again level off for large fh, the Lamb modes are again strongly excited. Each higher order mode is first strongly excited at $V/c \simeq 3.7$ and then in the limit of a thick shell at $V/c \simeq 2.0$. Therefore it should be possible to find an intermediate frequency range at which a Lamb type mode is generated with $V/c \simeq 3.7$ simultaneously with a lower order mode which has reached its high frequency limit $V/c \simeq 2.0$. Such a situation is seen in Fig. 41. Here a schlieren visualization is made at an fh value of 11.2. The target is a 3.4-cm-diameter-aluminum cylinder with $b/a = 0.9$. This corresponds to $ka = 476$ and $kh = 47.6$. Simultaneous observation of circumferential waves with $c_g^*/c = 3.7$ and $c_g^*/c = 2.0$ are seen. Figures 37 and 41 explain what has been referred to in the literature [12] as rare occurrences when slow and fast circumferential waves are observed simultaneously.

CONCLUSION

The steady state acoustic response of infinitely long solid elastic cylinders and cylindrical shells can be exactly computed in terms of a normal mode series. For rigid cylinders the backscattered form function can be described in terms of the interference of a specularly reflected wave and a Franz-type circumferential wave whose speed and attenuation are related to the peaks and nulls in the form function. Solid metal cylinders in water exhibit this purely rigid behavior in the low ka region, after which region the form function is dominated by minima related to resonances in the individual normal modes. For cylinders made of metals whose shear and compressional wave speeds

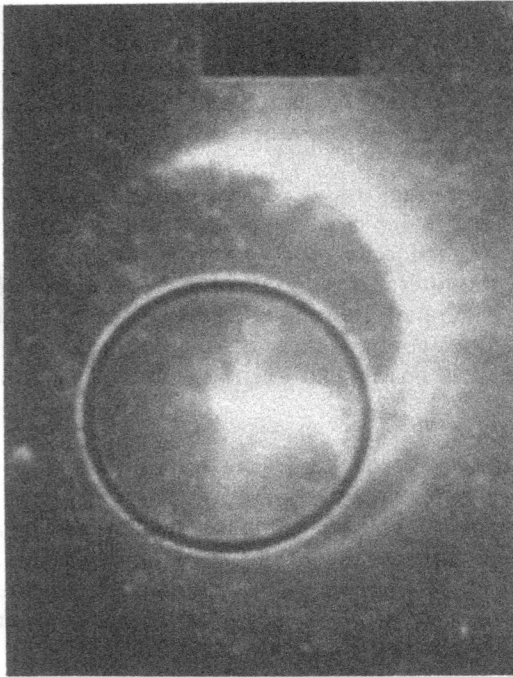

Fig. 41 — Schlieren visulization of circumferential waves generated on an aluminum shell, with $b/a = 0.9$ at $k_o a = 476$.

are greater than the wave speed of sound in water, the first resonance minimum observed is the $(2,1)$ resonance. This occurs at $ka = 4.78$ for aluminum, and the ka value at which it occurs for other metals can be computed using aluminum as a reference.

The normal mode resonances are related to circumferential waves predicted by creeping wave theory. A mode resonates when its modal velocity is matched by the velocity of a circumferential wave. A single circumferential wave generates a given eigenfrequency in successive modes. The $(n, 1)$ resonances are related to the Rayleigh wave, and the $(2,1)$, $(3,1)$, and $(4,1)$ modes are generated at ka values at which the cylinder circumference is 2, 3, or 4 Rayleigh wavelengths. Similarly the $(n, 2)$ resonances are related to the R_2 type whispering gallery mode and so on. The predominant circumferential waves in a given ka region can be predicted from the dominant resonance minima in $|f_\infty (\pi)|$. The "Rayleigh" wave was experimentally observed on aluminum in the predicted region.

The region of oscillations in $|f_\infty(\pi)|$ which begins at the position of the $(2,1)$ resonance persists as $ka \to \infty$; however, in reality a frequency will be reached after which absorption must be included in the theory. In the resonance region $(ka > 4.78$ for aluminum$)$ the scattering from the cylinder is made up of a rigid background part, on which the numerous resonances are superimposed. The resonance formalism of nuclear reaction theory is used to separate the exact normal mode series solution into rigid background and resonance terms, and resonance widths can be calculated.

For a thin cylindrical shell, the Franz wave does not measurably affect $|f_\infty(\pi)|$ even at low ka. This is consistent with soft rather than rigid scattering behavior, and it is demonstrated that as ka increases thin shells pass through three background regions. In the soft-background region at low ka the specular reflection is 180° out of phase with an incident pulse. This is followed by a region of intermediate background, and then a rigid-background region at which the specular and incident pulses are in phase and remain in phase as ka is further increased. Circumferential waves are isolated theoretically by applying fast Fourier transform techniques to the Fourier integral representing the echoes scattered, when a short acoustic pulse is incident on a shell. The relationship between the observed circumferential waves and the steady state form function shows that, for thin shells, the number of circumferential waves present, their velocity, and their relative significance can be obtained directly from the form function.

Lamb theory for plates is utilized to predict the ka range of possible excitation of specific circumferential waves. A circumferential wave related to the first symmetric mode is generated for $ka > 0$, for all thin shells. A circumferential wave related to the first antisymmetric mode is generated at ka values which vary with thickness in a predictable way. As shell thickness is increased, circumferential waves related to all high order Lamb modes are strongly generated with the same group velocity. This accounts for observations reported previously in the literature and thought to be the observation of a single circumferential wave.

REFERENCES

1. Lord Rayleigh, *Theory of Sound*, Dover (1945).

2. P. M. Morse, *Vibration of Sound*, McGraw-Hill (1936).

3. J. J. Faran, J. Acoust. Soc. Am. **23**, 405-418 (1951).

4.   R. Hickling, J. Acoust. Soc. Am. **34**, 1582-1592 (1962).

5.   R. Hickling, J. Acoust. Soc. Am. **36**, 1124-1137 (1964).

6.   L. D. Hampton and C.M. McKinney, J. Acoust. Soc. Am. **33**, 664-673 (1961).

7.   K. J. Diercks and R. Hickling, J. Acoust. Soc. Am. **41**, 380-393 (1967).

8.   W. G. Neubauer, R. H. Vogt, and L. R. Dragonette, J. Acoust. Soc. Am. **55**, 1123-1129 (1974).

9.   L. R. Dragonette, R. H. Vogt, L. Flax, and W. G. Neubauer, J. Acoust. Soc. Am. **55**, 1130-1137 (1974).

10. H. D. Dardy, J. A. Bucaro, L. S. Scheutz, and L. R. Dragonette, J. Acoust. Soc. Am. **62**, 1373-1376 (1977).

11. G. R. Barnard and C. M. McKinney, J. Acoust. Soc. Am. **33**, 226-238 (1961).

12. K. J. Diercks, T. G. Goldsberry, and C. W. Horton, J. Acoust. Soc. Am. **35**, 59-64 (1963).

13. C. W. Horton, W. R. King, and K. J. Diercks, J. Acoust. Soc. Am. **34**, 1929-1932 (1962).

14. T. G. Goldsberry, J. Acoust. Soc. Am., **42**, 1298-1305 (1967).

15. C. W. Horton and M. V. Mechler, J. Acoust. Soc. Am. **51**, 295-303 (1972).

16. W. Franz, Z. Naturforsch **9a**, 705-716 (1954).

17. G. N. Watson, Proc. Roy. Soc. (London) **A95**, 83-99 (1919).

18. W. Franz and K. Deppermann, Ann. Phys. **10**, 361-373 (1952).

19. L. Flax, J. Acoust. Soc. Am. **62**, 1502-1503 (1977).

20. H. Überall, R. D. Doolittle, and J. V. McNicholas, J. Acoust. Soc. Am. **39**, 564-578 (1966).

21. R. D. Doolittle, H. Überall, and P. Uginčius, J. Acoust. Soc. Am. **43**, 1-14 (1968).

22. O. D. Grace and R. R. Goodman, J. Acoust. Soc. Am. **39**, 173-174 (1966).

23. W. G. Neubauer, J. Acoust. Soc. Am. **44**, 298-299 (1968).

24. M. L. Harbold and B. N. Steinberg, J. Acoust. Soc. Am. **45**, 592-603 (1969).

25. R. E. Bunney, R. R. Goodman, and S. W. Marshall, J. Acoust. Soc. Am. **46**, 1223-1233 (1969).

26. W. G. Neubauer, J. Acoust. Soc. Am. **45**, 1134-1144 (1969).

27. W. G. Neubauer and L. R. Dragonette, J. Acoust. Soc. Am. **48**, 1135-1149 (1970).

28. D. Brill and H. Überall, J. Acoust. Soc. Am. **50**, 921-939, (1971).

29. K. Dransfeld and E. Salzmann, in *Physical Acoustics, Vol. VII*, edited by Warren P. Mason and R. N. Thruston, Academic Press (1970).

30. W. G. Neubauer and L. R. Dragonette, J. Appl. Phys., **45**, 618-622 (1974).

31. L. R. Dragonette, NRL Report 8216 (1978).

32. L. A. Scheutz and W. G. Neubauer, J. Acoust. Soc. Am. **62**, 513-517 (1977).

33. L. Flax and W. G. Neubauer, J. Acoust. Soc. Am. **61**, 307-312 (1977).

34. R. E. Bunney and R. R. Goodman, J. Acoust. Soc. Am. **53**, 1658-1662 (1973).

35. R. H. Vogt, L. Flax, L. R. Dragonette, and W. G. Neubauer, J. Acoust. Soc. Am. **57**, 558-561 (1975).

36. G. V. Frisk and H. Überall, J. Acoust. Soc. Am. **59**, 46-54 (1976).

37. W. G. Neubauer, J. Appl. Phys. **44**, 48-55 (1973).

38. H. L. Bertoni and T. Tamir, Appl. Phys. **2**, 157-172 (1973).

39. E. K. Sittig and C. A. Coquin, J. Acoust, Soc. Am. **48**, 1150-1159 (1970).

40. I. A. Viktorov, *Rayleigh and Lamb Waves*, Plenum Press (1967).

41. M. C. Junger and D. Feit, *Sound Structures and Their Interactions*, M.I.T. Press (1972).

42. R. H. Vogt and W. G. Neubauer, J. Acoust. Soc. Am. **60**, 15-22 (1976).

43. S. deBenedetti, *Nuclear Interactions*, Wiley (1964).

44. D. G. Tucker and N. J. Barnickle, J. Sound Vib. **9**, 393-397 (1969).

45. P. Uginčius and H. Überall, J. Acoust. Soc. Am. **43**, 1025-1035 (1968).

46. L. R. Dragonette, J. Acoust. Soc. Am. **51**, 920-935 (1972).

47. D. J. Shirley and K. J. Diercks, J. Acoust. Soc. Am. **48**, 1275-1282 (1970).

48. J. D. Murphy, E. D. Breitenback, and H. Überall, J. Acoust. Soc. Am. **64**, 677-683 (1978).

49. H. Lamb, Proc. Roy. Soc. (London) **93**, 114-128 (1917).

50. D. C. Worlton, J. Appl. Phys. **32**, 967-971 (1961).

51. M. F. Osborne and S. D. Hart, J. Acoust. Soc. Am. **17**, 1-18 (1945).

52. T. N. Grigsby and E. J. Tajchman, IRE Trans. **UE-8**, 26-33 (1961).

Appendix A

LIST OF SYMBOLS

a — the radius of the target

a_{ij} — matrix elements defined in Ref. 32

$\mathbf{A}$ — the vector potential

b — the inner radius of a cylindrical shell

B_n — a coefficient in Eq. 50b

c — the velocity of sound in water

c_p^F — the phase velocity of the Franz wave

c_g^F — the group velocity of the Franz wave

$c_n(ka)$ — the modal phase velocity for the nth normal mode

$c_n^g(ka)$ — the modal group velocity for the nth normal mode

$c_l(ka)$ — the phase velocity of the lth R-type circumferential wave

$c_l^g(ka)$ — the group velocity of the lth R-type circumferential wave

c_R — the phase velocity of the Rayleigh or R_1 circumferential

c_R^g — the group velocity of the Rayleigh or R_1 circumferential wave

c_L — the longitudinal wave velocity in a material

c_T — the shear wave velocity in a material

c_p^* — the phase velocity of a circumferential wave in a cylindrical shell

c_g^* — the group velocity of a circumferential wave in a cylindrical shell

C_n — a coefficient in Eq. 50a

d — distance

$D_n^{(1)}(ka)$ — A 2-by-2 matrix defined in Eq. 1a

$D_n^{(2)}(ka)$ — A 2-by-2 matrix defined in Eq. 1a

f_∞ — the far field form function

f_∞^R — the far field form function for a rigid cylinder

f_n — the nth partial wave or n^{th} modal contribution to the form function

f — frequency

f_o — the center frequency of an incident pulse

fh — the frequency thickness product

$(fh)_n$ — the dimensionless frequency thickness parameter, $(fh)_n = fh/c_T$

$g_i(ka)$	-	the spectrum of an incident pulse
$g_s(ka)$	-	the spectrum of a scattered echo
$G_n(Z)$	-	defined in Eq. 3
$G_n^R(Z)$	-	the expression to which $G_n(Z)$ reduces when the target is a rigid cylinder
h	-	the thickness of a cylindrical shell
$H_n(Z)$	-	the Hankel function of the first kind (order n, argument Z)
$H_n^{(2)}(Z)$	-	the Hankel function of the second kind (order n, argument Z)
$H_n'(Z)$	-	the derivative of the Hankel function of the first kind with respect to its argument
$H_n^{(2)'}(Z)$	-	the derivative of the Hankel function of the second kind with respect to its argument
$J_n(Z)$	-	the Bessel function (order n, argument Z)
$J_n'(Z)$	-	the derivative of the Bessel function with respect to its argument
k	-	the wavenumber in water given by $k = 2\pi/\lambda$
k_L	-	the longitudinal wavenumber in a material
k_T	-	the shear wavenumber in a material
ka	-	the dimensionless frequency variable, $ka = 2\pi a/\lambda$
$(ka)_o$	-	the center dimensionless frequency of a pulse in water
$(ka)_{peak}$	-	the ka value at which a peak in $|f_\infty|$ occurs
l	-	an integer, $l = 1, 2, ...$, used to number the eigenfrequencies of a given mode
L_n	-	defined in Eq. (1a)
n	-	an integer, $n = 1, 2, ...$, used to number the normal modes
$p_i(\tau)$	-	an incident acoustic pulse
p_o	-	the incident plane-wave amplitude
p_1	-	the pressure amplitude of the first backscattered circumferential echo
$p_s(\theta)$	-	the steady state scattered acoustic pressure at the bistatic angle θ
p_{spec}	-	the pressure amplitude of the specular reflection
q_l	-	the zeroes of the first derivatives of the Airy function
Q_n	-	defined in Eq. 54
r	-	the range or distance between the scatterer and the field point
R_l	-	the lth order Rayleigh-type circumferential wave
s_n	-	defined in Eq. 40
S_n	-	the elastic scattering function defined by $S_n \equiv \exp(2i\,\delta_n)$
$S_n^{(R)}$	-	the rigid body scattering function defined as $S_n^{(R)} \equiv \exp(2\,i\,\xi_n)$

t	-	time
u	-	the particle displacement
V	-	the phase velocity of a Lamb wave
V_g	-	the group velocity of a Lamb wave
V_n	-	dimensionless Lamb phase velocity given by $V_n = V/c^T$
V_{gn}	-	dimensionless Lamb group velocity givey by $V_{gn} = V_g/c_T$
$Y_n(Z)$	-	the Neumann function (argument Z, order n)
$Y_n'(Z)$	-	the derivative of the Neumann function with respect to its argument
Z	-	short-hand form for the dimensionless frequency variable $Z \equiv ka$
Z_n	-	the dimensionless frequency at a resonance, $Z \equiv (ka)_n$
Z_{pole}	-	the Z value at which a resonance pole in the scattering function S_n occurs
Z_{zero}	-	the Z value at which a resonance zero in the scattering function S_n occurs
z_1	-	defined in Eq. 37a
z_2	-	defined in Eq. 37b
α_i^F	-	the attenuation coefficient for the Franz wave
α_R	-	the attenuation coefficient for the Rayleigh or R_1 circumferential wave
α_R'	-	the dimensionless attenuation coefficient for the R_1 circumferential wave
α^*	-	the attenuation coefficient for a circumferential wave in a shell
β_n	-	coefficient in the Taylor series expansion (Eq. 42b)
Γ_n	-	the width of a resonance, given by $\Gamma_n = -2s_n/\beta_n$
δ_n	-	scattering phase shift for the elastic scattering function
Δd	-	change in distance
Δka	-	change in ka
Δt	-	change in time
$\Delta \tau$	-	change in the dimensionless time parameter
Δn	-	defined in Eq. 39
ϵ_n	-	the Neumann factor $\epsilon_n = 2$, $n = 0$; $\epsilon_n = 1$, $n > 0$
θ	-	the polar angle
θ_i	-	the incidence of angle of a plane wave
λ	-	the wavelength of sound in water
λ_R	-	the wavelength of the Rayleigh wave on a flat surface
λ^*	-	the wavelength of a circumferential wave
ν	-	a complex variable
ζ_n	-	the phase shifts for the rigid scattering function
ρ	-	the density of water
ρ_s	-	the density of the target material

τ - the dimensionless time parameter $\tau \equiv \dfrac{ct - a}{r}$

ϕ - the azimuthal angle

Ψ - a scalar potential

ω - the angular frequency, $\omega = 2\pi f$

Chapter 5

LAYERED ELASTIC ABSORPTIVE CYLINDERS*

INTRODUCTION

This scattering of sound by elastic hollow cylinders has been discussed by several authors [1-4]. In these papers, and in others mentioned in their references, the absorption of acoustic energy in the shell was assumed negligible. Although this assumption is valid in many cases, it is not, for certain cylindrical shells composed of nonmetallic materials. This was demonstrated [5] for spheres by showing the solution to be altered significantly when absorption was present. Excellent agreement between experiment and theory was obtained for that case only after absorption was accounted for. Acoustic reflection from two-layered cylindrical shells is formulated in which either layer of the shell may absorb wave energy, and different media may be assumed to be inside and outside the shell.

A mathematical model is developed to predict the scattering of a plane acoustic wave from a two-layered absorptive cylindrical shell with different fluid media inside and outside. Either layer or the shell may absorb wave energy. The hollow space within the cylinder is assumed to be occupied by a fluid or a vacuum. Solutions to the elastic problems considered here are constructed using scalar and vector potentials. The resulting equations are solved in terms of Bessel functions of complex argument. Matrix methods are used throughout.

DESCRIPTION OF PROBLEM

Figure 1 shows the cylindrical coordinate orientation and the direction of a plane wave incident on an infinitely long cylindrical shell in a fluid medium. The two-solid-elastic-absorptive layers are denoted by 1 and 2. The axis of the cylindrical shell is taken to be the z-axis of the cylindrical coordinate system (r, ϕ, z). There is a plane sound

---

*These results first appeared in: Lawrence Flax and Werner G. Neubauer, J. Acoust. Soc. Am. **61**, 307-312 (1977).

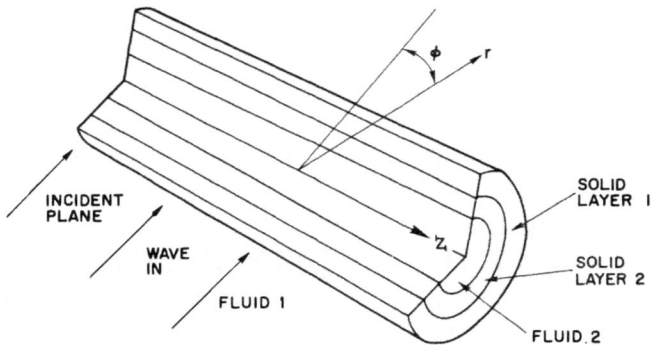

Fig. 1 — Geometry used for formulating the sound scattering
from an infinite circular two-layered cylindrical shell.

wave of circular frequency ω incident along the negative x-axis. The
fluid (2) inside the shell has a density of ρ_f and wave propagation
speed c_f. In general, the outer fluid (1) will be different and is
described by parameters ρ and c_ω.

In an effort to facilitate comparison with previous single-layered
computations, it is convenient to define cylinder parameters which
represent thickness ratios, b/a and c/a. The radius of the inner core is
b, a is the radius to the outside of layer 2, and c is the outside radius.

MATHEMATICAL ANALYSIS

The two fluids outside and inside the shell are labeled by w and f,
respectively. The two layers are described by the densities
$\rho_i(i = 1, 2)$, and longitudinal and shear speeds V_{Li} and V_{Si}, which are
complex.

In the outside fluid medium, the excess acoustic pressure is the
sum of the incident plane wave

$$p_i = p_0 \sum_{n=0}^{\infty} \epsilon_n (- i)^n J_n \ (kr) \cos \ (n\phi), \tag{1a}$$

and a scattered (outgoing) wave

$$p_s = p_0 \sum_{n=0}^{\infty} \epsilon_n C_n H_n^{(2)} \ (kr) \cos \ (n\phi). \tag{1b}$$

The incident pressure amplitude is p_0, ϵ_n is the Neumann factor, $k = \omega/c$, $H_n^{(2)} = J_n - iY_n$, where J_n and Y_n are Bessel functions of the first
and second kind, $H_n^{(2)}$ are Hankel functions, and C_n are the scattering

coefficients. Throughout, the time dependence $(e^{-i\omega t})$ has been suppressed.

The solution for the scalar and vector potentials in layer 1 are

$$\Phi = p_0 \sum_{n=0}^{\infty} \epsilon_n \, i^{-n} \, [A_n J_n(k_{L1}r) + B_n Y_n(k_{L1}r)] \cos(n\phi), \quad (2a)$$

$$A_z = p_0 \sum_{n=0}^{\infty} \epsilon_n \, i^{-n} \, [D_n J_n(k_{S1}r) + E_n Y_n(k_{S1}r)] \sin(n\phi), \quad (2b)$$

and in layer 2

$$\psi = p_0 \sum_{n=0}^{\infty} \epsilon_n i^{-n} \, [G_n J_n(k_{L2}r) + K_n Y_n(k_{L2}r)] \cos(n\phi), \quad (3a)$$

$$B_z = p_0 \sum_{n=0}^{\infty} \epsilon_n \, i^{-n} \, [I_n J_n(k_{S2}r) + L_n Y_n(k_{S2}r)] \sin(n\phi). \quad (3b)$$

In the core, one has again a compressional wave,

$$p_f = p_0 \sum_{n=0}^{\infty} \epsilon_n \, i^{-n} \, M_n J_n(k_f r) \cos(n\phi). \quad (4)$$

Since the core contains the origin, the solution must be regular at $r = 0$. Thus, the coefficients associated with $Y_n(k_f r)$ must be zero.

At the outside boundary of the shell $r = c$, the displacements and normal stresses must be continuous and the tangential stresses must be zero. Four boundary conditions are prescribed at the interface between the outer layer and inner layer $r = a$:

(i)   Radial displacements are continuous.

(ii)   Tangential displacements are continuous.

(iii)   Radial stresses of adjoining material are equal.

(iv)   Tangential stresses of adjoining material are equal.

Finally, three conditions are established between the inner layer and the internal fluid $r = b$. These are similar to the conditions present at $r = c$.

The coefficients C_n are determined from the boundary conditions and are of the form

$$C_n = \frac{J_n(Z)L_n - Z\,J_n^{(2)'}(Z)}{H_n^{(2)}(Z)L_n - ZH_n^{(2)'}(Z)'} \tag{5}$$

where the primes denote the derivative with respect to the argument $Z = kc$. The functions L_n are given by a division of two ninth-order determinants of the form

$$L_n = \frac{\rho}{\rho_1}\frac{P_n}{Q_n}, \tag{6}$$

where

$$P_n = \begin{vmatrix} a_{21} & a_{22} & a_{23} & a_{24} & 0 & 0 & 0 & 0 & 0 \\ a_{31} & a_{32} & a_{33} & a_{34} & 0 & 0 & 0 & 0 & 0 \\ a_{41} & a_{42} & a_{43} & a_{44} & a_{45} & a_{46} & a_{47} & a_{48} & 0 \\ a_{51} & a_{52} & a_{53} & a_{54} & a_{55} & a_{56} & a_{57} & a_{58} & 0 \\ a_{61} & a_{62} & a_{63} & a_{64} & a_{65} & a_{66} & a_{67} & a_{68} & 0 \\ a_{71} & a_{72} & a_{73} & a_{74} & a_{75} & a_{76} & a_{77} & a_{78} & 0 \\ 0 & 0 & 0 & 0 & a_{85} & a_{86} & a_{87} & a_{88} & a_{89} \\ 0 & 0 & 0 & 0 & a_{95} & a_{96} & a_{97} & a_{98} & 0 \\ 0 & 0 & 0 & 0 & a_{10,5} & a_{10,6} & a_{10,7} & a_{10,8} & a_{10,9} \end{vmatrix}$$

and

$$Q_n = \begin{vmatrix} a_{11} & a_{12} & a_{13} & a_{14} & 0 & 0 & 0 & 0 & 0 \\ a_{31} & a_{32} & a_{33} & a_{34} & a_{35} & 0 & 0 & 0 & 0 \\ a_{41} & a_{42} & a_{43} & a_{44} & a_{45} & a_{46} & a_{47} & a_{48} & 0 \\ a_{51} & a_{52} & a_{53} & a_{54} & a_{55} & a_{56} & a_{57} & a_{58} & 0 \\ a_{61} & a_{62} & a_{63} & a_{64} & a_{65} & a_{66} & a_{67} & a_{68} & 0 \\ a_{71} & a_{72} & a_{73} & a_{74} & a_{75} & a_{76} & a_{77} & a_{78} & 0 \\ 0 & 0 & 0 & 0 & a_{85} & a_{86} & a_{87} & a_{88} & a_{89} \\ 0 & 0 & 0 & 0 & a_{95} & a_{96} & a_{97} & a_{98} & 0 \\ 0 & 0 & 0 & 0 & a_{10,5} & a_{10,6} & a_{10,7} & a_{10,8} & a_{10,9} \end{vmatrix}$$

The elements of the determinants are given in the Appendix. The arguments of the function appearing in these expressions are determined from the following relationships:

$$r = c,$$

$$Z_L = k_{L1}c - i\beta_{L1}k_{L1}c, \quad Z_s = k_{S1}c - i\beta_{S1}k_{S1}c;$$

$$r = a,$$

$$t_L = k_{L1}a - i\beta_{L1}k_{L1}a, \quad t_s = k_{S1}a - i\beta_{S1}k_{S1}a,$$

$$q_L = k_{L2}a - i\beta_{L2}k_{L2}a, \quad q_s = k_{S2}a - i\beta_{S2}k_{S2}a;$$

$$r = b,$$

$$u_L = k_{L2}b - i\beta_{L2}k_{L2}b, \quad u_s = k_{S2}b - i\beta_{S2}k_{S2}b, \qquad (7)$$

where β's are absorption factors given in Np.

In addition, ratios of complex wave numbers appear in some of the matrix elements and are given by

$$G_1 = Z_L^2/(Z_S^2 - 2Z_L^2), \quad G_2 = q_L^2/(q_S^2 - 2q_L^2),$$

$$G_3 = u_L^2/(u_S^2 - 2u_L^2), \quad U = (\rho/\rho_1)Z_S^2/t_S^2. \qquad (8)$$

As a result of using complex wave numbers in the theory, the elements of the determinant defining L_n contain Bessel functions with complex arguments. Techniques [6] for calculating these functions are available.

The scattered pressure is obtained using Eqs. (1b), (5), and (6).

$$p_s = p_0 \sum_{n=0}^{\infty} \epsilon_n \left\{ \frac{[J_n(Z)L_n - Z J_n'(Z)] H_n^{(2)}(Z)}{H_n^{(2)}(Z)L_n - Z H_N^{(2)'}(Z)} \right\} \cos(n\phi). \qquad (9)$$

For large distances from the reflector, the asymptotic expressions for the $H_n^{(2)}(Z)$ can be used. For monostatic reflection $\phi = \pi$, making $\cos(n\pi) = (-1)^n$ and the scattered pressure becomes

$$p_s = p_0 \frac{(2c/r)^{1/2}\sqrt{2}}{(\pi Z)^{1/2}} \sum_{n=0}^{\infty} \epsilon_n (-1)^n \left| \frac{J_n(Z)L_n - Z_{J_n}'(Z)}{H_n^{(2)}(Z)L_n - Z H_n^{(2)'}(Z)} \right|. \qquad (10)$$

This may be rewritten as

$$p_s = p_0(2c/r)^{1/2}f_\infty(Z). \qquad (11)$$

From these results it is possible to consider several limiting cases. If $L_n \rightarrow 0$, the solution would apply to scattering by a rigid cylinder. If $L_n \rightarrow \infty$, the solution for a soft cylinder is obtained. The single layered fluid filled shell scattering coefficients are obtained if $c/a = 1.00$. The solid cylinder scattering coefficients are found by first taking $c/a = 1.00$ and letting b/a approach zero. One must make sure to take the limits of $Y_n(Z_L)$ and $Y_n(Z_S)$ each going to 0. The limiting process must be used since simply setting $Z_L = Z_S = 0$ gives 0 in both numerator and denominator of L_n.

CALCULATED RESULTS AND CONCLUSIONS

The calculations were performed for two-layered hollow (vacuum in the core) cylindrical shells in water. The layers consisted of Lucite (outer), an absorptive material, and iron (inner) considered to be lossless. Physical parameters for the materials assumed in the calculations are given in Table 1. Absorption is included in the theory by using complex wave numbers. Absorption factors expressed in Np were derived from Ref. 7 where they are given in dB.

Table 1 — Material properties

	$\rho \times 10^3$ (kg/m^3)	$V_L \times 10^3$ (m/sec)	$V_S \times 10^3$ (m/sec)	β_L (Np)	β (Np)
Iron	7.700	5.950	3.240	$\cdots$	$\cdot\cdot$
Lucite	1.180	2.680	1.380	0.00340	0.00531
Water	0.998	1.482^a	$\cdots$	$\cdots$	$\cdots$

^{a}Referred to as c_ω in the text.

Figures 2-8 show the computation of $|f_\infty(ka)|$, called MODULUS in the plots, for various shells. Computation for shells of three thicknesses ($b/a = 0.99$, 0.97, and 0.95) without an outer covering ($c/a = 1.00$) is shown in Figs. 2(a)-2(c), respectively. The functions were computed for intervals of ka of 0.05. Cases of shells having iron inner layers of the same thicknesses as the above-mentioned single-layer shells but with a thin Lucite covering layer ($c/a = 1.01$) are shown in Fig. 3(a)-3(c), and with a thicker Lucite covering layer ($c/a = 1.10'$) are shown in Fig. 4(a)-4(c). In all plots the abscissa is ka where $k = 2\pi/\lambda$ and λ is the wavelength of sound in water. In all cases as the iron thickness is increased, the curves exhibit similar behavior at low ka ($ka < 5$). The first dip becomes less pronounced and moves to a higher ka. Comparing Figs. 2(a) and 3(a), Figs. 2(b) and 3(b), or Figs. 2(c) and 3(c) would lead one to predict a decrease in variation of the modulus as ka is increased. This is however, fallacious. It can be noted that with a thicker Lucite outer layer, even for the case of the thinnest shell whose reflection was computed [$c/a = 1.10$ and $b/a = 0.99$, Fig. 4(a)], the modulus variation increased at higher

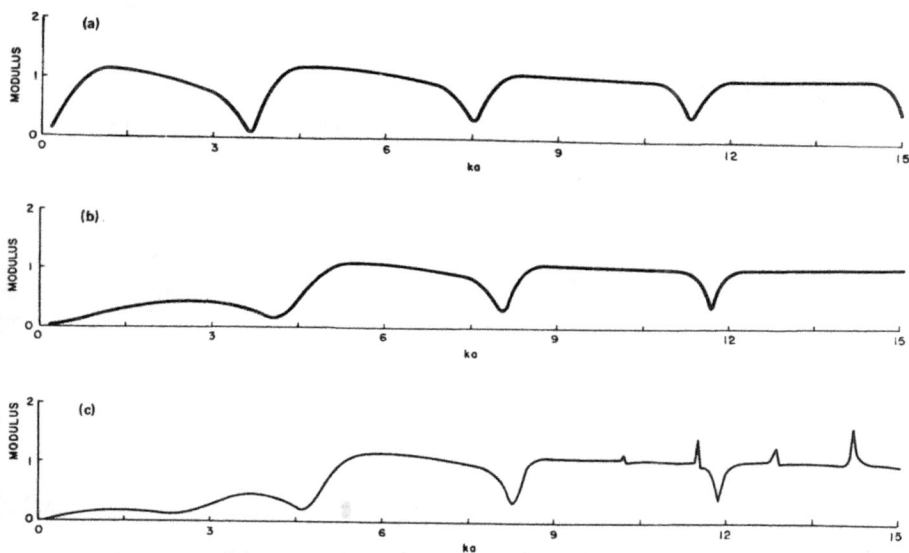

Fig. 2 — Theoretical calculations of the monostatic reflection expressed as MODULUS ($|f_\infty|$) from a two-layered hollow cylindrical shell with vacuum inside (a) $c/a = 1.00$, $b/a = 0.99$; (b) $c/a = 1.00$, $b/a = 0.97$; and (c) $c/a = 1.00$, $b/a = 0.95$.

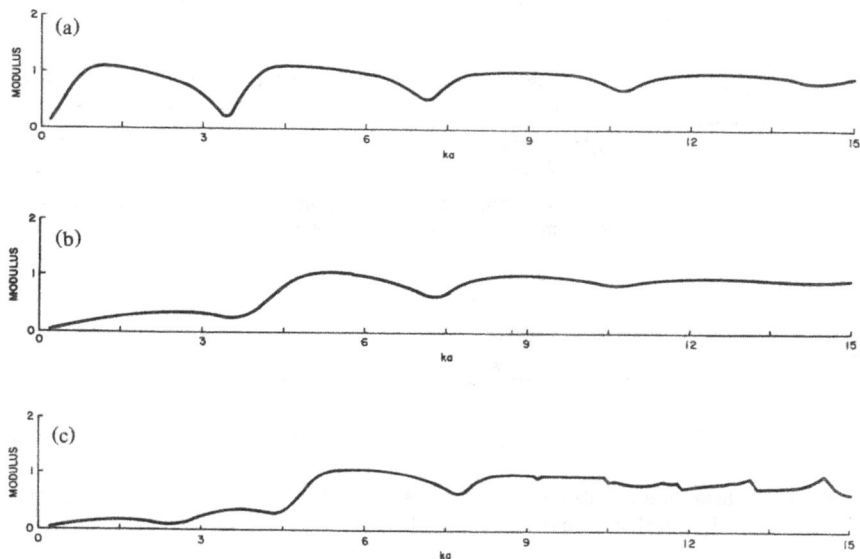

Fig. 3 — Theoretical calculation of the monostatic reflection expressed as MODULUS ($|f_\infty|$) from a two-layered hollow cylindrical shell with vacuum inside (a) $c/a = 1.01$, $b/a = 0.99$; (b) $c/a = 1.01$, $b/a = 0.97$; and (c) $c/a = 1.01$, $b/a = 0.95$.

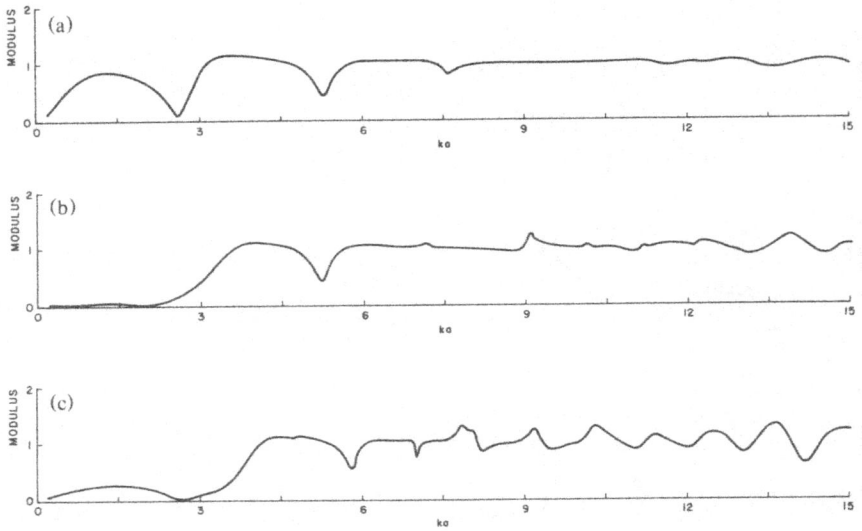

Fig. 4 — Theoretical calculation of the monostatic reflection expressed as MODULUS ($|f_\infty|$) from a two-layered hollow cylindrical shell with vacuum inside (a) $c/a = 1.10$, $b/a = 0.99$; (b) $c/a = 1.10$, $b/a = 0.97$; and (c) $c/a = 1.10$, $b/a = 0.95$.

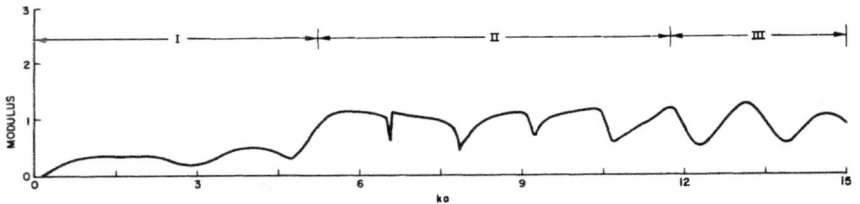

Fig. 5 — Delineation of three regions for the calculation of the reflection from a two-layered hollow cylindrical shell with a vacuum inside ($c/a = 1.01$, $b/a = 0.93$).

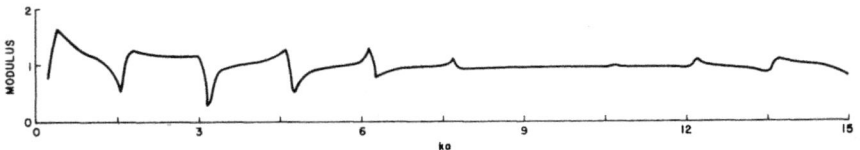

Fig. 6 — Theoretical calculation of the monostatic reflection from a hollow cylindrical Lucite shell with a vacuum inside ($c/a = 1.10$, $b/a = 1.00$).

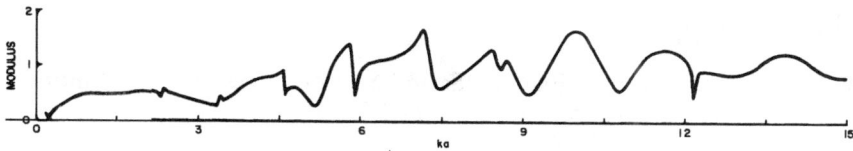

Fig. 7 — Theoretical calculation of the monostatic reflection from a hollow cylindrical iron shell with a vacuum inside ($c/a = 1.10$, $b/a = 0.90$).

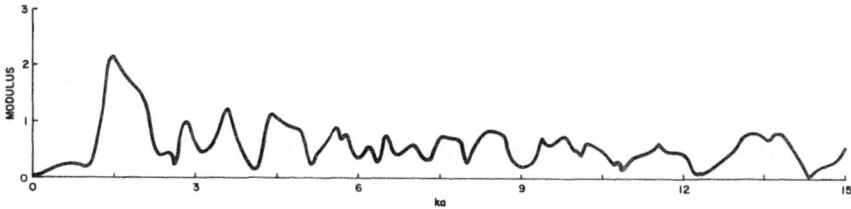

Fig. 8 — Reflection by a solid Lucite cylinder.

ka ($ka > 12$) in the same range where only a small variation was observed for a thinner Lucite covering [$c/a = 1.01$, Fig. 3(a)]. For a thick outer Lucite layer (layer 1 in Fig. 1), increased variation of the modulus resulted [Figs. 3(b) and 3(c)] over the same range of the ka values ($12 < ka < 15$) for which little variation occurred for an outer layer only one tenth as thick [Figs. 3(b) and 3(c)].

For many cases we can delineate three regions of the function $|f_\infty(ka)|$. Figure 5 illustrates these regions Region I is characterized by a fairly regular slow variation with ka. Region II is an irregular or mixed region. Region III tends to be regular or periodic and varying faster with ka than region I. These regions appear to be bounded within a certain ka range that can be identified in many cases. Varying the thickness ratios or complex moduli will change the ka delineation of these regions and in some cases can eliminate one or more of the regions.

By taking the appropriate limits in the matrix equations, these solutions reduce to the single-layered shell or solid cylinder case. The plot in Fig. 6 shows $|f_\infty(ka)|$ for a Lucite shell without any other material layer computed by taking $b/a = 1.00$ and $c/a = 1.10$. Similarly, a simple iron shell of the same thickness is considered by taking $b/a = 0.90$ and $c/a = 1.00$. The resulting computation is shown in Fig. 7. Allowing $b/a \to 0$ and $c/a = 1.00$, a solid cylinder case is approached. Figure 8 shows the computations for a solid Lucite cylinder.

REFERENCES

R. D. Doolittle and H. Überall, "Sound Scattering by Elastic Cylindrical Shells," J. Acoust Soc Am. **39**, 272-275 (1966).

2. P. Uginčius and H. Überall, "Creeping-Wave Analysis of Acoustic Scattering by Elastic Cylindrical Shells," J. Acoust. Soc. Am. **43**, 1025-1035 (1968).

3. Miquel C. Junger, "Sound Scattering by Thin Elastic Shells," J. Acoust Soc. Am. **24**, 366-373 (1952).

4. C. W. Horton, W. R. King, and K. J. Diercks, "Theoretical Analysis of the Scattering of Short Acoustic Pulses by a Thin-Walled Metallic Cylinder in Water," J. Acoust. Soc. Am. **34**, 1929-1932 (1962).

5. Richard H. Vogt, Lawrence Flax, Louis R. Dragonette, and Werner G. Neubauer, "Monostatic Reflection of a Plane Wave From an Absorbing Sphere," J. Acoust. Soc. Am. **57**, 558-561 (1975).

6. Lawrence Flax and Janet P. Mason, "Fortran Subroutines to Evaluate, in Single or Double Precision, Bessel Functions of The First and Second Kinds for Complex Arguments," Naval Research Laboratory Report No. 7997, Washington, DC, 1976.

7. Bruce Hartman and Jacek Jarzynski, "Ultrasonic Hysteresis Absorption in Polymers," J. Appl. Phys. **43**, 4304-4312 (1972).

APPENDIX: MATRIX ELEMENTS

The following are the expressions for the elements of the determinants appearing in Eqs. (5)-(11) in the text.

$$a_{11} = [J_n(Z_L) - 2G_1 J_n''(Z_L)]/(1 + 2G_1),$$

$$a_{12} = [Y_n(Z_L) - 2G_1 Y_n''(Z_L)]/(1 + 2G_1),$$

$$a_{13} = -2n[Z_s J_n'(Z_s) - J_n(Z_s)]/Z_S^2,$$

$$a_{14} = -2n[Z_s Y_n'(Z_s) - Y_n(Z_s)]/Z_S^2,$$

$$a_{21} = Z_L J_n'(Z_L),$$

$$a_{22} = Z_L Y_n'(Z_L),$$

$$a_{23} = nJ_n(Z_S),$$

$$a_{24} = nY_n(Z_S),$$

$$a_{31} = 2n[Z_L J_n'(Z_L) - J_n(Z_L)],$$

$$a_{32} = 2n[Z_L Y_n'(Z_L) - Y_n(Z_L)],$$

$$a_{33} = (Z_S)^2 J_n''(Z_S) + n^2 J_n(Z_S) - Z_S J_n'(Z_S),$$

$$a_{34} = (Z_S)^2 Y_n''(Z_S) + n^2 Y_n(Z_S) - Z_S Y_n'(Z_S),$$

$$a_{41} = [J_n(t_L) - 2G_1 J_n''(t_L)]/(1 + 2G_1),$$

$$a_{42} = [Y_n(t_L) - 2G_1 Y_n''(t_L)]/(1 + 2G_1),$$

$$a_{43} = 2n[t_s J_n'(t_s) - J_n(t_S)]/t_S^2,$$

$$a_{44} = -2n[t_s Y_n'(t_s) - Y_n(t_S)]/t_S^2,$$

$$a_{45} = (\rho_2/\rho_1)[J_n(q_L) - 2G_2 J_n''(q_L)]/(1 + 2G_2),$$

$$a_{46} = (\rho_2/\rho_1)[J_n(q_L) - 2G_L Y_n''(q_L)]/(1 + 2G_2),$$

$$a_{47} = -2n(\rho_2/\rho_1)[q_s J_n'(q_s) - J_n(q_S)]/q_S^2,$$

$$a_{48} = -2n(\rho_2/\rho_1)[q_s Y_n'(q_s) - Y_n(q_S)]/q_S^2,$$

$$a_{51} = t_L J_n'(t_L),$$

$$a_{52} = t_L Y_n'(t_L),$$

$$5_{53} = nJ_n(t_S),$$

$$5_{54} = nY_n(t_S),$$

$$a_{55} = q_L J_n'(q_L),$$

$$a_{56} = q_L Y_n'(q_L),$$

$$a_{57} = nJ_n(q_S),$$

$$a_{58} = nY_n(q_S),$$

$$a_{61} = 2n[t_L J_n'(t_L) - J_n(t_L)],$$

$$a_{62} = 2n[t_L Y_n'(t_L) - Y_n(t_L)],$$

$$a_{63} = [t_S^2 J_n''(t_S) + n^2 J_n(t_S) - t_S J_n'(t_S)],$$

$$a_{64} = [t_S^2 Y_n''(t_S) + n^2 Y_n(t_S) - t_S Y_n'(t_S)],$$

$$a_{65} = 2Un[q_L J_n'(q_L) - J_n(q_L)],$$

$$a_{66} = 2Un[q_L Y_n'(q_L) - Y_n(q_L)],$$

$$a_{67} = 2U[q_S^2 J_n''(q_S) + n^2 J_n(q_S) - q_S J_n'(q_S)],$$

$$a_{68} = 2U[q_S^2 Y_n''(q_S) + n^2 Y_n(q_S) - q_S Y_n'(q_S)],$$

$$a_{71} = nJ_N(t_L),$$

$$a_{72} = nY_N(t_L),$$

$$a_{73} = t_S J_n'(t_S),$$

$$a_{74} = t_S Y_n'(t_S),$$

$$a_{75} = nJ(q_L),$$

$$a_{76} = nY_n(q_L),$$

$$a_{77} = q_S J_n'(q_S),$$

$$a_{78} = q_S Y_n'(q_S),$$

$$a_{85} = [J_n(u_L) - 2G_2 J_n''(u_L)]/(1 + 2G_2),$$

$$a_{86} = [Y_n(u_L) - 2G_2 Y_n''(u_L)]/(1 + 2G_2),$$

$$a_{87} = -2n[(u_S J_n'(u_S) - J_n(u_S)]/u_S^2,$$

$$a_{88} = -2n[(u_S Y_n'(u_S) - Y_n(u_S)]/u_S^2,$$

$$a_{89} = -J_n(u)/\rho_f,$$

$$a_{95} = 2n[(u_L J_n'(u_L) - J_n(u_L)],$$

$$a_{96} = 2n[(u_L Y_n'(u_L) - Y_n(u_L)],$$

$$a_{97} = -u_S J_n'(u_S) + n^2 J_n(u_S) + u_S^2 J_n''(u_S),$$

$$a_{98} = -u_S Y_n'(u_S) + n^2 Y_n(u_S) + u_S^2 Y_n''(u_S),$$

$$a_{10,5} = u_L J_n'(u_L),$$
$$a_{10,6} = u_L Y_n'(u_L),$$
$$a_{10,7} = n J_n(u_S),$$
$$a_{10,8} = n Y_n(u_S),$$
$$a_{10,9} = u J_n'(u),$$

The argument of the functions are given in Eq. (7) and, as usual, the prime denotes derivative with respect to the argument.

Chapter 6

NONABSORBING AND ABSORBING CYLINDERS*

INTRODUCTION

The scattering of a plane sound wave by an infinite elastic cylinder has been examined by several investigators [1-3]. In those papers, the absorption of acoustic energy in the cylinder is assumed negligible. Although this assumption is undoubtedly valid in many real situations, there are, however, cases where absorption plays an important role. Here an infinitely long cylinder is considered analytically that is composed of a material in which wave absorption may be ignored and also a material in which both longitudinal and shear waves are significantly attenuated by absorption.

The material of the cylinder is assumed to be solid and isotropic, supporting compressional and shear waves with speeds c_L and c_S, respectively, and immersed in a fluid of sound speed c_f.

The pressure p_s reflected by a cylinder is given by Eq. (9) of Chapter 5 [4].

$$p_s = p_0 \sum_{n=0}^{\infty} \epsilon_n \left[\frac{J_n(ka)L_n - kaJ_n'(ka)H_n^{(2)}(kr)}{H_n^{(2)}(ka)L_n - kaH_n^{(2)'}(ka)} \right] \cos(n\theta), \qquad (1)$$

where p_0 is the incident acoustic pressure, J_n and $H_n^{(2)}$ are the Bessel function of the first kind and a Hankel function of the second kind, respectively, and ϵ_n is the Neumann factor. Primes indicate a derivative with respect to the argument. The polar coordinates of the field point are r and θ, a is the cylinder radius and k is the incident wave number. Equation (1) can be specialized to the farfield by substituting an asymptotic form for $H_n^{(2)}(kr)$ for large kr. The farfield form function magnitude is then

$$|F_\infty(ka, \theta)|$$
$$= \frac{2}{(\pi ka)^{1/2}} \sum_{n=0}^{\infty} \epsilon_n \left[\frac{J_n(ka)L_n - kaJ_n'(ka)}{H_n^{(2)}(ka)L_n - kaH_n^{(2)'}(ka)} \right] \cos(n\theta),$$

where the L_n are defined by Eq. (6) of Chapter 5.

---

*These results first appeared in: Luise S. Schuetz and Werner G. Neubauer, J. Acoust. Soc. Am., **62**, 513-517 (1977)

139

Absorption has been included, by introducing complex shear and compressional wave numbers in the solid as

$$k_s \equiv k(c_f/c_S)(1 - i\beta_S), \text{ and } k_L \equiv k(c_f/c_L)(1 - i\beta_L).$$

The β's are absorption factors, with the dimensions of Np, derived from the measurements of Hartmann and Jarzynski, [5] who reported their measurements as absorption coefficients expressed as $\alpha_L\lambda_L$ and $\alpha_S\lambda_S$ in dB. Conversion between these factors is accomplished according to the relations

$$\beta_L = \alpha_L\lambda_L/40\pi \log e \text{ and } \beta_s = \alpha_S\lambda_S/40\pi \log e.$$

The nonabsorptive case is given by $\beta_S = \beta_L = 0$.

The chief difficulty in this solution lies in the evaluation. Unexpected difficulties are encountered in calculating the Bessel functions over the required wide range of complex arguments [6]. Comparisons were made with experimental monostatic and bistatic form functions measured using lucite (absorbing) and aluminum (essentially nonabsorbing) cylinders.

EXPERIMENTS

The theory is two-dimensional, i.e., an infinite cylinder is assumed which, of course, can not be realized in a real so-called free field. However, reasonable agreement between two-dimensional theory and experiments was shown in Chapter 4 using finite metallic cylinders if reasonable care is exercised in performing the experiments. Plane-wave incidence demanded by the theoretical solution was approximated using directional sources. Schlieren photographs were taken to allow a preliminary examination of the entire acoustic field. Figure 1 shows the dark-field schlieren observation for an aluminum cylinder, with the magnitude of the theoretical bistatic form function, $|f(ka, \theta)|$, superimposed. The cylinder is seen end-on as a dark disk in the center and the source is visible as a black rectangular area at the top of the picture. A long incident pulse was used, and the photograph is taken at the instant when the pulse is just leaving the cylinder. The theoretical curve in the forward-scatter direction is not shown, since the theoretical results do *not* include the incident field.

This photograph indicates good agreement between theory and experiment at a distance of two or three diameters from the scatterer, implying that the cylinder is placed in the plane-wave field of the source.

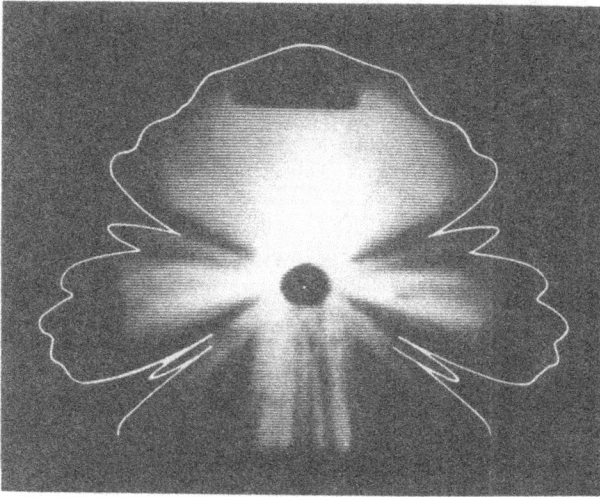

Fig. 1 — Schlieren photograph of the field scattered by a aluminum cylinder (ka = 24.58) with the calculated bistatic form
function superimposed.

A water-tank facility similar to the one described by Neubauer [7]
was used for the quantitative experiments. Cylinders with a large ratio
of length to radius were insonified using directional sources. Conventional piezoelectric transducers with square or circular active areas were
used. In fact, some of the results at 200 kHz were obtained using a
"fish-finder" source available at boating equipment stores for little cost.

In Fig. 1, the cylinder was placed in the nearfield, i.e., in the
plane-wave field, of the source. This was impossible in the quantitative
experiments. Instead the cylinder was placed at a large enough distance
so that the center of the main lobe was a sufficient approximation to a
plane wave. Figure 2 shows typical beam patterns of sources that were
used. Their use resulted in an incident field which was approximately
plane over a section of the cylinder between 2.5 and 5 cm in length.

Transducers were suspended in the water from two movable
tracks. One of these held the source, on the other the receiver was
mounted from a rotating arm, with the cylinder suspended at its axis of
rotation. Source-to-cylinder distances ranged from 15 to 30 cylinder
diameters. A fine nichrome wire mounted in the upper end cap was
used to hang the cylinder.

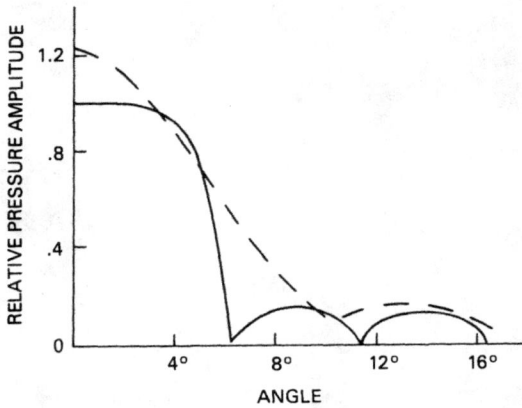

Fig. 2 — Typical measured beam patterns of
sources used in quantitative experiments.

The incident field was pulsed to reduce interfering reflections. The steady-state results were in some cases derived from the pulsed data by Fourier transforming short-pulse returns [8]; in others they were measured directly by approximating the steady state by a very long pulse. A repeated harmonic pulse was generated by gating the output of a synthesizer. A counter operating from the signal frequency was used to trigger the gate, and the same trigger, with a delay, initiated display of the received pulse. The resulting return was stable enough to permit signal averaging over long periods of time.

None of the following data are absolute. All are normalized in such a way that the averages of the theoretical and empirical curves are equal.

NONABSORBING CYLINDERS

The facility permitted the taking of bistatic data in the backscatter half-space. Figures 3-6 are plots of the scattering form function versus scattering angle at various ka, for cylinders composed of 6061 aluminum. Data are taken using cylinders ranging in diameter from 0.90 to 2.54 cm with acoustic frequencies of from 0.2 to 1.2 MHz. Previous work using aluminum spheres [8] at similar frequencies gave good agreement between theory and experiment when the theory neglected absorption. Therefore, it is reasonable to assume that at these frequencies aluminum is virtually nonabsorbing. For the theoretical curves shown here that assumption is made. The theoretical curves are calculated taking the longitudinal and shear wave speeds of 6370 and 3120

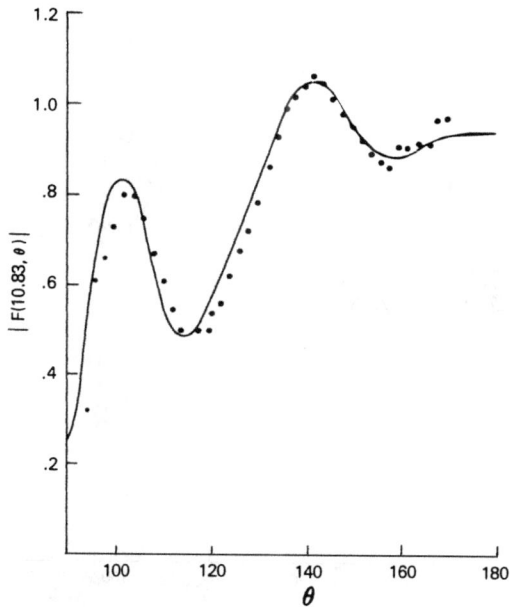

Fig. 3 — Bistatic form function of a 6061 aluminum cylinder, $ka = 10.83$, $c_f = 1473.39$ m/sec. Theory ——; experiment, ●.

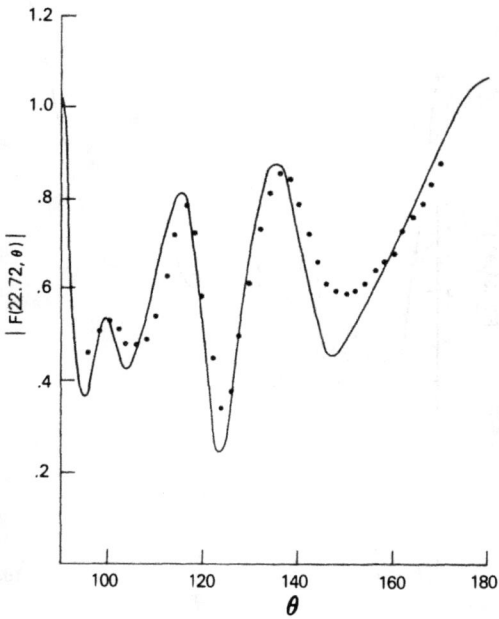

Fig. 4 — Bistatic form function of a 6061 aluminum cylinder, $ka = 22.72$, $c_f = 1470.00$ m/sec. Theory,——; experiment, ●.

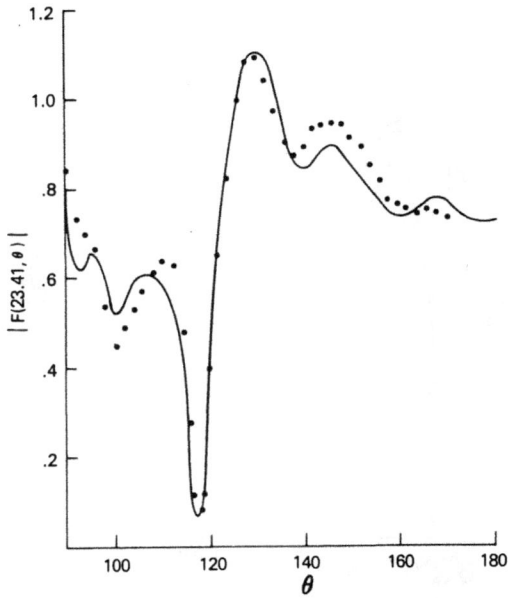

Fig. 5 — Bistatic form function of a 6061 aluminum cylinder, $ka = 23.41$, $c_f = 1477.65$ m/sec. Theory, ——; experiment, ●.

Fig. 6 — Bistatic form function of a 6061 aluminum cylinder, $ka = 31.31$, $c_f = 1468.96$ m/sec. Theory, ——; experiment 1, ×; experiment 2, ●.

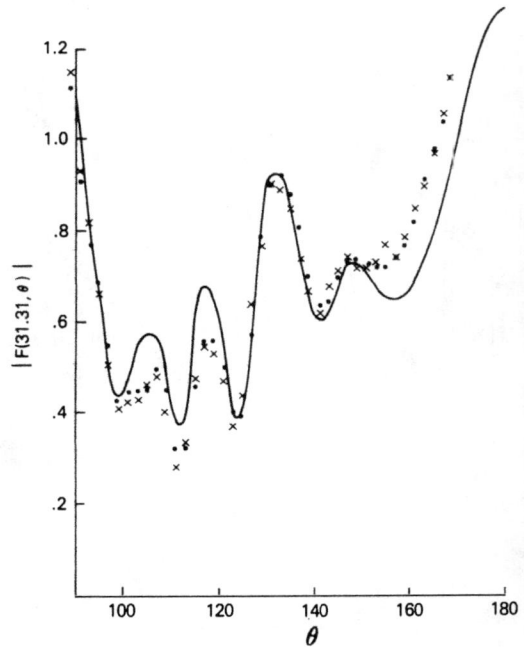

m/sec, respectively. These wave speeds are believed to be accurate to within one percent. The sound speed in water was derived from Ref. 9.

As indicated above, the section of the cylinder which is approximately plane-wave illuminated is only about three diameters in length. In view of this fact, the agreement between theory and experiment is surprisingly good. The ratio of plane-wave illuminated length to cylinder diameter decreases as *ka* increases in these experiments. It is, therefore, not surprising to see a decrease in agreement with increasing *ka*.

Figure 7 is a plot of the backscatter form function versus *ka*, also using a cylinder of 6061 aluminum. The experimental curve is the result of analysis of transient data processed as described in Chapter 8. Essentially, the form function versus frequency is calculated by fast Fourier transforming both a short incident pulse and the reflected return. The form-function magnitude is then given by the normalized ratio of the reflected to the incident spectrum.

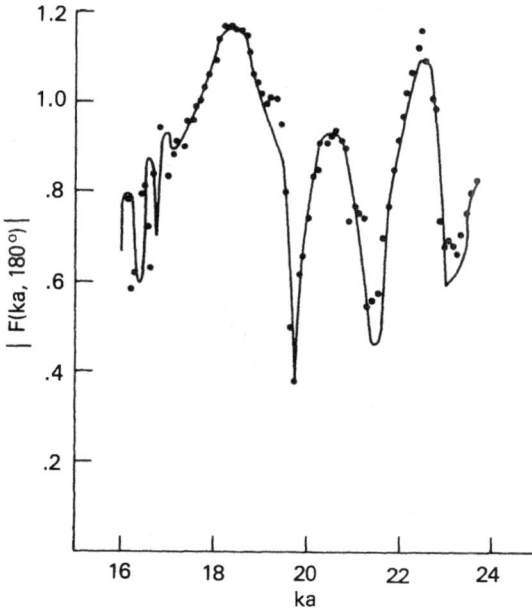

Fig. 7 — Backscatter form function of a 6061 aluminum cylinder. Theory, ———; experiment, •.

ABSORBING CYLINDERS

In view of the agreement obtained in the above cases, it was assumed that the two-dimensional geometry of the theory had been sufficiently approximated. Under the same experimental conditions similar observations were made using lucite cylinders at ka of 4.30 and 10.90. In this case, longitudinal and shear wave absorption was included for calculating the theoretical curves. The absorption factors β_L and β_S were taken equal to -0.00348 and -0.00531, respectively.

By far the greatest uncertainty in this case lies in the values of wave speeds, which are difficult to measure, and which, furthermore, have been found to vary greatly from one sample of plastic to the next. For example, Refs. 5 and 6 arrive at shear wave speeds of 1380 m/sec and 1340 m/sec, respectively, in two different samples of lucite. As shown in Fig. 8, the theoretical form function is highly sensitive to these velocities, so that an error of 1% or 2% completely changes the character of the curve. With only a limited knowledge of the wave speeds (within 3%) it was nevertheless possible to test the model in the least-squares sense. Highly overdetermined systems were used, with the wave speeds as parameters. In every case, the result was an extremely good fit, with narrow confidence limits, and a small standard deviation. Calculated wave speeds fell well within the expected range.

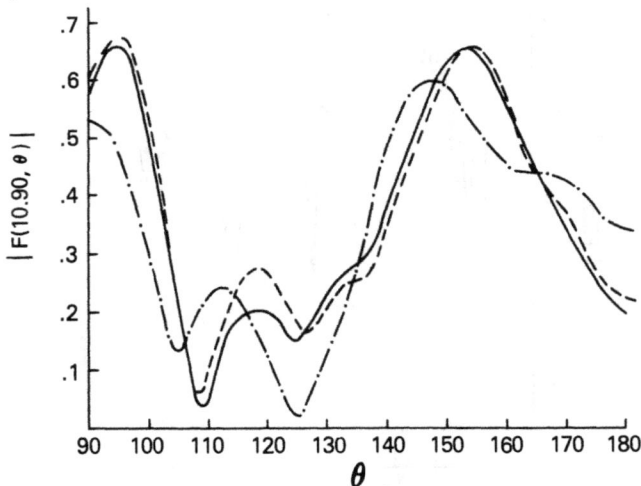

Fig. 8 — Variation of theoretical form function with small differences in wave speeds for a lucite cylinder, $ka = 10.90$, $c_f = 1469.64$ m/sec; $c_L = 2680.00$ m/sec, $c_S = 1380.00$ m/sec ———; $c_L = 2680.00$ m/sec, $c_S = 1350.00$ m/sec — · — · — ·; and $c_L = 2710.00$ m/sec, c_S 1380.00 m/sec.

Figures 9 and 10 are the bistatic form functions of two different lucite cylinders. Wave speeds obtained from the least-squares fit are indicated in the captions. Both cases resulted in an excellent fit, with a standard deviation of 0.02 in the first case, and 0.01 in the second case.

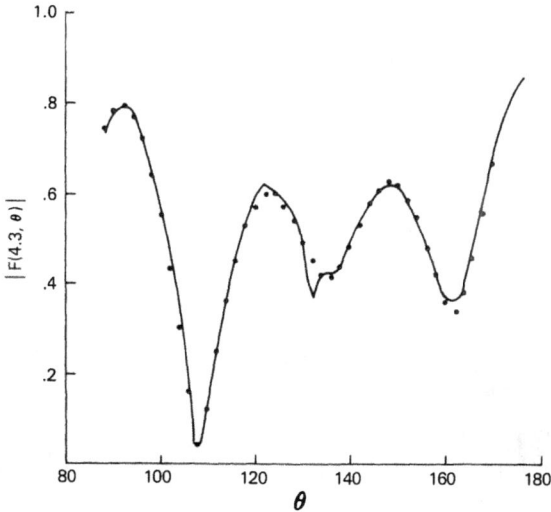

Fig. 9 — Bistatic form function for a lucite cylinder, $ka = 4.30$, $c_L = 2717.30$ m/sec, $c_S = 1312.90$ m/sec, and $c_f = 1468.27$ m/sec. Experiment ——; theory ●.

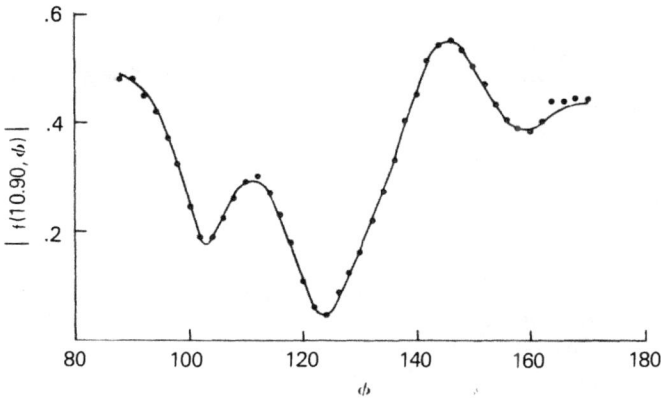

Fig. 10 — Bistatic form function for a lucite cylinder, $ka = 10.90$, $c_L = 2688.50$ m/sec, $c_S = 1340.00$ m/sec, and $c_f = 1468.72$ m/sec. Experiment ——; theory ●.

The importance of the absorption in the theory is indicated by Fig. 11. The two curves are identically calculated backscatter form functions, except that one allows the absorption coefficient to go to zero, and the other does not. In the latter case (solid curve) $\alpha_L = -0.00348$ and $\alpha_S = -0.00531$.

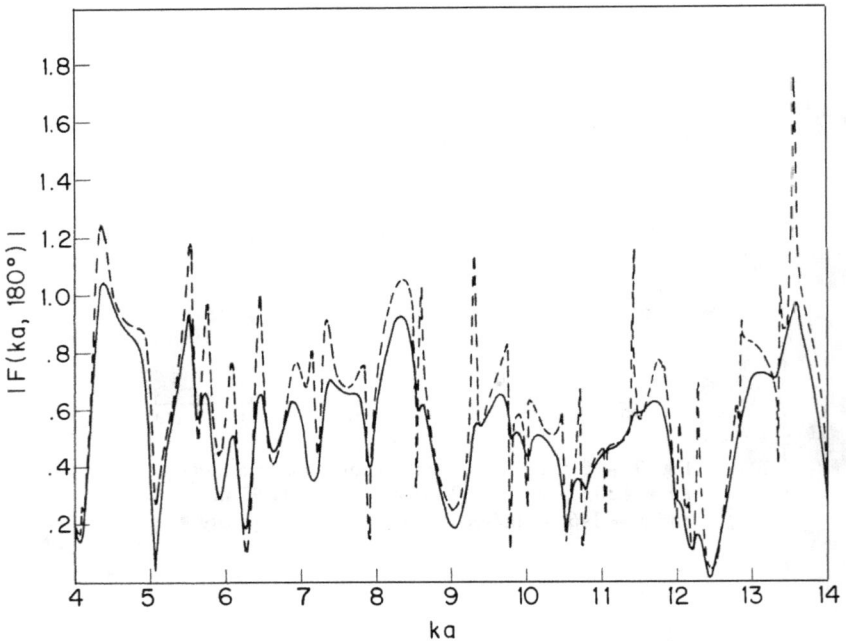

Fig. 11 — Effects of absorption on backscatter form function of a lucite cylinder; – – – neglecting absorption ($\alpha_S = \alpha_L = 0$) and ——— including absorption ($\alpha_S = -0.00531$ and $\alpha_L = -0.00348$).

CONCLUSIONS

In summary, we find that very good predictions can be made theoretically of scattering from absorbing cylinders, provided the acoustic wave speeds and absorption factors are known to the required accuracy. In the case of lucite, in particular, these wave speeds are difficult to measure, and widely variable from one sample to the next.

Furthermore, these results indicate that experiments using conventional sources can give good approximations of two-dimensional calculations.

REFERENCES

1. J. J. Faran, Jr., J. Acoust. Soc. Am. **23**, 405-418 (1951).

2. C. W. Horton, W. R. King, and K. J. Diercks, J. Acoust. Soc. Am. **34**, 1929-1932 (1962).

3. William W. Ryan, Jr., Applied Research Laboratory, University of Texas Report ARL TR-73-20, Austin, TX (August 1973) (unpublished).

4. Lawrence Flax and Werner G. Neubauer, J. Acoust. Soc. Am. **61**, 307-312 (1977).

5. Bruce Hartmann and Jacek Jarzynski, J. Appl. Phys. **43**, 4304 (1972).

6. Lawrence Flax and Janet P. Mason, NRL Report 7997, Washington, DC (April 1976) (unpublished).

7. W. G. Neubauer, NRL Report 5688, Washington, DC (October 1961) (unpublished).

8. Louis R. Dragonette, Richard H. Vogt, Lawrence Flax, and Werner G. Neubauer, J. Acoust. Soc. Am. **55**, 1130-1137 (1974).

9. Werner G. Neubauer and Louis R. Dragonette, J. Acoust. Soc. Am. **36**, 1685-1690 (1964).

Chapter 7

A CYLINDRICAL CAVITY IN AN
ABSORPTIVE MEDIUM*

INTRODUCTION

Formulation and computation of elastic wave reflections by a cylindrical cavity in a solid homogeneous isotropic medium with no losses has been given by Lewis et al., [1] among others [2-4]. Several materials have been considered as the wave-supporting medium. Among these is polyethylene, a material known [5] to have the property of significant attenuation of both compressional and shear waves. In the range of low megahertz frequencies most applications would indeed appropriately ignore absorption in the ambient medium for the case of most metals. However, grainy metals such as brass can have extraordinary losses and thus can cause problems in some cases. If polymers and other nonmetals are considered, wave attenuations become significant and should be included in the theory. Although a very large number of materials were not considered, a sufficient number were considered by Lewis et al. that it seems that materials fall into two rather arbitrary but convenient categories depending on the behavior of the total scattering cross section (SCS) versus ka curves. In this paper k will designate the wave number of the longitudinal wave, κ that of the shear wave, and a is the cavity radius. In their work some materials approach a limit of SCS with increasing wave parameter $(ka, \kappa a)$ from above and some from below when no losses are considered. The plots of Fig. 1 distinguish those two categories. The ordinate labeled SCS is the total scattering cross section following, but deviating slightly from, the definition of White [2]. He defines the total scattering cross section as the total scattered power per unit length of cylinder divided by the intensity. We imposed the additional modification used by Lewis et al. [1] of normalizing to the cylinder radius by simply dividing by $2a$. The upper curve in Fig. 1 is computed for polyethylene, here considered lossless, approaching a limit from above with increasing ka of SCS slightly above 2. This case also has a peak at

*These computations first appear in: Lawrence Flax and Werner G. Neubauer, J. Acoust. Soc. Am. **63**, 675-680 (1978).

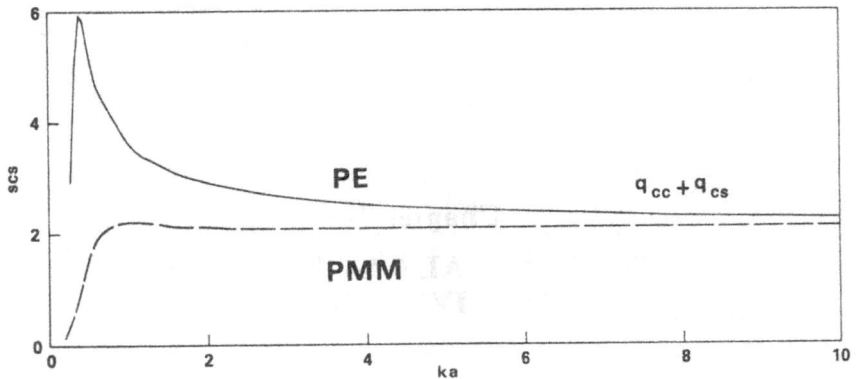

Fig. 1 — A comparison of the scattering cross section (SCS) of a cylindrical cavity for an incident compressional wave as a function of ka for PMM and PE without attenuation.

low ka. Lucite is the material for which the lower curve was computed and approaches the limiting value of SCS from below with increasing ka. The parameter that characterizes the scattering-cross-section-function behavior is the ratio of compressional-wave speed c_c to shear-wave speed c_s. The "critical value" for that ratio (c_c/c_s) which determines to which of the two classes a particular material belongs is very close to 2. Examples of metals [1] that belong to the same class as lucite are beryllium $(c_c/c_s = 1.45)$, silver (1.64), and tungsten (1.82) and ones that fall in the same category as polyethylene are aluminum (2.1) and gold (2.7). A thorough parametric study of all possible case was not attempted; rather in this study samples of Lucite (a brand of polymethyl methacrylate PMM) were chosen as an example to calculate as a member of one category and having significant, but small, wave attenuations. Polyethylene (PE) was chosen as the member of the other category but is known to have significantly greater wave attenuations than PMM.

NOTATION

The elastic material containing the cylindrical cavity will support shear and compressional waves and both will be considered as incident waves and mode conversion on reflection will produce shear from compressional as well as shear, and vice versa. Also, computations ignoring and including absorption will be presented. Hence, a word about notation is in order. Scattering cross sections of separate waves will be designated with a q. A first subscript will designate the kind of incident wave, c for compressional and s for shear. A second subscript, again c or s, will designate the reflected wave type. Whereas an unprimed q designates a computation ignoring wave attenuations, primes designate the same computations including attenuations. The physical constants used in the computations are given in Table 1.

Table 1 — Material Properties

	PMM[a]	PE
c_c (m/s)	2680	1950[b]
c_s (m/s)	1350	540[b]
c_c/c_s	1.94	3.61[b]
β_c (Np)	0.0035	0.0070[c]
β_s (Np)	0.0053	0.0212[c]

[a]See Ref. 7
[b]See Ref. 1
[c]See Ref. 5

THEORY

Consider a cylindrical cavity of radius a embedded in an isotropic absorptive medium characterized by density ρ and speeds c_c and c_s as determined earlier. The axis of the cylindrical cavity is taken to be the z axis of the cylindrical coordinate system (r, ϕ, z). Both types of incident waves of circular frequency ω are considered, i.e., compressional waves and shear waves polarized in a direction normal to the cylindrical axis.

Compressional-Wave Incidence

In the solid medium the incident plane acoustic wave $p_i = p_0 \exp(-ik_c x)$ can be expanded in cylindrical coordinates as

$$p_i = p_0 \sum_{n=0}^{\infty} \epsilon_n (-i)^n J_n (k_c r) \cos (n\phi). \tag{1a}$$

The scattered pressure is

$$p_s = p_0 \sum_{n=0}^{\infty} \epsilon_n C_n H_n^2 (k_c r) \cos (n\phi) \tag{1b}$$

$$+ p_0 \sum_{n} \epsilon_n B_n H_n^{(2)} (k_s r) \cos (n\phi).$$

The incident pressure amplitude is p_0, ϵ_n is the Neumann factor, $k_c = \omega/c_c$, $k_s = \omega/c_s$, $H_n^{(2)} = J_n - iY_n$, where J_n and Y_n are Bessel functions of the first and second kind. The $H_n^{(2)}$ are Hankel functions, and C_n

and B_n are the scattering coefficients. Throughout the time dependence $\exp(i\omega t)$ has been suppressed.

At the boundary of the cavity $r = a$, the tangential stresses must be zero. The coefficients C_n and B_n are determined from the boundary conditions and are given by a division of two second-order determinates of the form

$$C_n = \frac{\begin{vmatrix} a_{11} & a_{13} \\ a_{31} & a_{33} \end{vmatrix}}{\begin{vmatrix} a_{12} & -a_{13} \\ a_{32} & -a_{33} \end{vmatrix}},$$

(2)

$$B_n = \frac{\begin{vmatrix} a_{12} & -a_{11} \\ a_{32} & -a_{31} \end{vmatrix}}{\begin{vmatrix} a_{12} & -a_{13} \\ a_{32} & -a_{31} \end{vmatrix}}.$$

The elements of the determinants are

$$a_{11} = [J_n(Z_c) - 2GJ_n''(Z_c)]/(1 + 2G),$$
$$a_{12} = [H_n^{(2)}(Z_c) - 2GH_n''^{(2)}(Z_c)]/(1 + 2G),$$
$$a_{13} = -(2n/Z_s^2)[Z_s H_n'^{(2)}(Z_s) - H_n^{(2)}(Z_s)],$$
$$a_{14} = -(2n/Z_s^2)[Z_s J_n'(Z_s) - J_n(Z_s)],$$

(3)

$$a_{31} = 2n[Z_c J_n'(Z_c) - J_n(Z_c)],$$
$$a_{32} = 2n[Z_c H_n'^{(2)}(Z_c) - H_n^{(2)}(Z_c)],$$
$$a_{33} = Z_s^2 H_s''^{(2)}(Z_s) - Z_s H_n'^{(2)}(Z_s) + n^2 H_n^{(2)}(Z_s),$$
$$a_{34} = Z_s^2 J_n''(Z_s) - Z_s J_n'(Z_s) + n^2 J_n(Z_s).$$

The arguments of the function appearing in these expressions are determined from the following relationships:

$$Z_c = k_c a - i\beta_c k_c a,$$

(4)

$$Z_s = k_s a - i\beta_s k_s a,$$

where β's are absorption factors given in nepers.

In addition, ratios of complex wave numbers appear in some of the matrix elements and are given by

$$G = Z_c^2/(Z_s^2 - 2Z_c^2).$$

(5)

As a result of using complex wave numbers in the theory, the elements of the determinants defining C_n and B_n contain Bessel functions with complex arguments. Techniques developed by Flax and Mason [6] for calculating these functions were used.

Dimensionless scattering cross sections, as defined in the Introduction are

$$q_{cc} = \sum_{n=0}^{\infty} \epsilon_n |C_n|^2 |H_n^{(2)}(Z_c r/a)|^2,$$

$$q_{cs} = \sum_{n=0}^{\infty} \epsilon_n |B_n|^2 |H_n^{(2)}(Z_s r/a)|^2. \tag{6}$$

The nondimensional parameter r/a is the distance from the cavity axis. For large ratios of r/a, the asymptotic expression for $H_n^{(2)}$ is used:

$$H_n^{(2)}(Z) = (2/\pi Z)^{1/2} \exp[-i(Z - n\pi/2 - \pi/4)], \tag{7}$$

where $Z = x + iy$.

Shear-Wave Incidence

For the case of an incident shear wave the same procedure was used as for the compressional case. The coefficients D_n and E_n are again determined from the boundary conditions and are given by the quotient of two second-order determinants of the form:

$$D_n = \frac{\begin{vmatrix} a_{13} & a_{14} \\ a_{33} & a_{34} \end{vmatrix}}{\begin{vmatrix} a_{12} & -a_{13} \\ a_{32} & -a_{33} \end{vmatrix}},$$

$$E_n = \frac{\begin{vmatrix} a_{12} & a_{14} \\ a_{32} & a_{34} \end{vmatrix}}{\begin{vmatrix} a_{12} & -a_{13} \\ a_{32} & -a_{33} \end{vmatrix}}. \tag{8}$$

Similarly, the normalized cross sections for monostatic reflectors are found to be

$$q_{sc} = \sum_{n=0}^{\infty} \epsilon_n |D_n|^2 |H_n^{(2)}(Z_c r/a)|^2,$$

$$q_{ss} = \sum_{n=0}^{\infty} \epsilon_n |E_n|^2 |H_n^{(2)}(Z_s r/a)|^2. \tag{9}$$

COMPUTATIONAL RESULTS

In all calculations an assumed value of the range at which the field is computed is taken to be 10 radii of the cylindrical hole, i.e., $r/a = 10$. This was taken to be the farfield as defined by a comparison of a computation using an asymptotic form of the Hankel function for that r and an equivalent explicit evaluation.

The scattering cross sections for compressional-wave incident on a cavity in PMM and a resultant scattered compressional wave is shown versus ka in Fig. 2 for calculations ignoring and including attenuation. In all plots, the wave number plotted on the abscissa is that of the incident wave. It can be seen in Fig. 2 that for a hole diameter slightly larger than one incident wavelength $(ka = 10)$, the SCS attributed to a compressional wave with attenuation is approximately one-half of the SCS in PMM when attenuation is ignored. The comparable diminution in SCS because of shear reflection, generated by mode conversion from the incident compressional wave is shown in Fig. 3. The SCS for a compressional-wave incident and a combined result of compressional and shear reflection from the cylindrical hole is shown in Fig. 4 again for PMM. An interesting behavior of the computation near a ka of unity is apparent when the curves labeled q'_{cc} and q'_{cs} are plotted together as in Fig. 5. A decreased value or "dip" in the curve for q'_{cc} is compensated for by an increased value or "bump" in the curve for q'_{cs} to result in a combined smooth curve in that region in the plot of $q'_{cc} + q'_{cs}$.

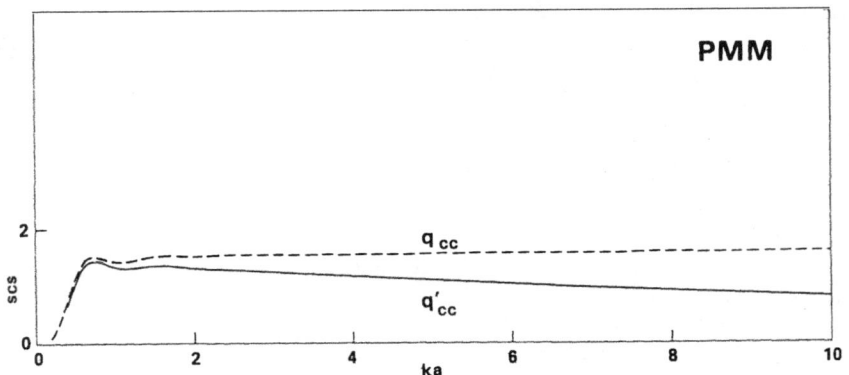

Fig. 2 — A comparison of the compressional component of the SCS for PMM as a function of ka with and without wave attenuation for compressional-wave incidence.

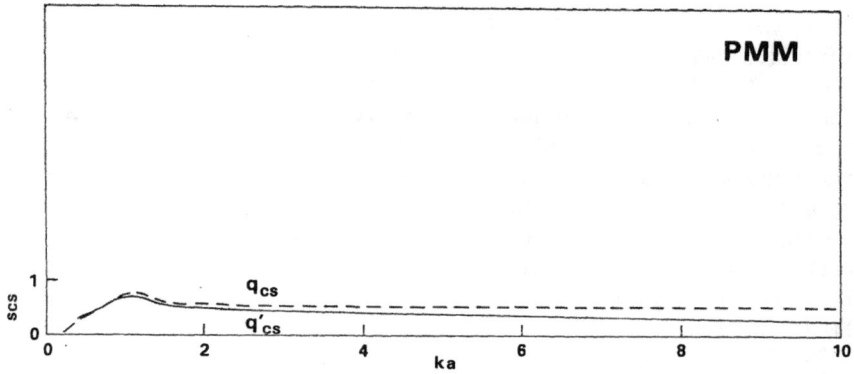

Fig. 3 — A comparison of the shear component of the SCS for PMM as a function of ka with and without wave attenuation for compressional-wave incidence.

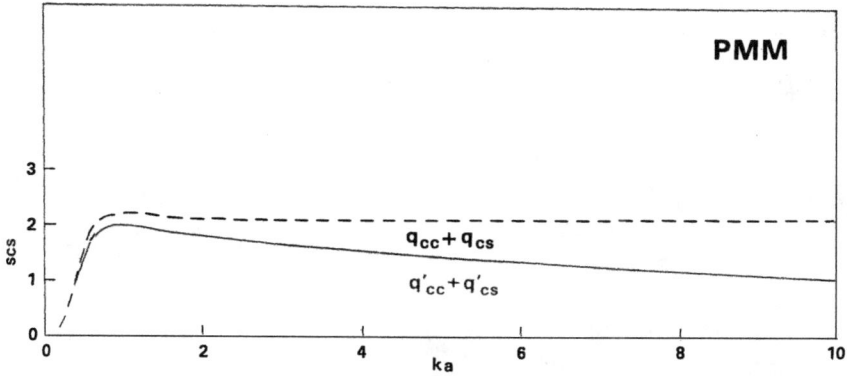

Fig. 4 — A comparison of the net SCS for PMM as a function of ka with and without attenuation for compressional-wave incidence.

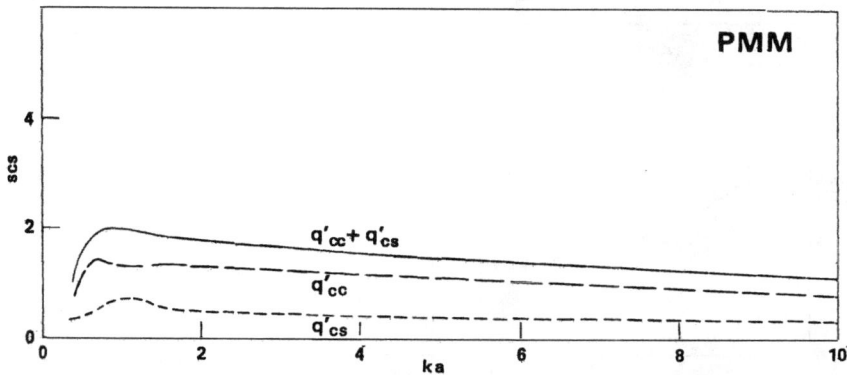

Fig. 5 — The net and components of the SCS for PMM as a function of ka with wave attenuation for compressional-wave incidence.

The results of similar computations are shown in Figs. 6 and 7 for PE which has, as can be seen in Table 1, somewhat lower wave speeds, double the compressional-wave attenuation and approximately four times the shear-wave attenuation. The contribution to the total SCS by the shear wave generated from mode conversion of a compressional wave incident on the cavity is very small for all values of ka. This is apparent from the similarities of the differences between each pair of curves in Figs. 6 and 7. The importance of including absorption in the calculations is apparent from the difference in magnitude of SCS between the computations with and without attenuation. Curves similar to those shown in Fig. 1 which seem to approach a limiting value of SCS when no attenuation was included are shown in Fig. 8 for computations including attenuation. The waves in the two categories referred to previously do not approach a single value. They cross over at a value of ka between 3 and 4 for these materials.

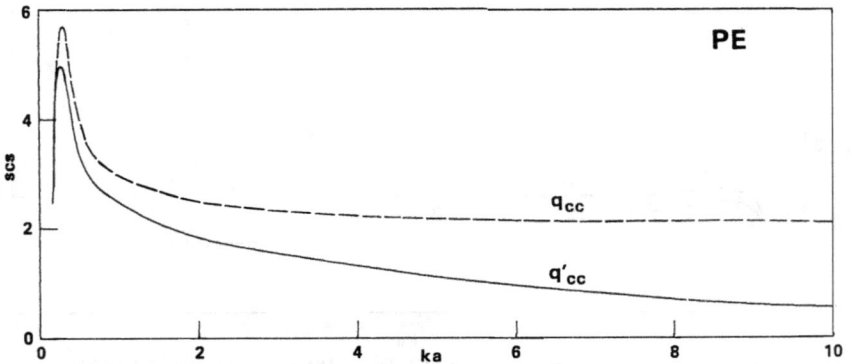

Fig. 6 — A comparison of the compressional component of the SCS and PE as a function of ka with and without wave attenuation for compressional-wave incidence.

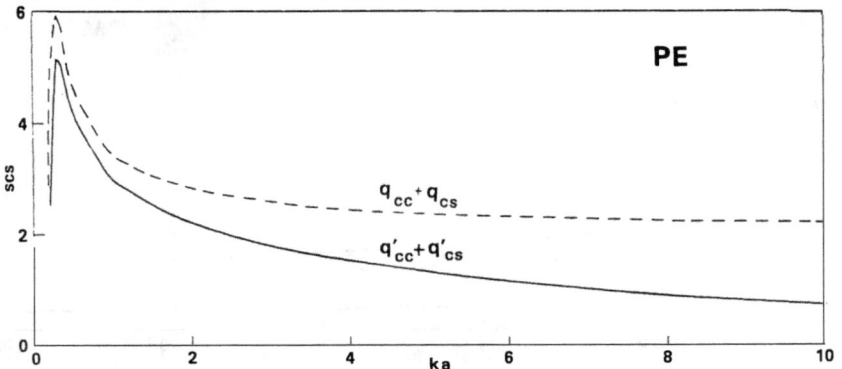

Fig. 7 — A comparison of the net SCS for PE as a function of ka with and without wave attenuations for compressional-wave incidence.

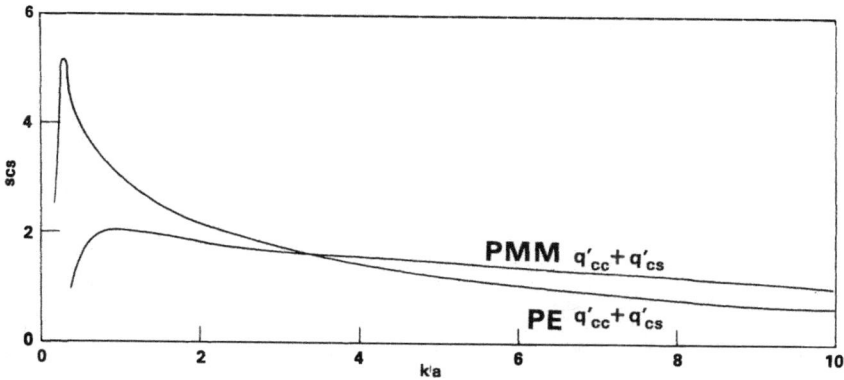

Fig. 8 — A comparison of the SCS of a cylindrical cavity for an incident compressional
wave as a function of *ka* for PMM and PE with attenuation.

Computations of SCS similar to those for an incident compressional wave on the cavity were carried out for an incident shear wave.
The results of those computations are plotted in Figs. 9 and 10 for
PMM and PE, respectively. The plots of Fig. 9 show a curious result.
The computations of SCS including wave attenuation results in a magnitude which is, over a small *ka* range near unity, larger than that not
including attenuation. No explanation is known for the behavior at
present. It should be mentioned that for PE not only is the mode
conversion a negligible contribution to the SCS, as suggested by others,
[1] but the total shear-wave contribution is negligible for κa above 8.

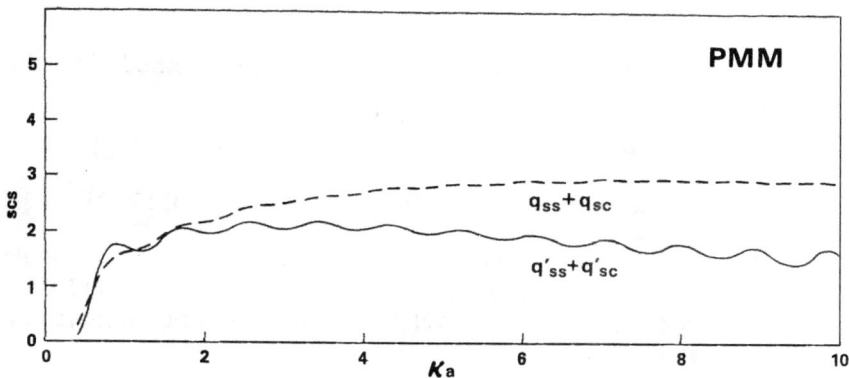

Fig. 9 — A comparison of the net SCS for PMM as a function of κa with
and without wave attenuations for shear-wave incidence.

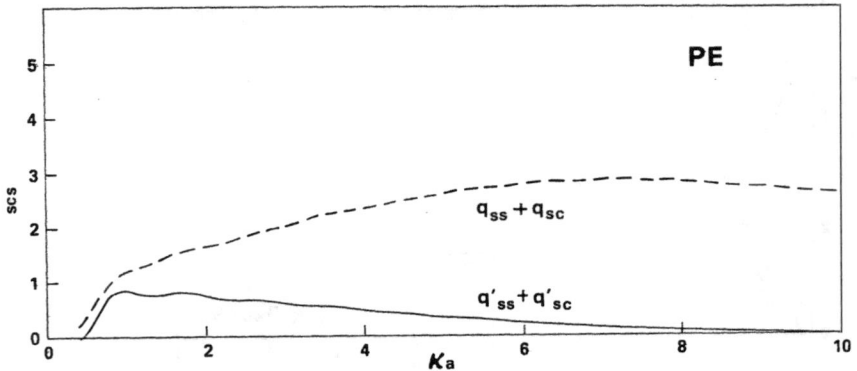

Fig. 10 — A comparison of the net SCS for PE as a function of κa with and without wave attenuations for shear-wave incidence.

A caution in conceptualization of the problem treated here is in order. The scattering of waves by a cavity in a solid does not really change locally at the cavity. The apparent scattering behavior appears different as expressed by the function of SCS versus ka or κa because the waves, compressional and shear, are attenuated in the course of propagation to a point in the field. Therefore, it is probably more useful to consider the hole and a surrounding elastic medium out to a given radius as a total system whose characteristics are sought.

REFERENCES

1. T.S. Lewis, D.W. Kraft, and N. Hom, *J. Appl. Phys.* **47**, 1795-1798 (1976).

2. R.M. White, *J. Acoust. Soc. Am.* **30**, 771-785 (1958).

3. R. Truell, C. Elbaum, and B. Chick, Ultrasonic Methods in Solid State Physics (Academic, New York, 1969).

4. C.F. Ying and R. Truell, *J. Appl. Phys.* **27**, 1086-1097 (1956).

5. B. Hartman and J. Jarzynski, *J. Appl. Phys.* **43**, 4304-4312 (1972).

6. L. Flax and J.P. Mason, "Fortran Subroutines to Evaluate, in Single or Double Precision, Bessel Functions of the First and Second Kind for Complex Arguments," Naval Research Laboratory Report No. 7997, Washington, DC (1976).

7. L. Flax and W.G. Neubauer, *J. Acoust. Soc. Am.* **61**, 307-312 (1977).

Chapter 8

ELASTIC SPHERES —
STEADY-STATE SIGNALS

INTRODUCTION

The closest approach of experiment to theory in a reflection situation would be expected to be for a problem of finite elastic geometry for which the wave equation is separable and for which all conditions are known. In geometries that are not finite, such as cylindrical geometry, the approximation of the experiment would be in question, i.e., the lack of experimental satisfaction of the infinite dimension inherent in the solution of cylindrical geometry, or the approximations of the theory would be in question, i.e., the lack of theoretical validity of approximating the finite length of the cylinder. The problem of reflection from a sphere is the simplest problem of finite geometry suitable for defining the methods and accuracy of the system and methods used for measuring reflection. Meaningful definitive measurements of reflected pressure cannot be expected to be any better for other reflectors than the correspondence that can be achieved for such a well-defined analytically calculable problem.

The free-field backscattered acoustic pressure amplitude was measured in the farfield for aluminum and tungsten-carbide spheres immersed in water. These were compared with the wave harmonic steady-state computer solutions incorporating the conditions of the experiments. A close correspondence between theory and experiment was sought for the relatively simple case of the sphere as well as a determination of the effect caused by specific experimental parameters, such as temperature and material wave speeds, on achieving agreement. This was necessary so that the experimental facilities and methods could be used to obtain measurements having known limits of error or reasonably estimated ones for cases in which little or no guiding theory is available.

The scattering of acoustic waves by a solid elastic sphere in water was described theoretically and experimentally by others [1-3]. Acoustic scattering by a solid, homogeneous, isotropic, elastic sphere immersed in a uniform, inviscid, elastic fluid was described quantitatively by Faran [1] before the advent of digital computers for the case of a plane-harmonic-incident wave. Hampton and McKinney [2] made an experimental study in the absence of theoretical computation of the scattering from solid metal spheres in water, for pulses of sine waves which were both of long and short duration compared to the size of the sphere. Hickling [3] obtained expressions describing the backscattered field by the use of a digital computer for incident sinusoidal pulses, either spherical or plane, and obtained computer solutions for several sphere materials. Numerical evaluations of the formulas describing monostatic scattering of harmonic waves used to compare the measurements discussed below with the theory were made by Rudgers [4].

The elements of the theory for the reflection by a sphere are reproduced here for the purposes of definition. Let the pressure of a plane wave traveling in the direction of the positive z-axis and incident on the sphere centered at the origin be

$$p_i = p_0 \exp [i(\omega t - kz)],$$

where k is the wavenumber and ω the angular frequency. The pressure at a point outside the sphere is the sum of the incident pressure field p_i and a scattered pressure field p_s,

$$p = p_i + p_s,$$

where the scattered pressure amplitude may be written

$$p_s = (p_0 a/2r) \exp [i(\omega t - kr)]f_r,$$

where a is the radius of the sphere, r the distance from the origin to the field point and f_r is called the reflection "form-function." That function is a complex quantity whose modulus is related to the experimental observations by the expression

$$|f_r| = (2r/a)(|p_s|/p_0),$$

where p_0 is the measured incident pressure amplitude at the origin in the absence of the sphere and $|p_s|$ the observed scattered pressure amplitude.

A sound pulse was used for the measurements, an approximation to the steady-state amplitude being achieved by making the length of the pulse train many sphere diameters and appropriately interpreting a steady-state amplitude at a specific position within the scattered pulse.

In the monostatic case (scattering at 180°, or backscattering) the form function is given by [3]

$$f_r = -2(kr/ka)e^{ikr}\sum_{n=0}^{\infty}(-i)^{n+1}h_n^{(2)}(kr)(-1)^n(2n+1)\sin\eta_n e^{i\eta_n},$$

where $h_n^{(2)}(kr)$ is a spherical Hankel function and η_n the phase angle of the n th partial wave. Expressions for η_n are given in Refs. 1 and 4.

For large distances from the sphere, use of the asymptotic expressions for $h_n^{(2)}(kr)$ gives the farfield form function in the monostatic case,

$$f_\infty = -(2/ka)\sum_{n=0}^{\infty}(-1)^n(2n+1)\sin\eta_n e^{i\eta_n}$$

In order to compare the measurements with the theoretical calculations, the reflected pressure amplitude was observed at a distance of 10 or more sphere diameters from the sphere. At that distance the measured results correspond to the farfield form function in the range of frequency parameter ka from 0 to 30.

EXPERIMENTAL METHOD

The experimentally determined modulus of the farfield form function $|f_\infty|$ was obtained from the relative pressure amplitudes p_0 and $|p_s|$ of the incident and scattered waves, respectively, measured with the same equipment. The incident pressure was determined at the frequencies of interest before and after each run with the receiver at the position normally occupied by the center of the sphere, and the sphere removed. Usually the size of the sphere was held constant and the frequency was varied during a run over a range of ka of approximately unity. Then the sphere was changed to be a different size and the same range of frequency was covered to achieve a different range of ka. The steady-state condition was approximated by a burst of sine waves 300 to 800 μsec in duration.

It was found necessary to enhance the signal-to-noise ratio of the reflected signal. For that purpose, the box-car integrator was used to scan and record on a strip chart recorder, five cycles of the sine wave located at a position in the pulse judged to be a true representation of the steady state. The average of the 10 peak values so obtained gave the peak pressure amplitude. The sphere was then removed and the box-car integrator continued scanning to record background. If a coherent background was present, a correction for its effect was estimated from the relationship between its amplitude and phase and that of the total pressure wave measured with the sphere in place.

The pressure wavefront incident on the sphere was slightly distorted by the presence of the receiver situated on the straight line between the transmitter and the sphere. The result was to reduce the incident pressure at the sphere by a few percent. When the receiver replaced the sphere to measure the incident pressure p_0 a dummy receiver was placed in the position normally occupied by the receiver when the scattered pressure $|p_s|$ was measured and the result was compared with that obtained with no dummy probe. In practice, the pressure pulse incident on the sphere was usually observed with no dummy probe and the experimentally determined appropriate correction (of the order of 2%-4%) was applied.

EXPERIMENTAL EQUIPMENT

The water-tank facility is similar to the one described by Neubauer [5]. A cypress-wood tank containing a volume of water 12 ft × 6 ft × 6 ft is surrounded by a platform which supports the electronic equipment and serves as a working area. The temperature of the water in the tank was measured at six points at different depths and was held constant to within 0.1°C for the duration of each measurement. A residual chlorine level between 1 and 4 ppm was maintained by the addition of sodium hypochlorite solution to the water to keep its optical clarity at a level required to measure distances optically in the water.

The source, receiver, and scattering sphere were placed along a horizontal line parallel to the long dimension of the tank and held in position by a system of tracks, rails,, and carriages mounted over the tank. The placement system made longitudinal, lateral, rotational, and elevation movements possible. All positioning of the elements was done by hand. Final determinations of placement and all lateral and longitudinal positions of objects in the tank were made by optical squares which moved on precision flat surfaces located on support beams above two adjacent sides of the tank. The vertical location was fixed by placing all objects of interest in the same horizontal plane, determined by a level telescope sighting through a porthole at one end of the tank.

The aluminum spheres used for the scattering measurements were precision machined in a range of sizes from 0.25 to 2.5 in. in diameter, known to within 0.0005 in., and were suspended in the water from one of the carriages by nichrome wire 0.005 in. in diameter attached to the sphere with epoxy resin in a radial hole at the surface about $\frac{1}{32}$ in. deep and 0.009 in. in diameter. The tungsten-carbide spheres were commercially available sizing balls suspended in a similar manner.

The hydrophone was similar to the piezoceramic device described by Neubauer [5]. The disk, 0.125 in. in diameter by 0.010 in. thick, was contained in a waterproof housing, covered by a unicellular rubber jacket and the whole assembly held in place by a titanium tube 0.063 in. in diameter extending from the receiver assembly to a connector above the surface of the water. The tube also acted as an electrical shield for the signal lead. The projectors used were circular pistons from 1 to 3 in. in diameter.

Sine waves were generated by a synthesizer and monitored by a counter. The frequency was divided by a preset counter determining the repetition rate of the acoustical pulse by synchronizing the pulse generator which provides the rectangular modulating pulse for the gate and also the timing pulse for the receiving oscilloscope. The acoustical signal received by the pressure sensing probe was amplified, filtered, and displayed on an oscilloscope. A box-car integrator was operated in the scan mode to extract the repetitive signal from the background noise and its output was recorded with a strip chart recorder.

RESULTS AND DISCUSSION

Aluminum

The curve in Fig. 1 shows the modulus of the farfield reflection form function as a function of *ka* for an aluminum sphere in water. The curve starts at the upper part of the figure and is continued below. The experimentally determined points, shown by open circles, are back-scattered relative pressure amplitudes normalized in amplitude. It was of greatest interest to make the measurements in regions where the form function is rapidly varying, especially near the sharp minima in the curve. Calculations showed close correspondence between the frequencies of free vibration of the elastic sphere and the frequencies at which the minima occur in the form function.

Faran [6] noted and verified experimentally for cylinders that in some materials nulls in the backscattered pressure amplitude are close to the frequencies of free vibration of the elastic cylinders and spheres. In Faran's work, the distribution in angle of the scattering was observed and computed for relatively few selected values of the frequency. In

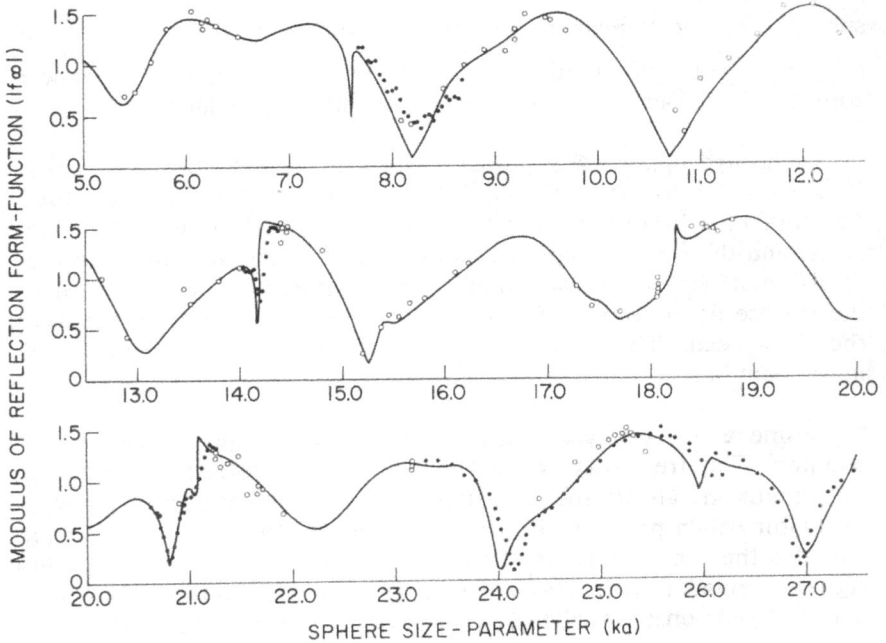

Fig. 1 — Comparison between the exact theoretical calculation, of the form function vs *ka* (solid curve) for an aluminum (1100) sphere, and experimental measurement (points)

this work, only backscattering was observed and the monostatic form function was computed for a range of frequency parameters from 0 to 27.5 for aluminum and to 22.5 for tungsten-carbide, thereby showing in more detail the proximity of the resonant frequencies of the free elastic sphere to changes in the shape of the monostatic form function.

Faran noted that the *n* th term in the partial wave series expansion for the scattered pressure has a term with infinites at the frequencies of the free vibrations of the scattering body, which satisfy the conditions of symmetry of the problem. In the case of the cylinder, the infinities coincide with the zeros of the secular equation giving the vibrational frequencies. The same conclusion holds in the spherical case for a class of vibrations called the spheroidal modes. The resonant frequencies of the spheriodal oscillations were computed by solving numerically the equation given by Sato and Usami [7]. The spheroidal modes are herein designated S_{ni} (there are various conventions), where *n* corresponds to the *n* th term in the partial wave series expansion of the pressure and *l* is the overtone with 1 corresponding to the fundamental.

Tables 1 and 2 list, respectively, the computed frequency parameters of the spheroidal modes, S_{ni}, of interest for the measurements for aluminum spheres in water at 19.3°C and tungsten-carbide spheres in water at 20°C. The frequency parameter is given by $2\pi\nu a/c$, where ν is the frequency of the spheroidal vibration, c is the speed of sound in water at the temperature of the measurements, and a is the radius of the sphere. The resonant mode of lowest frequency is seen to be the oblate-prolate mode S_{21}, as is well known. Mode charts showing the nondimensional frequency of vibration of the modes as a function of Poisson's ratio are given by Faran, and more extensively by Fraser and

Table 1 — Frequencies Expressed in ka of Spherical Modes, S_{n1}, of an Aluminum Sphere at 19.3°C

$n\backslash l$	1	2	3	4	5
0	11.994	26.355	...	...	...
1	...	7.616	15.320	18.380	22.629
2	5.588	10.792	18.229	23.559	...
3	8.334	14.187	21.090	27.536	...
4	10.696	17.564	23.949	...	...
5	12.907	20.831	26.832	...	...
6	15.043	23.954	...	...	...
7	17.137	26.930	...	...	...
8	19.204	...	...	...	...
9	21.254	...	...	...	...
10	23.290	...	...	...	...
11	25.316	...	...	...	...
12	27.335	...	...	...	...

Table 2 — Frequencies Expressed in ka of Spherical Modes, S_{n1}, of a Tungsten Carbide Sphere at 20.0°C

$n\backslash l$	1	2	3	4
0	11.213	27.604	...	...
1	...	9.312	18.019	21.800
2	7.431	13.266	22.752	...
3	10.983	17.708	...	...
4	14.015	22.235	...	...
5	16.857	...	...	...
6	19.611	...	...	...
7	22.317	...	...	...

LeCraw [8]. In the references, the frequencies are normalized to either the longitudinal or shear sound speed in the material of the scattering body. Tables of vibrational frequencies for the incompressible case and for Poisson's ratio of 0.25 are given by Sato and Usami [7]. The vibrations of the free elastic sphere are discussed in Refs. 7-12.

The free vibration of the sphere and the form-function minima are sensitive to changes in the shear wave speed of the sphere material and not very sensitive to changes in the longitudinal wave speed. Shear and longitudinal wave speeds were measured on cylindrical samples of the same piece of aluminum of which the spheres were fabricated. The measured values were 3100 ±30 and 6370 ±40 m/sec for shear and longitudinal waves, respectively.

Both shear and longitudinal wave speeds were varied within the region of experimental error, until the computed frequencies of free oscillation of the sphere best corresponded to the experimental minima in Fig. 1, and these velocities were used in computing the form function. As noted by Faran [6], the position of a minimum or other prominent feature in the form function would be expected to show a shift from the frequency of free vibration of the sphere, but in the case of solids with low losses and with densities significantly greater than that of the surrounding fluid, the frequency shift is small.

The open circles in Fig. 1 are the results of amplitude measurements made with an oscilloscope, and the solid circles were obtained from recordings with the box-car integrator. Both are absolute mea- surements and not normalized in amplitude. The points near the minima at ka equal to 14.1 and 20.8 are based on the most complete set of measurements and correspond to resonant frequencies of oscillation of the free sphere to within about 0.5%. The measured shear speed in aluminum was increased by 20 m/sec, within the known possible error of its independent measurement, to bring the calculated curve into the agreement with the observed results shown at the two minima near ka equal to 14.1 and 20.8. There is still some disagreement at ka of 24. It was found that a 1% change in longitudinal sound speed had less than one-tenth of the effect of a 1% change in shear wave speed in shifting the calculated position of the vibration frequencies at ka values 14.1 and 20.8. The wave speeds used for the calculated curve of Fig. 1 were 3120 and 6370 m/sec, respectively, for shear and longitudinal waves in aluminum and 1480.08 m/sec for the wave speed in water at 19.3°C.

Figure 2 shows an incident pulse, of 500-μsec duration, and pulses reflected by an aluminum sphere recorded by the scanning box-car integrator. The time scale is increasing from left to right. At a *ka* of value of 20.78 (Fig.2(b)), the form function has a minimum from which it rises steeply to a maximum at *ka* of 21.21 (Fig. 2(c)). In this case, both reflected pulses show the effect of the elastic vibrations and the pulses in Figs. 2(a) and 2(b) have a duration considerably greater than that of the incident pulse.

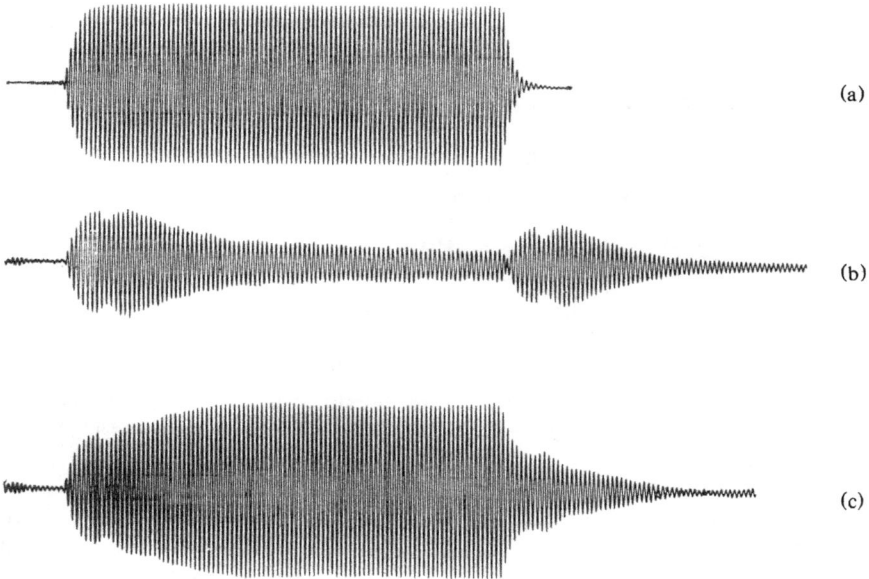

(a)

(b)

(c)

Fig. 2 — (a) A 500-μsec-long incident pulse. The pulse reflected by an aluminum sphere when the pulse in Fig. 2(a) is incident for (b) *ka* = 20.78, and (c) *ka* = 21.21.

In recording the data, only three to five cycles were recorded for the measurement of the pressure amplitude instead of the complete pulse as shown. In the case shown, the five cycles would have been located about 300 μsec from the beginning of the pulse.

Tungsten-Carbide

The curve in Fig. 3 shows the computed farfield form function of a tungsten-carbide sphere in water at 20°C. The values of the shear and longitudinal wave speeds in tungsten-carbide used for the computation were 4150 and 6860 m/sec, respectively, as estimated from the

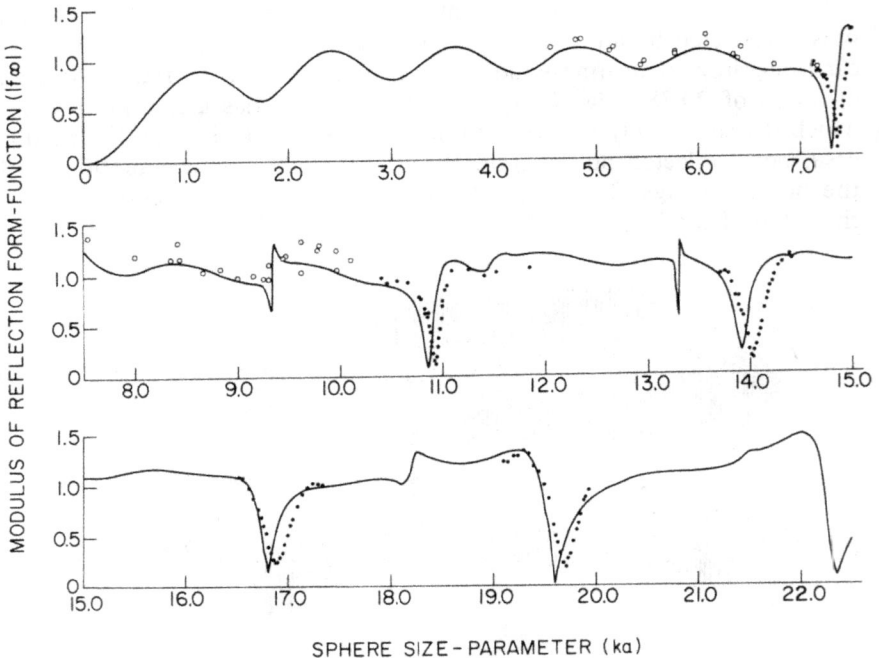

SPHERE SIZE-PARAMETER (ka)

Fig. 3 — Comparison between exact theoretical calculation, of the form-function vs
ka (solid curve) for a tungsten carbide sphere, and experimental measurement
(points)

results of Balashov and Voronov [13]. No separate sample of the identical material of which the sizing balls were made was available for
independent measurement of wave speeds in tungsten-carbide. The
measured points are given as open and solid circles and were obtained
as described for the case of aluminum. Except for the presence of
sharp minima and narrow regions where it rises abruptly, the curve
resembles the form function of the rigid sphere. The minima and
sharp rises in the curve correspond closely to the calculated modes of
oscillation of the free tungsten-carbide sphere, usually to within a few
tenths of a percent in ka. Figure 3 shows the experimentally determined minima slightly to the right of the computed ones. Since they
are close to the frequencies of free oscillation, which are most strongly
dependent on the shear speed of waves in the tungsten-carbide, the
shear speed was adjusted to bring the computed curve into better agreement with the measurements. The effect on the calculation of increasing the shear speed by 35 m/sec in the sphere material is shown in Fig.
4. This measurement is a reasonable value that is within the limits of
probable error for shear wave speed in tungsten-carbide.

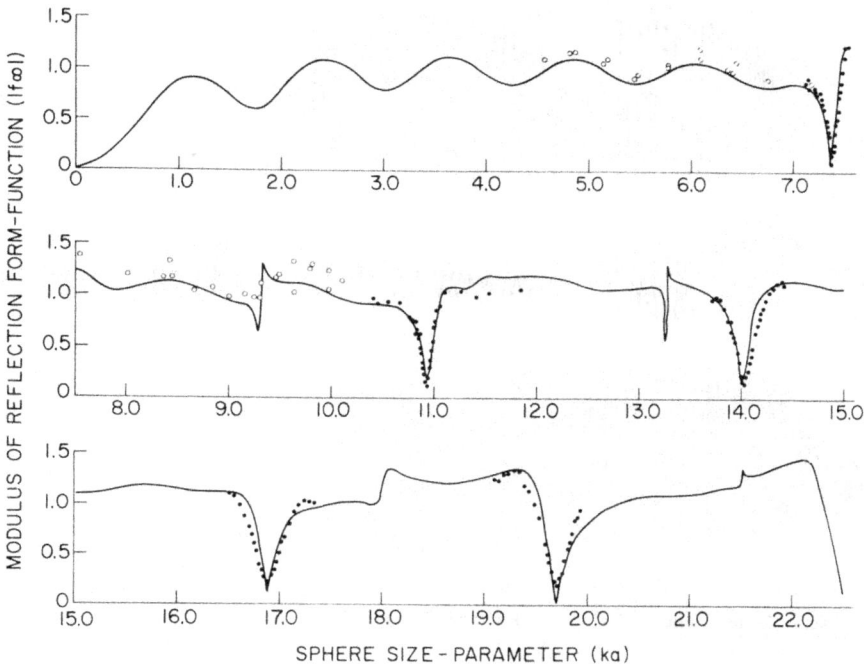

Fig. 4 — The same comparison given in Fig. 3, where the theoretical curve has been shifted, due to a 35 m/sec upward adjustment in the shear velocity in tungsten-carbide

In Fig. 5 the scattered pulse with carrier frequency, corresponding to a resonance at *ka* equal to 7.3 (Fig. 5(b)), is compared with the scattered pulse corresponding to *ka* of 8.4 (Fig. 5(c)) and the incident pulse (Fig. 5(a)). The pulses were recorded by scanning the periodically recurring signal with the box-car integrator. The amplitude of the incident pulse, shown in Fig. 5(a), is not related to that of the reflected pulses. the length of the incident pulse shown in the figure is about 300 μsec and the diameters of the spheres used for scattering at *ka* of 7.36 and 8.46 were 22/32 and 25/32 in., respectively. The shape of the pulse scattered at a *ka* of 7.36 (Fig. 5(b)) is observed where the amplitude of the form function has a sharp minimum, corresponding in this case to a resonant frequency of vibration of the free elastic sphere. The form function is changing rapidly in amplitude and phase and it can be seen that the reflected pulse is considerably longer than the incident pulse as the sphere continues to oscillate and return energy to the detector.

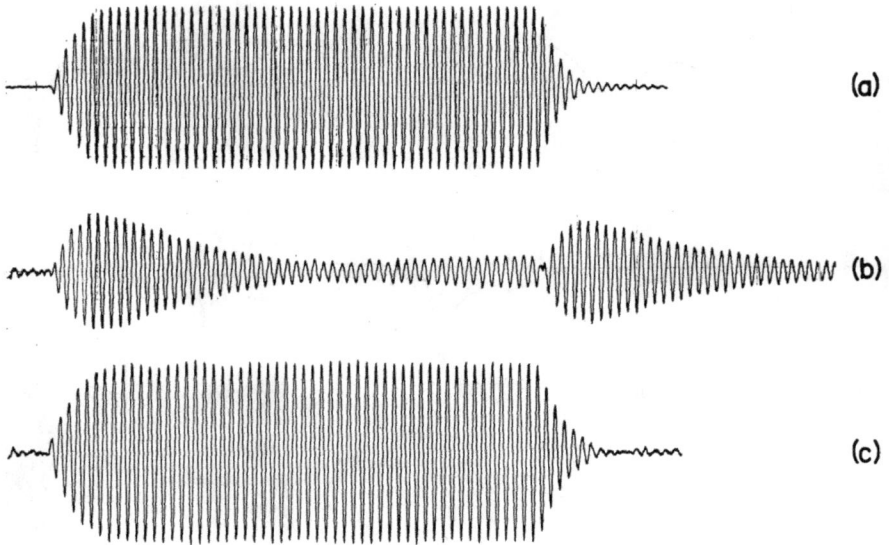

Fig. 5 — (a) A 300-μsec-long incident pulse. The pulse reflected by a tungsten-carbide sphere when the pulse in Fig. 5(a) is incident for (b) $ka = 7.36$, and (c) $ka = 8.46$.

The pulse scattered at a ka of 8.46 (Fig. 5(c)) corresponds to a region where the form function is not changing rapidly and is far from the nearest resonance. The shape of the scattered pulse is similar to that of the incident pulse with little energy going into the elastic vibrations of the sphere.

Effect of Temperature

Figure 6 shows the form function of aluminum measured near the minimum at ka of 20.8. The measurements, shown with open circles and error bars, were made at 17.9° and 20.9°C, and the solid curves show the form functions calculated at those two temperatures. The measurements show a shift of approximately -0.2% per °C in the abscissa of the minimum, in agreement with the shift in the computed curves. This is accounted for, as was expected, by the change in the speed of sound with the temperature of the water, the only parameter changed in the calculation.

From the experimental results in this case, it appears that the abscissa of the minimum near ka of 20.8 can be estimated to the nearest 0.05. As previously mentioned, it was found that the values of

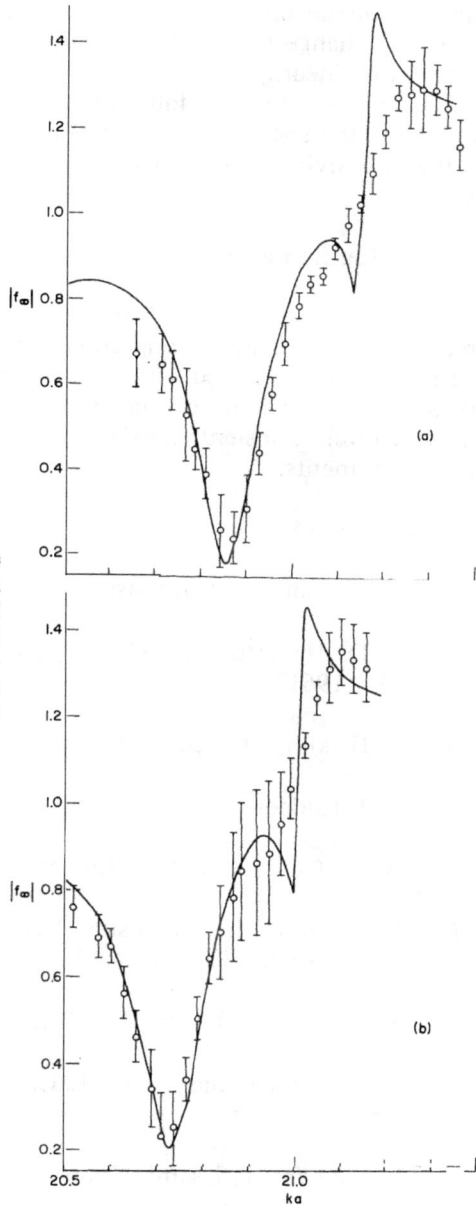

Fig. 6 — The shift of a minimum in the curve of the form function vs *ka* for an aluminum (1100) sphere to a 3°C change in temperature. (a) 17.9°C and (b) 20.9°C. The solid curve is theoretical; the points are experimental for temperature of (a) 17.9°C and (b) 20.9°C.

the minima in the calculated form function were sensitive to changes in the shear wave speed used in making the calculation. For example, the abscissa at the minimum in the calculated form function near 20.8 was found to change by 0.12 for a change in shear wave speed of 20 m/sec. Assuming linearity, it follows that for the location of the minimum in the computed curve to fall within the *ka* range of 0.05 found experimentally, the shear wave speed must be known to within 8 m/sec, i.e., ±0.3%, provided the other parameters used in the calculation are known.

Concluding Remarks

These results of the backscattered reflection from spheres used a pulsed sine wave signal long enough that the experimental result could be reasonably compared with calculation for reflection at a specific value of *ka*. The transient portion of the pulse was ignored. See Chapter 8 in which only transient signals (i.e., relatively short pulses) are used in the experiments.

REFERENCES

1. J. J. Faran, Jr., J. Acoust. Soc. Am. **23**, 405 (1951).

2. L. D. Hampton and C. M. McKinney, J. Acoust. Soc. Am. **33**, 694 (1961).

3. R. Hickling, J. Acoust. Soc. Am. **34**, 1582 (1962).

4. A. J. Rudgers, NRL Rep. 6551 (June 1967).

5. W. G. Neubauer, NRL Rep. 5688 (Oct. 1961).

6. J. J. Faran, Jr., Acoustics Research Laboratory, Harvard Univ. Tech. Mem. No 22 (Mar. 1951).

7. Y. Sato and T. Usami, Geophys. Mag. **31**, 15 (1962-63).

8. D. B. Fraser and R. C. LeCraw, Rev. Sci. Instrum. **35**, 1113 (1964).

9. Y. Sato and T. Usami, Geophys. Mag. **31**, 25 (1962-63).

10. A. E. H. Love, *The Mathematical Theory of Elasticity* (Dover, New York, 1944), 4th ed.

11. P. M. Morse and H. Feshback, *Methods of Theoretical Physics* (McGraw-Hill, New York, 1953), p. 1872.

12. B. A. Auld, *Acoustics Fields and Waves in Solids* (Wiley, New York, 1973), Vol. II.

13. D. B. Balashov and F. F. Voronov, Phys. Metals Metalog. 9, No. 4, 127 (1960).

Chapter 9

ELASTIC SPHERES AND RIGID SPHERES AND SPHEROIDS — TRANSIENT SIGNALS*

INTRODUCTION

In Chapter 8 previous comparison of theoretical calculation and experimental measurements [1], of the reflection by a sphere versus frequency, required the use of an acoustic pulse long enough so that a steady-state value of the reflected pressure amplitude was achieved at a well-defined single frequency. Such an experiment requires a number of reflection measurements: one at each frequency of interest. Reference to earlier work can be found in the references for Chapter 8.

The reflected pressure amplitude is expressed here as

$$p_r = (a/2r)p_0|f_r(ka)|, \tag{1}$$

where a is the sphere radius, r is the distance from the center of the reflecting sphere to the point of reception of p_r, $f_r(ka)$ is the so-called form function, and ka is the size parameter in the fluid surrounding the sphere. If reflection measurements are made at a far field point of the sphere, usually defined as a distance of at least 10 sphere diameters, $f_r(ka)$ becomes independent of r and is called the farfield form function $f_\infty(ka)$. For a long pulse experiment approximating steady state, $f_\infty(ka)$ is derived from the substitution of measured amplitudes p_r and p_0 in the expression

$$|f_\infty(ka)| = (2r/a)\ (p_r/p_0). \tag{2}$$

Measurements of p_r and p_0, made at several frequencies for a given sphere or for several spheres at a given frequency, are used in Eq. (2) to determine specific values of $|f_\infty(ka)|$.

The incident pulse may be regarded as a function of time, so that now the acoustic pressure measured at the position of the center of the reflecting sphere, in its absence, is

$$p_i = p_0 e^{ik_0ct},$$

*This work was first published in: Louis R. Dragonette, Richard H. Vogt, Lawrence Flax, and Werner G. Neubauer, J. Acoust. Soc. Am. **55**, 1130-1137 (1974).

177

where c is the acoustic wave speed in the liquid in which the sphere is immersed. Introducing the transformation $\tau = (ct - r)/a$ and rewriting Eq. (2) in terms of τ,

$$p_r e^{ik_0 a\tau} = \frac{a}{2r} p_0 e^{ik_0 a\tau} |f_\infty(k_0 a)|. \tag{3}$$

The quantity τ represents the nondimensionalized time by normalizing the time parameter to a/c and reducing it by the ratio r/a, i.e., the distance from sphere center to the reception point of p_r measured in units of the sphere radius. Using τ amounts to beginning time reckoning at the center of the reflecting sphere and measuring in units of wave travel time in water over the distance of a sphere radius a.

A pressure pulse containing frequencies expressed in ka space has a spectrum $g(ka)$ and its instantaneous time history in τ is expressed according to the Fourier theorem by the integral

$$p_r(\tau) = \int_{-\infty}^{\infty} g(ka) e^{ika\tau} d(ka),$$

and the inverse transform defining the pulse spectrum is

$$g(ka) = \frac{1}{2\pi} \int_{-\infty}^{\infty} p_r(\tau) e^{-ika\tau} d\tau.$$

The quantities $p_i(\tau)$ and $p_r(\tau)$ are now instantaneous pressures and the relationship between them in terms of $|f_\infty(ka)|$, similar to the amplitude equation for a single frequency given by Eq. (3), is expressed by

$$\int_{-\infty}^{\infty} p_r(\tau) e^{-ika\tau} d\tau = \frac{a}{2r} f_\infty(ka) \int_{-\infty}^{\infty} p_i(\tau) e^{-ika\tau} d\tau.$$

or further,

$$|f_\infty(ka)| = \frac{2r}{a} \left| \int_{-\infty}^{\infty} p_r(\tau) e^{-ika\tau} d\tau \right| / \left| \int_{-\infty}^{\infty} p_i(\tau) e^{-ika\tau} d\tau \right|. \tag{4}$$

Rather than analytically examining the effects of computational procedures for performing the integrals in Eq. (4), a well-defined case of reflection was carried through in the mathematical operation indicated by that equation. The response of a rigid sphere to a two-cycle incident pulse at a ka of 4.328 was computed by exact theory as described in Ref. 2. The resulting calculated reflected pulse was sampled at intervals $\Delta\tau = 0.16$. These values were used as a sampled representative of $p_r(\tau)$. A theoretical two-cycle unit-amplitude incident pulse, of dimensionless frequency $k_0 a = 4.238$, was also sampled at the same $\Delta\tau$. This sampling rate corresponds to about 10 samples per cycle. These values were used as a sampled representative of $p_i(\tau)$. Figure 1(a) shows the spectrum $g_i(ka)$ computed from $p_i(\tau)$ and Fig. 1(b) shows the spectrum of $g_r(ka)$ of the reflected pulse. Fast Fourier

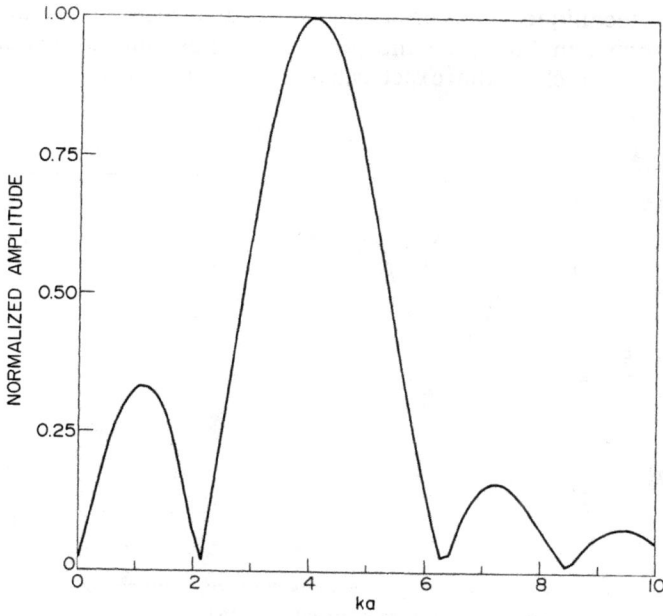

(a)

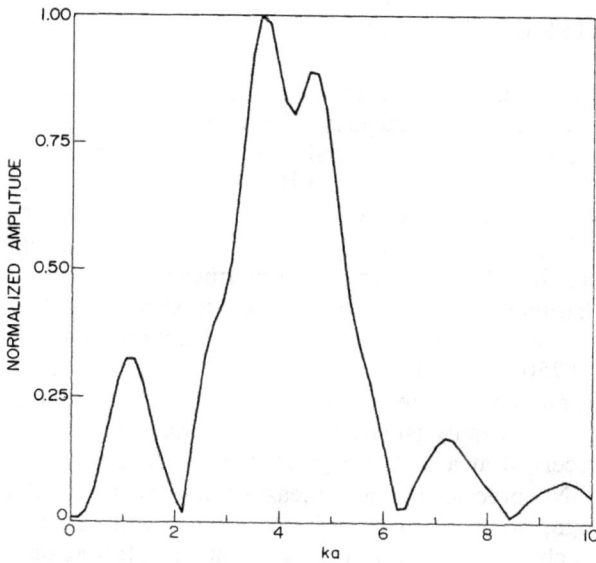

(b)

Fig. 1 — (a) Spectrum of an ideal two-cycle pulse. (b) Spectrum of the computed reflected pulse from a rigid sphere.

transform techniques were used and the results of the evaluation of Eq.
(4) are shown in Fig. 2 by the points and these are compared to the
solid curve, which is the exact calculation of the form function for a
rigid sphere.

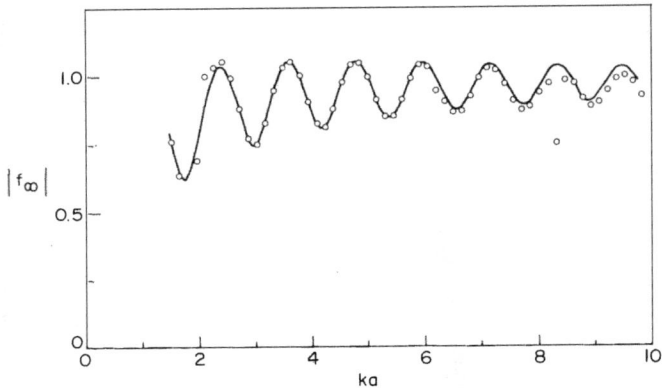

Fig. 2 — Calculated $|f_\infty|$ vs ka (solid curve) and computed
points derived from sampled ideal two-cycle pulse.

RIGID SPHERE AND SPHEROID

Monostatic reflection from metal shapes was measured in air at a
pulse center frequency of 26 kHz. Rigid boundary conditions are very
nearly satisfied by any solid metal object for airborne acoustic waves.
The acoustic source that generated the pulse was a capacitance speaker
6 in. in diameter. The pulse received at the position later occupied by
the center of the reflector is shown in Fig. 3(a). Throughout the air
experiments a Brüel and Kjaer capacitor microphone was used that was
1/4 in. in diameter. All pulseforms that are shown are the actual sampled outputs used in computation. The reflection of the pulse in Fig.
3(a) by a 0.750-in.-diam metal sphere is shown in Fig. 3(b). The
apparent tail on the reflected pulse is the result of a creeping wave contribution that is partially isolated in time. This monostatically reflected
pulse was received at a distance greater than 10 sphere diameters from
the sphere. No precise distance measurement was available in the air-
acoustic range, so the pulses are not shown with amplitudes correct
relative to each other. The normalized spectra plotted on the abscissa,
expressed in units of ka for the pulses in Figs. 3(a) and 3(b), are given
in Figs. 4(a) and 4(b), respectively.

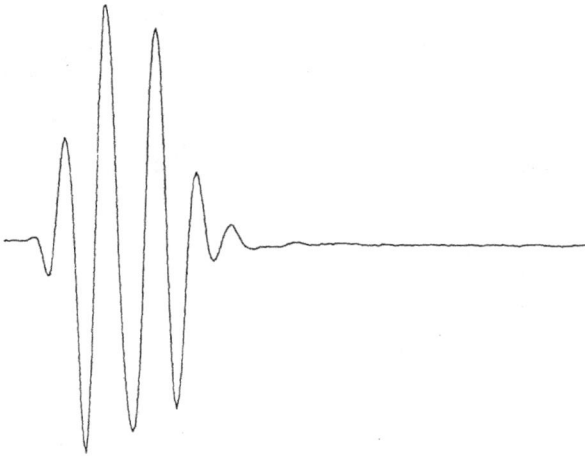

(a)

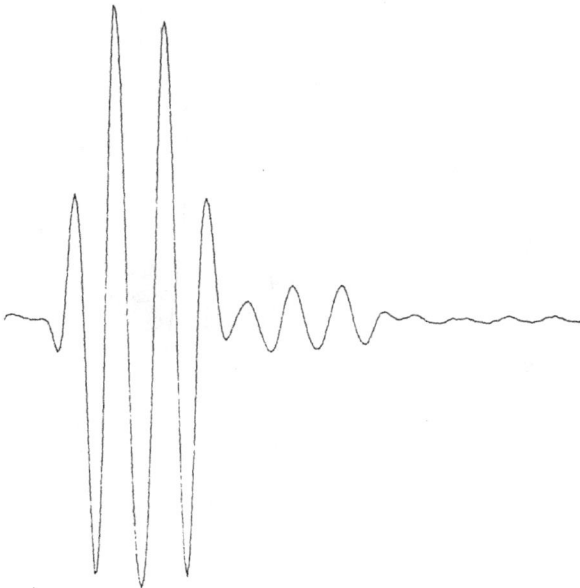

(b)

Fig. 3 – (a) Incident acoustic pulse used to reflect from sphere and
spheroid in air. (b) Reflected pulse from a metal sphere in air.

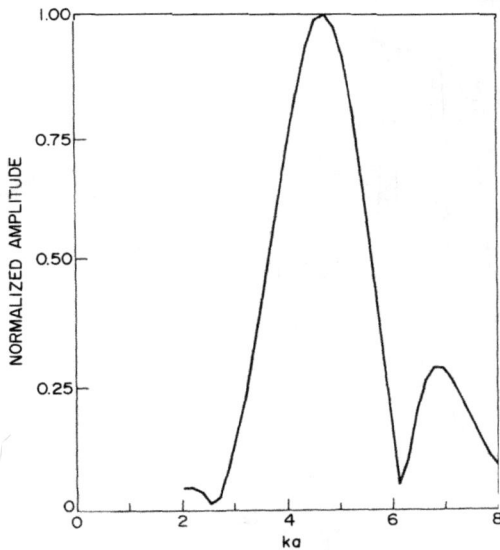

(a)

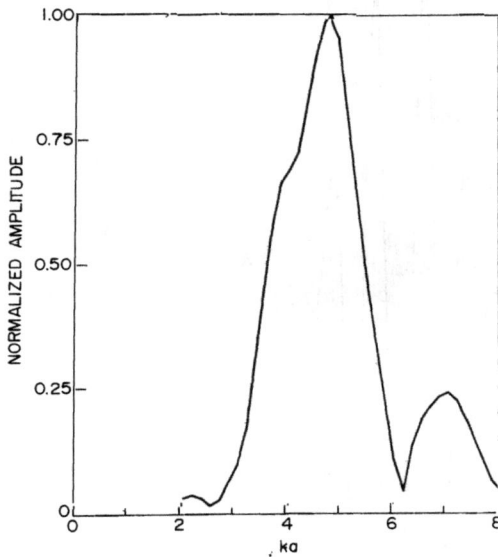

(b)

Fig. 4 — Spectrum of (a) the incident pulse shown in Fig. 3(a), and (b) the reflected pulse shown in Fig. 3(b).

Calculation of $|f_\infty|$ by using the 20-samples-per-cycle output plotted in Fig. 3(a) in the denominator and the similarly sampled output in Fig. 3(b) in the numerator of the right side of Eq. (4) results in the plot of Fig. 5. The experimental values shown by points are normalized to the exact calculation shown as a solid curve to the center of the spectral band. The values of ka at which the spectral values are 10 dB below the peak amplitude on both sides of the peak are shown in Fig. 5 by vertical lines. A description of the experimental apparatus is available in Ref. 3.

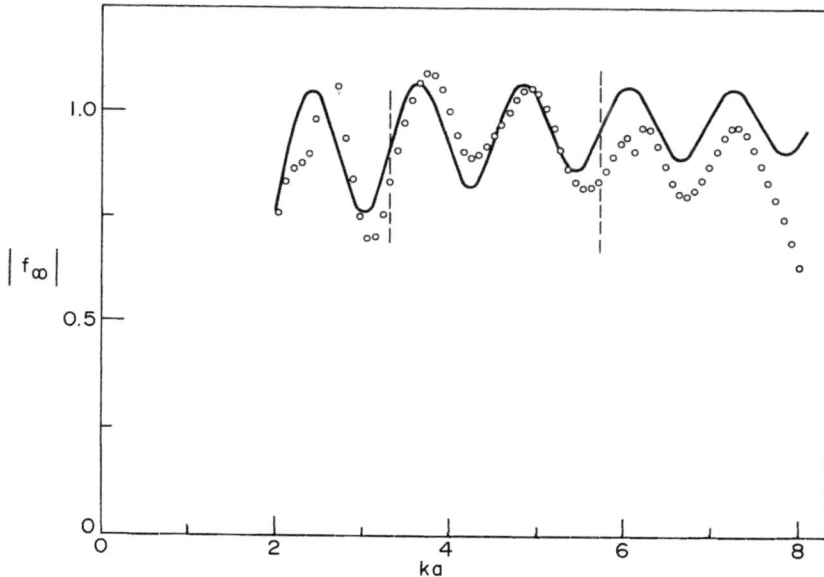

Fig. 5 — Theoretical reflection form-function $|f_\infty|$ vs ka (solid curve) and experimental points derived from the single reflected pulse shown in Fig. 3(b).

The same incident pulse (Fig. 3(a)) was reflected from a metal prolate spheroid with a 5/3 fineness ratio for which half the interfocal distance was 1.041 cm. The pulse reflected from the end of the spheroid is given in Fig. 6(a). This is the aspect of the spheroid for which the major axis is pointing in the monostatic direction. It is interesting to observe the creeping wave pulse that is distinctly separated from and smaller than the initial specularly reflected pulse. The spheroid was rotated to a position 90° from the position at which the pulse in Fig. 6(a) was measured, so that the pulse was reflected from its side. Figure 6(b) shows a plot of that pulse. Received pulse amplitudes are set to optimize the electronic measurement of the pulseforms so the plotted relative amplitudes are not meaningful.

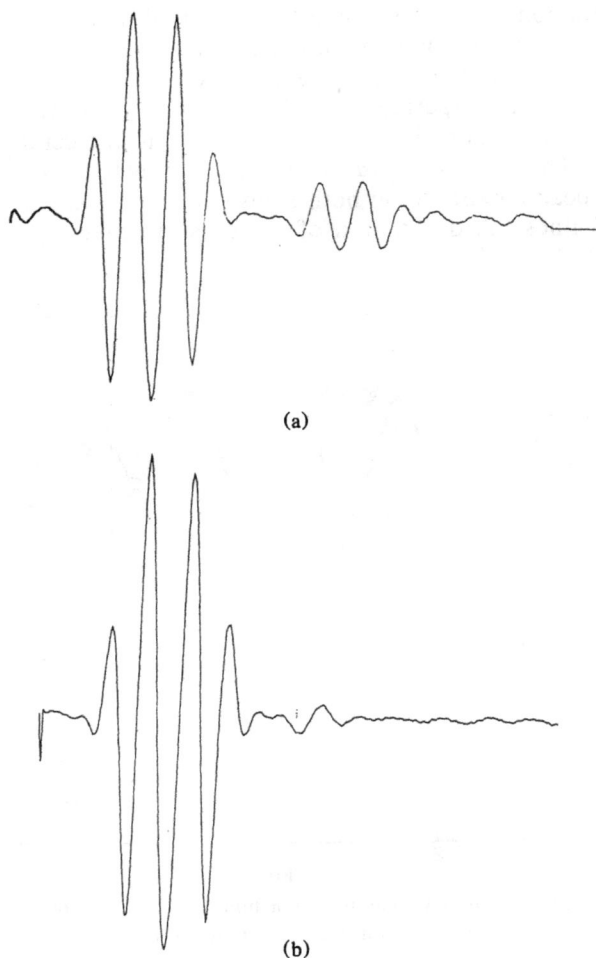

(a)

(b)

Fig. 6 — Pulse reflected from (a) the end of the spheroid in air (along major axis), and (b) the side of the spheroid in air (along minor axis).

In Fig. 6(b), a small creeping wave pulse can be observed at the end of the specularly reflected pulse. The path around the small dimension of the spheroid is, however, insufficient to cause separation of the specular and creeping wave pulses, which is evident in Fig. 6(a). Calculation of $|f_\infty|$ by means of Eq. (4), using sampled incident and reflected pulse data, is plotted in Fig. 7. The x's are data for end reflection (incidence and reflection along the major axis) and the open circles are for side reflection (minor axis incidence) from a rigid prolate spheroid. Theoretical results [4] for the end reflection exist up to a ka

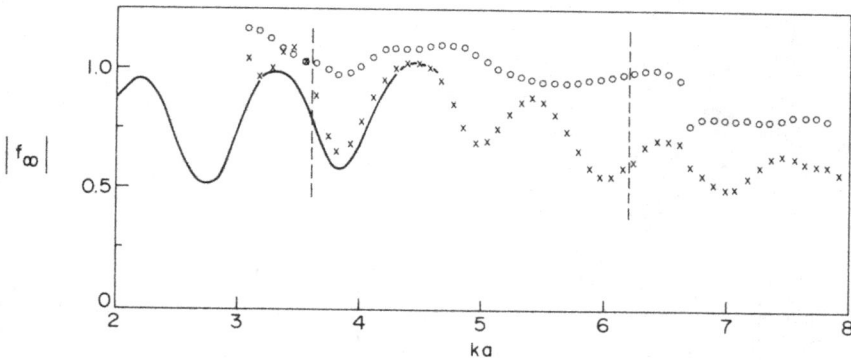

Fig. 7 — Theoretical reflection form-function $|f_\infty|$ vs ka (solid curve) for end reflection from the spheroid and experimental points for end reflection (crosses) and side reflection (circles).

of 4.6 and are plotted as the solid curve. The end reflection data are normalized to the peak in $|f_\infty|$ at $ka = 4.47$. The side reflection is arbitrarily normalized to 1.1 at the ka at the center of the incident pulse spectrum, since no theoretical values were available. This normalization was convenient so that the two curves would be separated and could be compared. The large-amplitude oscillations for the case of end reflection are consistent with the larger-amplitude creeping wave interfering with the specular reflection. For the end reflection case, all creeping wave paths are the same length and are symmetrical around the body. When the reflection is from the side, the creeping wave paths are not all the same length, and therefore the signals that take these paths do not act in concert to produce the successive large oscillations over the frequency range of the pulse, as in the end reflection case. Between the limits of ka of 3.6 and 6.2, the spectrum of the incident pulse is no more than 10 dB below its peak.

ELASTIC SPHERES IN WATER

Portions of the reflection form function were obtained by the use of a single reflected pulse for spheres made of three different materials. The materials were aluminum (1100), brass (70−30), and tungsten-carbide sizing balls and were the same spheres for which steady-state reflections were measured and described in Chapter 8.

The incident pulseform used for the aluminum and brass spheres over the same ka range is shown in Fig. 8. This pulse was received by a hydrophone at the position otherwise occupied by the center of the

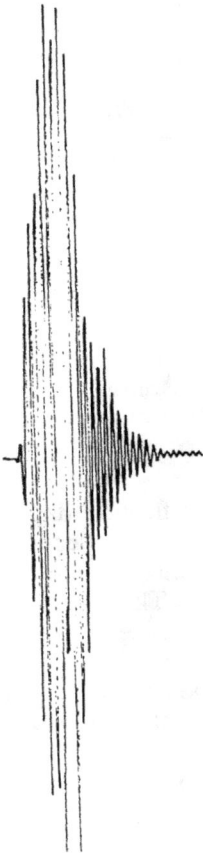

Fig. 8 — Incident acoustic pulse used for aluminum and brass sphere insonification, measured at the position of the center of the spheres in their absence.

sphere when the monostatically reflected pulse was received. The normalized computed spectrum of the pulse in Fig. 8 is plotted in Fig. 9. The peak of the spectrum occurs at a frequency of 212.8 kHz, which corresponds to a *ka* value of 20.05 for the 4.445-cm-diam sphere used in the experiment. The actual pulse reflected from the aluminum sphere is given in Fig. 10. Its computed spectrum is given in Fig. 11. Both incident and reflected pulseforms were digitized at an interval of 1 μsec, or about five samples per cycle with 8-bit accuracy.

The calculation resulting from the use of Eq. (4) for the incident and reflected pulses in Figs. 8 and 10 is plotted in Fig. 12 at specific values of *ka* on the same plot with the exact computer solution [4] shown by a solid line. Values of *ka* where the incident pulse spectrum was 10 dB below the peak are indicated by vertical broken lines. The experimental points are normalized to the value of $|f_\infty|$ at the peak frequency of the transmitted pulse, since the reflection measurements

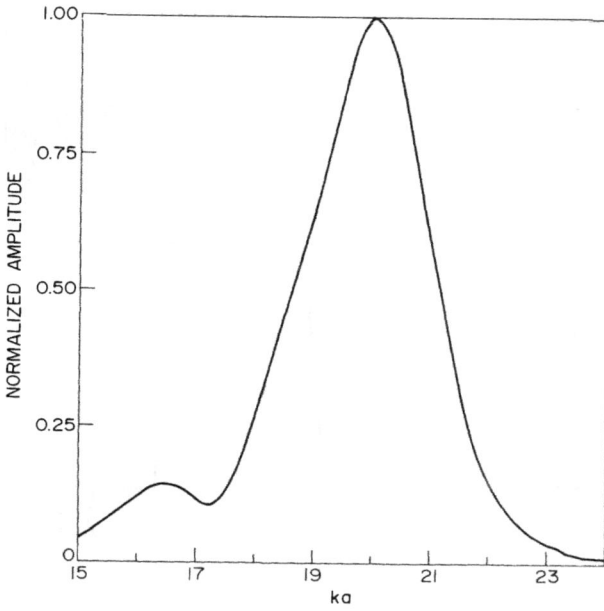

Fig. 9 — Frequency spectrum of incident acoustic pulse
in Fig. 8.

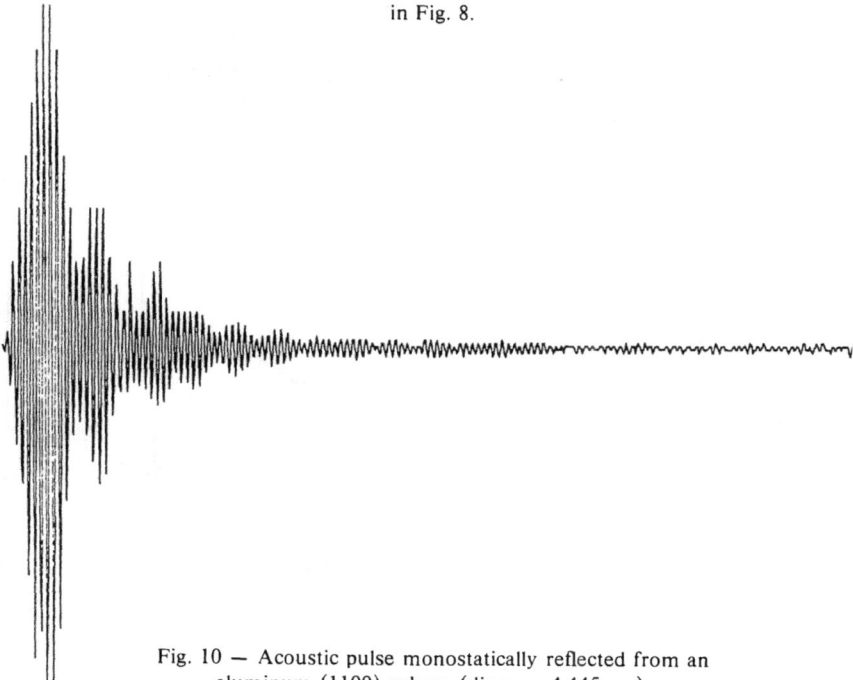

Fig. 10 — Acoustic pulse monostatically reflected from an
aluminum (1100) sphere (diam = 4.445 cm).

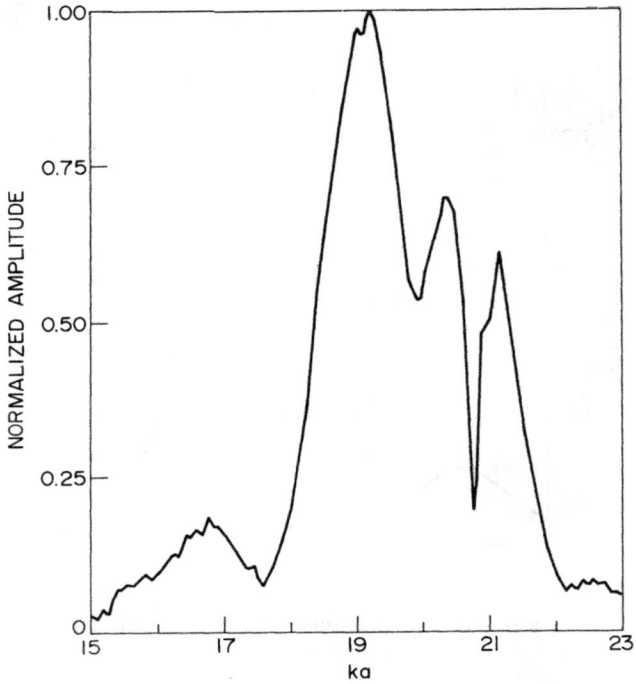

Fig. 11 — Frequency spectrum of the acoustic pulse reflected
from an aluminum sphere shown in Fig. 10.

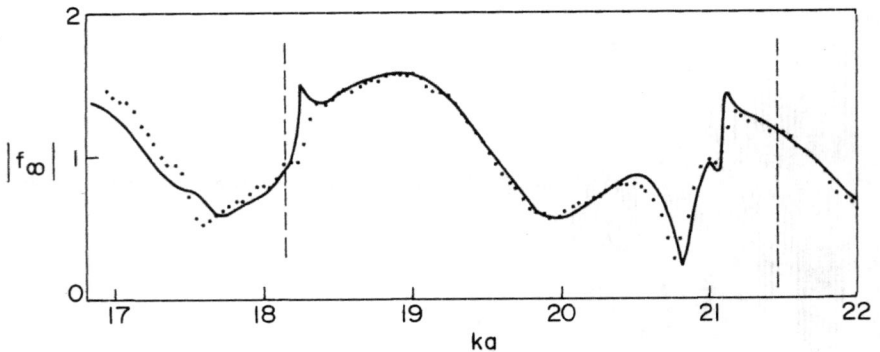

Fig. 12 — Theoretical reflection form-function $|f_\infty|$ vs ka (solid curve) for an aluminum
sphere and experimental points derived from the single reflected pulse shown in Fig. 10.

were made in a 150,000 gallon pool which was not yet instrumented to measure accurately distance between the acoustical elements in the water.

The acoustic wave speed in water at 20°C in the calculation of $|f_\infty|$ was 1482.25 m/sec. Longitudinal and shear wave speeds used in computation of $|f_\infty|$ for aluminum and tungsten carbide were those given in Chapter 8. The longitudinal wave speed used in the calculation for brass was measured, in a separate cylindrical sample (6.4-cm diam × 7.6 cm) by direct propagation between parallel faces, to be 4700 ± 47 m/sec. The cylindrical brass was a piece of the same bar from which the sphere was made. The measured shear wave speed in the same sample was 2110 ± 21 m/sec. The pulse reflected in the monostatic direction from a brass sphere the same size as the aluminum sphere is given in Fig. 13. This reflection results from the incident pulse shown in Fig. 8. The computed spectrum of the pulse in Fig. 13 is given in Fig. 14 with an abscissa in units of ka. A calculation similar

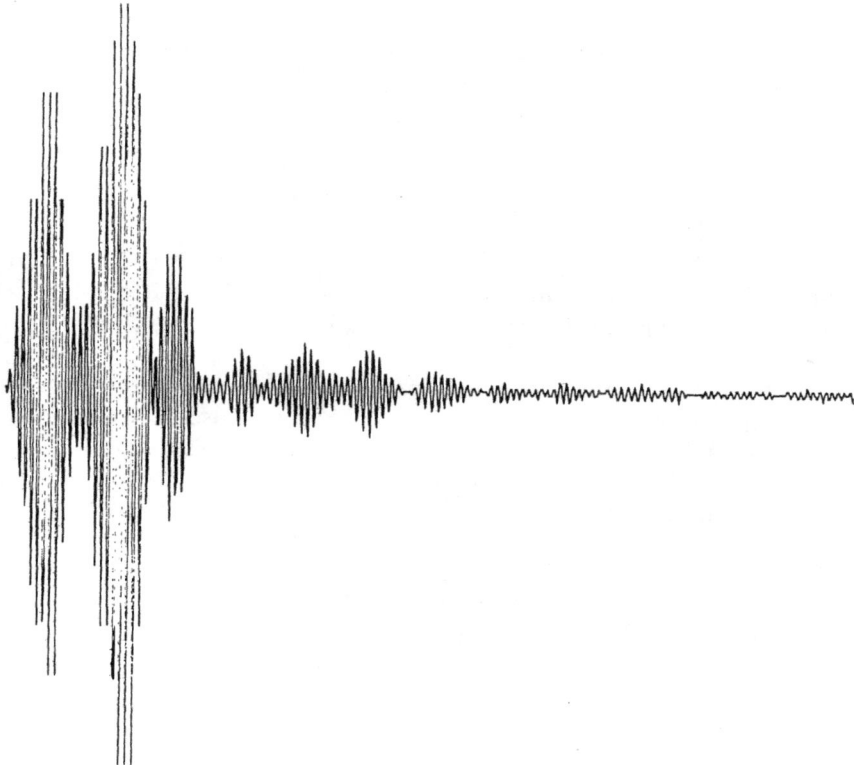

Fig. 13 — Acoustic pulse monostatically reflected from a brass sphere.

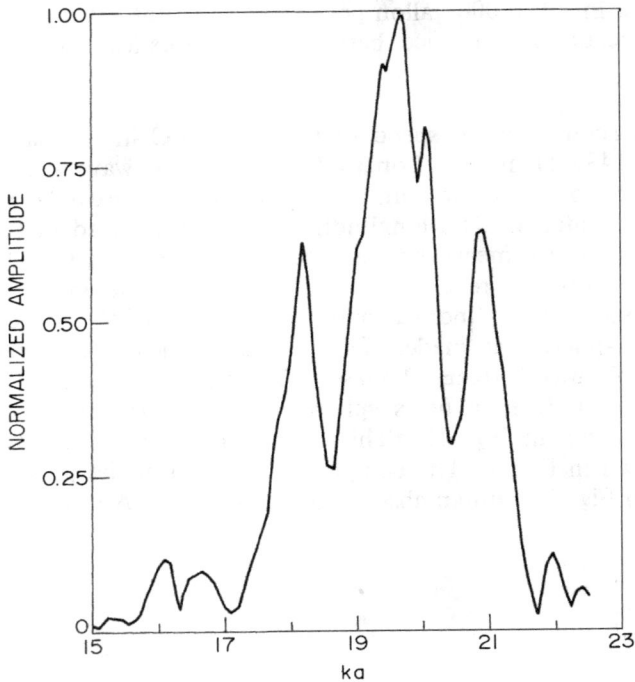

Fig. 14 — Frequency spectrum of the acoustic pulse
reflected from a brass sphere shown in Fig. 13.

to that for aluminum by the use of Eq. (4) is plotted in Fig. 15 for
brass. Vertical broken lines on the plot again indicate *ka* values at
which the incident pulse spectrum was 10 dB below the peak value.

The incident pulse and the reflected pulse from the tungsten-
carbide sphere for a different range of *ka* than that used for aluminum
and brass are shown in Figs. 16(a) and 16(b), respectively. The
incident pulse has the same center frequency as the pulse incident on
aluminum and brass but has a slightly different shape. The diameter of
the tungsten-carbide sphere used was 2.54 cm. The spectra of the
incident and reflected pulses are given in Figs. 17(a) and 17(b),
respectively. The theoretical calculation by straightforward wave harmonic means is plotted in Fig. 18 by the solid line and the result of use
of pulse data for tungsten carbide in Eq. (4) is shown by points with
10-dB spectral limits indicated.

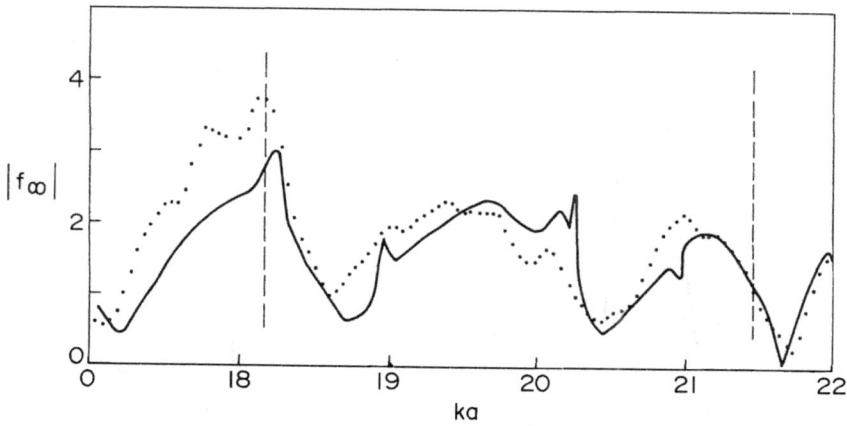

Fig. 15 — Theoretical reflection form-function $|f_\infty|$ vs ka (solid curve) for a brass sphere and experimental points derived from the single reflected pulse shown in Fig. 13.

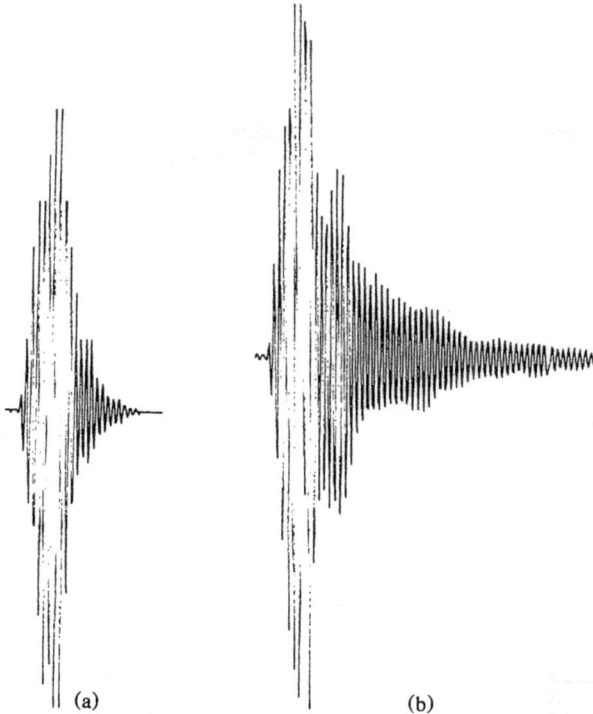

(a)   (b)

Fig. 16 — (a) Incident acoustic pulse used for tungsten-carbide sphere insonification, measured at the position of the center of the sphere in its absence. (b) Acoustic pulse monostatically reflected from a tungsten-carbide sphere (diam = 2.54 cm).

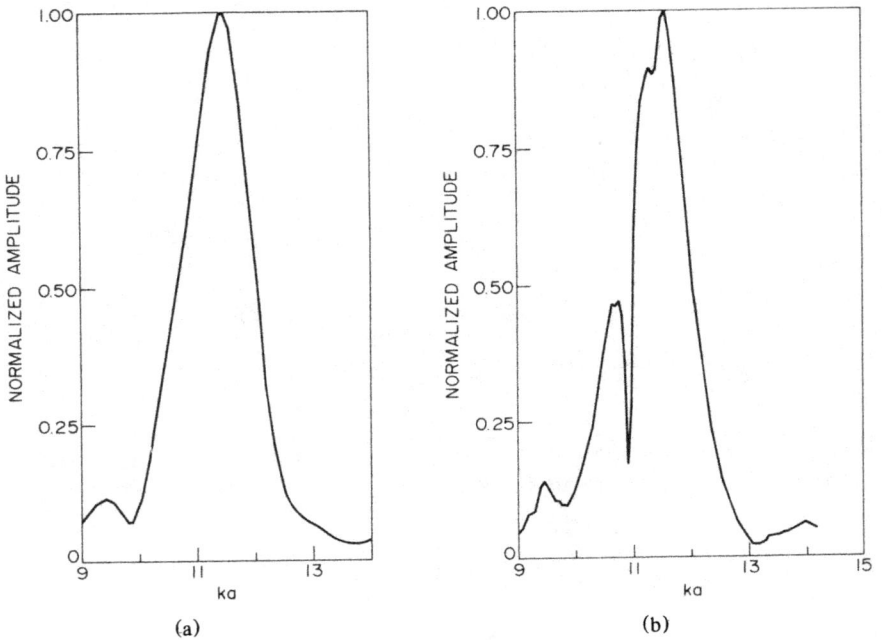

Fig. 17 — Frequency spectrum of (a) the acoustic pulse incident on a tungsten carbide-sphere and (b) the reflected acoustic pulse in Fig. 16(b).

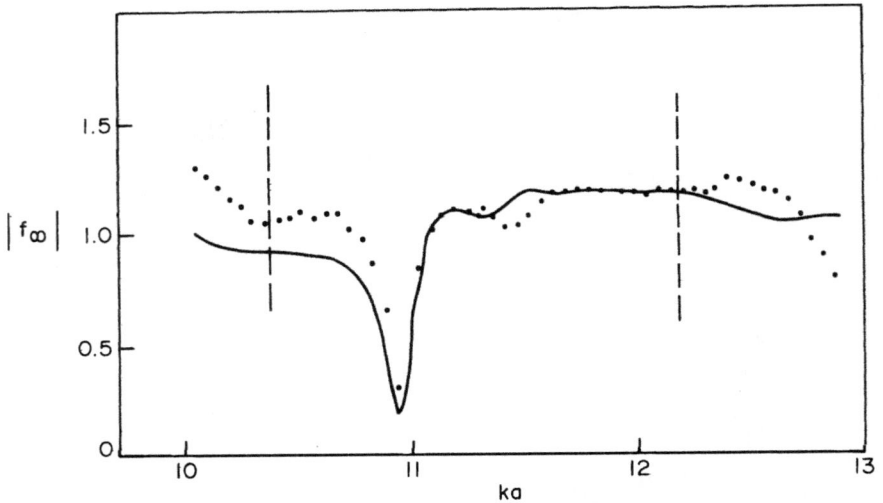

Fig. 18 — Theoretical reflection form-function $|f_\infty|$ vs ka (solid curve) for a tungsten-carbide sphere and experimental points derived from the single reflected pulse shown in Fig. 16(b).

It is interesting that the deviation of the experiment from theory in the case of tungsten carbide around a *ka* of 11.4 was also observed in other measurements carried out at a different time in a different method of using a long pulse approximating steady state. This indicates that some constants of the material of which the sphere is made are probably not sufficiently well known to define the theory accurately.

All previous comparisons of theory and experimental results using short pulses have been normalized to the value of $|f_\infty|$ at a *ka* value in the center of the frequency band of the incident pulse. One absolute measurement, for an aluminum sphere, was made for which all gains in the electronics were recorded. Also, the sphere-to-receiver distance *r* was measured to within 5 mm, so that the coefficient in Eq. (4) $a/2r$ could be evaluated. When all of these factors were taken into account, the absolute comparisons in a *ka* range from 9.7 to 12.5, shown in Fig. 19, resulted.

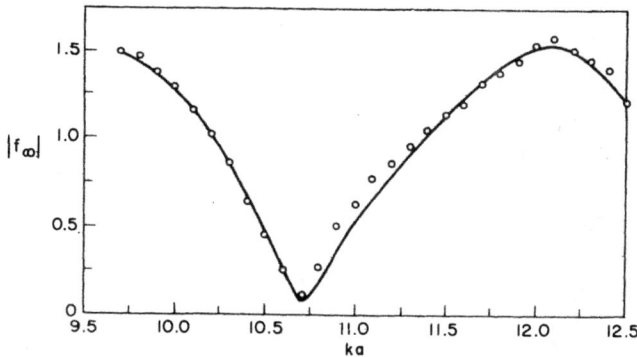

Fig. 19 — An absolute comparison between the theoretical reflection form-function $|f_\infty|$ vs *ka* (solid curve) and experimental points derived from a single pulse reflected from an aluminum sphere.

CONCLUSION

Accurate monostatic reflection characteristics, $|f_\infty|$, of rigid bodies in air and elastic spheres in water were obtained by using short tone bursts to insonify the targets. Agreement between exact elastic calculations and underwater experiments was excellent for spheres of aluminum, tungsten carbide, and brass. Comparisons between rigid theory and measurements in air on metal spheres and prolate spheroids also demonstrated excellent agreement. The use of a short pulse, rather than a steady-state acoustic wave or a pulse long enough to approximate steady-state, has major advantages because of the frequency dependence of $|f_\infty|$. The tone burst covers a broad frequency range, allowing a significant portion of the $|f_\infty|$ vs ka curve to be obtained from one experiment. Steady-state experiments give only the value of $|f_\infty|$ at a specific ka, and numerous such experiments must be performed to obtain information over a significant ka range. An added advantage of short pulse experiments, especially under laboratory conditions, is the ease with which the echoes of interest can be separated from echoes reflected at the test facility boundaries. The pulses used in air were shorter than those which could be obtained in water, because of the higher Q of the ceramic transducers used in the underwater measurements, as compared to the capacitive devices available for acoustic generation in air.

The nulls in the form function vs ka curves for elastic spheres are highly dependent on the material constants of the sphere material, especially the shear speed. Ordinary methods of measurement of the elastic constants would seldom yield sufficiently accurate theory to allow exact agreement with experiment. The wave speeds used here were measured to within 1% by direct propagation measurements, and the shear speed was varied within this tolerance until theoretical and experimental nulls agreed.

REFERENCES

1. W. G. Neubauer, R. H. Vogt, and L. R. Dragonette, J. Acoust. Soc. Am. **55**, 1123-1129 (1974).

2. A. J. Rudgers, J. Acoust. Soc. Am. **45**, 900-910 (1969).

3. W. E. Moore, "Recovery of Acoustic Pulse Waveforms Using Calculator Controlled Signal Processing," NRL Rep. No. 7658 (1974).

4. R. Hickling, J. Acoust. Soc. Am. **30**, 137-139 (1958).

5. A. J. Rudgers, "Techniques for Numerically Evaluating the Formulas Describing Monostatic Reflection of Acoustic Waves by Elastic Spheres," NRL Rep. No. 6551 (1967).

Chapter 10

ABSORBING SPHERES*

INTRODUCTION

The monostatic reflection from elastic spheres has been the subject of numerous theoretical [1-4] and experimental [5-7] treatments. Two methods demonstrated excellent agreement between exact elastic theory and experiment for the monostatic reflection from solid metal spheres in water. The steady-state technique described in Chapter 8, utilized long acoustic pulses to insonify the spheres, the transient technique described in Chapter 9 utilized short incident acoustic pulses. The latter method is generally more useful since a single experiment gives reflected pressure-vs-ka information over a range of ka values, whereas each steady-state measurement gives a single point on a reflected-pressure-vs-ka curve. The factors in the size parameter ka are the wave number k and the radius of the sphere a.

Figure 1 demonstrates quantitative agreement between the theoretical results obtained by means of a harmonic series analysis [2,3] (the solid curve) and both steady-state and transient experiments. Each of the points was obtained from a separate steady-state measurement in which frequency and/or sphere size was varied, while the dashed curve resulted from a single short pulse experiment with a single sphere. The theoretical analysis did not include effects due to absorption of shear and compressional waves in the material used to fabricate the spheres. The results, given in Fig. 1 and other similar comparisons, demonstrate that ignoring absorption effects is a very good assumption for spheres of tungsten carbide, aluminum, and brass, which were the three metals observed, [6,7] and probably for most other metals. In order to determine the possible effects of absorption on the reflection form function, a lucite sphere was investigated, and no agreement between the harmonic series analysis without absorption in the sphere and experiment was found. A modification of the theory to include the effects of the absorption of shear and compressional waves in lucite was demonstrated to be the major solution to the disparity.

---

*This work was reported earlier in: R. H. Vogt, L. Flax, and W. G. Neubauer, J. Acoust. Soc. Am. **57**, 558 (1975).

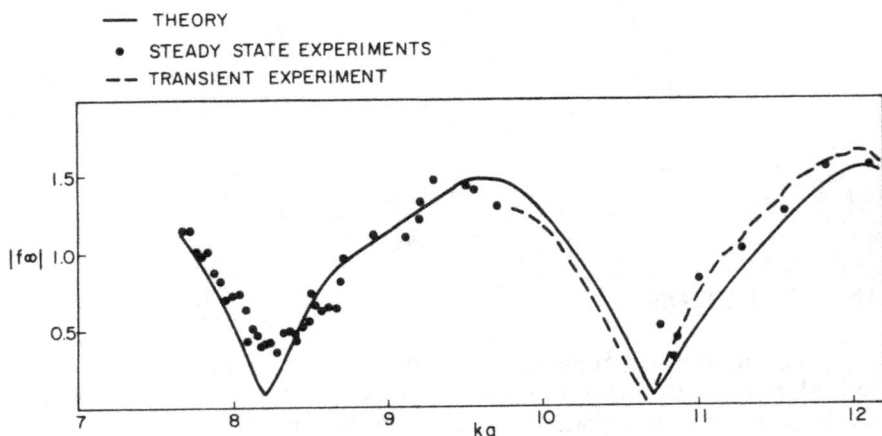

— THEORY
• STEADY STATE EXPERIMENTS
— — TRANSIENT EXPERIMENT

Fig. 1 — Theoretical calculation and two different types of experimental
measurement of the reflection from an aluminum sphere.

EXPERIMENTAL RESULTS AND COMPARISON WITH
HARMONIC SERIES ANALYSIS

Measurements were made on a Lucite sphere, employing the
technique described in Chapter 9. The form function $|f_\infty(ka)|$ is
defined in terms of experimentally measurable quantities by

$$|f_\infty(ka)| = \frac{2r}{a} \frac{|p_r|}{|p_i|} \qquad (1)$$

where r is the distance between the sphere center and the receiving
hydrophone, $|p_r|$ amplitude of the reflected pressure, and $|p_i|$ the
amplitude of the incident pressure measured, at the point occupied by
the center of the sphere, in the absence of the sphere. The pressure
amplitudes in Eq. (1) are steady-state values. The modified expression
for $|f_\infty(ka)|$, suitable for the transient measurements, is [7]

$$|f_\infty(ka)| = \frac{2r}{a} \frac{|g_r(ka)|}{|g_i(ka)|}, \qquad (2)$$

where g_r and g_i are, respectively, the Fourier transforms of the
reflected and incident pulses. The geometry of the measurement is the
same as described above for the steady-state case. The method of pulse
digitization and transformation is described in Ref. 8.

Figure 2(a) shows a pulse used to insonify a 3.175-cm-diam
Lucite sphere. The center frequency of the pulse is 210 kHz, and its
Fourier transform is given in Fig. 2(b). The abscissa of the curve in

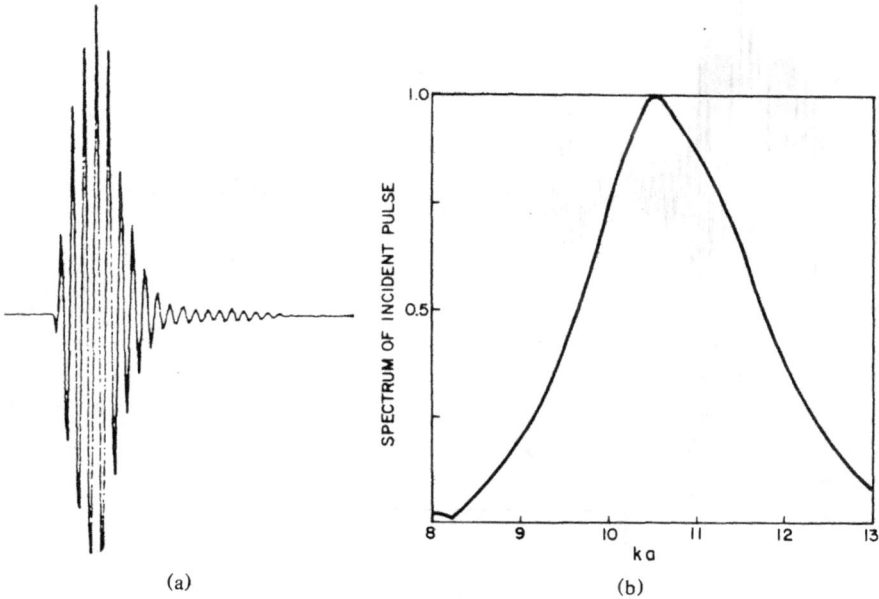

(a)   (b)

Fig. 2 — (a) The pulse incident on a Lucite sphere. (b) The normalized frequency
spectrum of the pulse in (a).

Fig. 2(b) has been converted from frequency space to ka, since the
final result of the analysis will be a form function vs ka plot. The pulse
reflected by the lucite sphere and its Fourier transform are given in
Figs. 3(a) and 3(b). The operations indicated by Eq. (2) were carried
out with a CDC 3800 computer and the resulting, experimentally
obtained, form function is given in Fig. 4, which also shows the
theoretical results obtained from a harmonic series analysis [2,3] which
neglects absorption. The limits on the abscissa in Fig. 4,
$9 < ka < 12.5$, were determined from the experiment as the ka values
at which the spectrum of the incident pulse is down 10 dB from its
maximum amplitude. This 10-dB-down value was empirically determined to be a reasonable choice of a cutoff value from extensive previous measurements [7]. Agreement between theory and experiment was
obtained [7] beyond the 10-dB points, so this is a conservative cutoff
value.

A lack of quantitative agreement between theory and experiment
is evident from Fig. 4. The accuracy of the experimental measurements made on metal spheres, [7] by the same technique, inferred that
the experimental curve in Fig. 4 was correct and that the theory was in
need of modification. The most obvious modification considered was

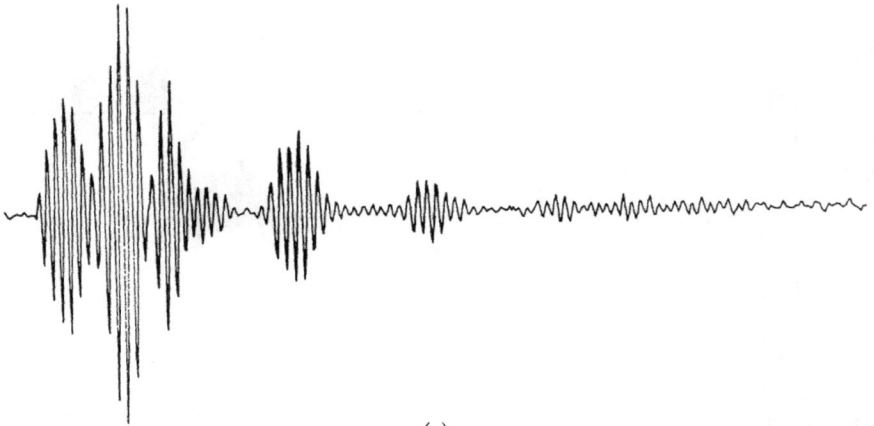

(a)

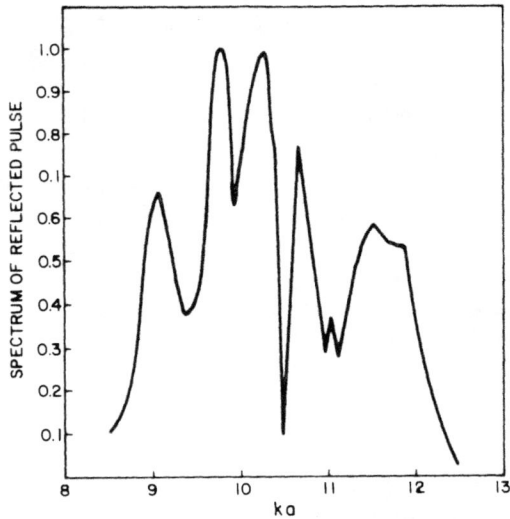

(b)

Fig. 3 — (a) The pulse monostatically reflected by a Lucite sphere when the pulse shown in Fig. 2 is incident. (b) The normalized frequency spectrum of the pulse seen in (a).

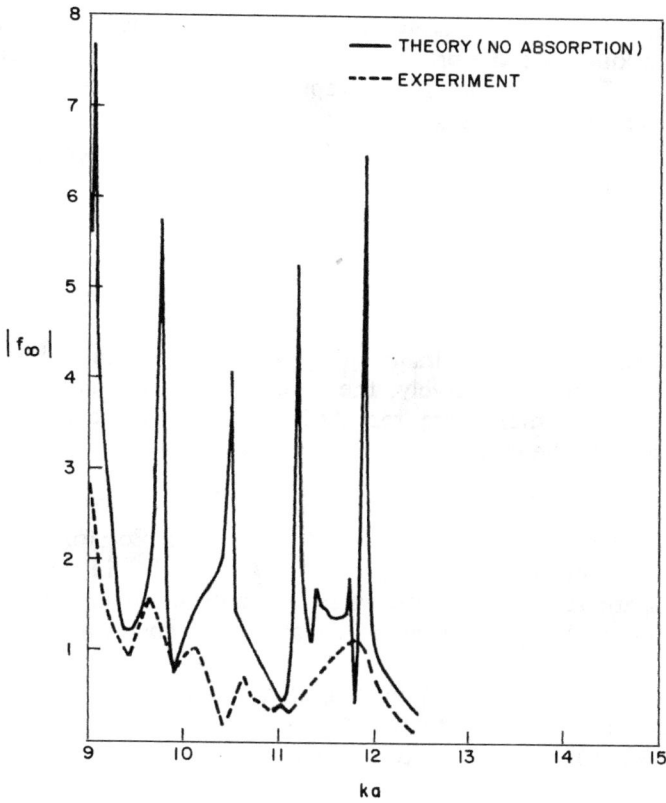

Fig. 4 — A comparison between theory which does not include absorption (−) and experiment (———) for a Lucite sphere.

the inclusion of the effects of the absorption of shear and compressional waves in lucite.

MODIFICATION OF THEORY TO INCLUDE ABSORPTION

For an elastic sphere in the absence of absorption, the exact series solution for $|f_\infty(ka)|$ is given by [2,3]

$$f_\infty(ka) = \left|\frac{-2}{ka}\right| \sum_{n=0}^{\infty} (-1)^n (2n + 1)\sin\eta_n e^{i\eta_n}, \tag{3}$$

with

$$\tan\eta_n = -\ [j_n(ka)L_n - (ka)j_n'(ka)]/[y_n(ka)L_n - (ka)y_n'(ka)]. \tag{4}$$

The j_n and y_n are, respectively, spherical Bessel functions of the first and second kind, and the primes denote derivatives with respect to the arguments. The function L_n is represented by a division of two second-order determinants of the form

$$L_n = \frac{\begin{vmatrix} a_{21} & a_{23} \\ a_{31} & a_{33} \end{vmatrix}}{\begin{vmatrix} a_{11} & a_{13} \\ a_{31} & a_{33} \end{vmatrix}} \cdot \rho/\rho_S. \tag{5}$$

The elements of the determinants are given by Goodman and Stern, [9] and ρ and ρ_S are, respectively, the densities of water and the sphere material. Two elements are repeated here for later comparisons in order to indicate the complication introduced by absorption:

$$a_{21} = k_L a j_n'(k_L a) \quad \text{and} \quad a_{23} = n(n + 1)j_n(k_S a). \tag{6}$$

The arguments of the spherical Bessel function in Eq. (6) include the compressional and shear wave numbers (k_L and k_S) in lucite. These arguments are real in the case where no absorption is included, and computation of the spherical Bessel functions is standard procedure.

Absorption is included by the standard method of introducing complex wave numbers into the theory. The complex compressional and shear wavenumbers $\tilde{k}_L$ and $\tilde{k}_S$ are given by

$$\tilde{k}_L a = (ka) \frac{c}{c_L} - ika \frac{c}{c_L} \beta_L, \tag{7}$$

$$\tilde{k}_S a = (ka) \frac{c}{c_S} - ika \frac{c}{c_S} \beta_S. \tag{8}$$

The subscripts L ans S represent compressional and shear in the Lucite; the nonsubscripted quantities refer to water. The c's are wave speeds and the β's are absorption factors derived from the measurements of Hartmann and Jarzynski [10]. In the frequency region of interest here, the absorption coefficients are given in Ref. 10 as $\alpha_L \cdot \lambda_L$ (compressional) $= 0.19$ dB and $\alpha_S \cdot \lambda_S$ (shear) $= 0.29$ dB. The wave speeds $c_S = 1380$ m/sec and $c_L = 2680$ m/sec were obtained from the Ocean Materials Criteria Branch of NRL. The velocities are measured to an accuracy of $\pm 1\%$, but were unfortunately not measured on a sample from the same stock used to fabricate the sphere, since no such sample existed. The speed of sound in water was 1482 m/sec [11]. As a result of introducing complex wave numbers into the theory, the elements of the determinant defining L_n contain spherical Bessel functions with complex arguments. No tabulations of Bessel functions with noninteger order and complex arguments were available, hence these functions were redefined in terms of the hypergeometric function using the relations [12]:

$$j_n(Z) = A {}_0F_1[\gamma + 1; - (Z)^2/4](Z)^{\gamma-1/2}, \tag{9}$$

where

$$A = \frac{\sqrt{\pi}}{\Gamma(\gamma + 1) 2^{\gamma+(1/2)}}, \quad \gamma = n + \frac{1}{2}, \tag{10}$$

Γ is the gamma function, ${}_0F_1$ is the hypergeometric function, and Z is a complex quantity, in this case either $(\tilde{k}_L a)$ or $(\tilde{k}_S a)$. The derivatives j_n and y_n are expressed in terms of hypergeometric functions using the relation

$$\frac{d^n}{dx^n} {}_0F_1(b; x) = \frac{1}{(b)_n} {}_0F_1(b + n; x). \tag{11}$$

The matrix elements describing L_n (Eq. 5) are given below in terms of the hypergeometric function. The following definitions are used in the expressions for the matrix elements:

$$Z_L \equiv \tilde{k}_L a; \qquad\qquad Z_S \equiv \tilde{k}_S a,$$
$$U \equiv - Z_L^2/4; \qquad\qquad l \equiv - Z_S^2/4;$$
$$F_L \equiv {}_0F_1(\gamma + 1; U); \qquad F_S \equiv {}_0F_1(\gamma + 1; l),$$
$$F_{L1} \equiv {}_0F_1(\gamma + 2; U); \qquad F_{S2} \equiv {}_0F_1(\gamma + 2; l),$$
$$F_{L2} \equiv {}_0F_1(\gamma + 3; U); \qquad F_{S2} \equiv {}_0F_1(\gamma + 3; l).$$

In the following expressions, μ and λ are the Lamé constants, and λ and A are defined by Eq. (9-10). With the above definitions, the matrix elements are

$$a_{21} = a \left[\left(\gamma + \frac{1}{2} \right) Z_L^{\gamma-1/2} F_L - \frac{Z_L^{\gamma+3/2}}{2(\gamma + 1)} F_{L1} \right], \tag{12}$$

$$a_{23} = A n (n + 1) Z_S^{\gamma-1/2} F_S, \tag{13}$$

$$a_{31} = 2[a_{21} - A Z_L^{\gamma-1/2} F_L]. \tag{14}$$

$$a_{33} = A \left[2(n^2 - 1) Z_S^{\gamma-1/2} F_S \right. \tag{15}$$
$$\left. - \frac{\gamma Z_S^{\gamma+(3/2)}}{(\gamma + 1)} F_{S1} + \frac{Z_S^{\gamma+(7/2)}}{4(\gamma + 1)(\gamma + 2)} F_{S2} \right],$$

$$a_{11} = A \left[Z_L^{\gamma-(1/2)} F_L - 2\mu/\lambda \left((\gamma - 1/2)(\gamma - 3/2) Z_L^{\gamma-(5/2)} F_L \right. \right.$$

$$\left. \left. - \frac{\gamma Z_L^{\gamma-(1/2)}}{(\gamma + 1)} F_{L1} + \frac{Z_L^{\gamma+(3/2)}}{4(\gamma + 1)(\gamma + 2)} F_{L2} \right) \right], \qquad (16)$$

$$a_{12} = - \frac{2(n + 1) nA}{Z_S^2} \left[(\gamma - 3/2) Z_S^{\gamma-(1/2)} F_S - \frac{Z_S^{\gamma+(3/2)}}{2(\gamma + 1)} F_{S1} \right]. (17)$$

The form function $|f_\infty(ka)|$ is obtained from Eqs. (3-5) using Eqs. (12-17). Calculations were made with a CDC 3800 computer.

A comparison of curves calculated with and without the inclusion of absorption effects is given in Fig. 5 while Fig. 6 shows a comparison between the predictions of absorption theory and two independent experiments. The significance of modifying the theory to include absorption is demonstrated by Figs. 4-6. Quantitative agreement between theory and experiment is found in Fig. 6.

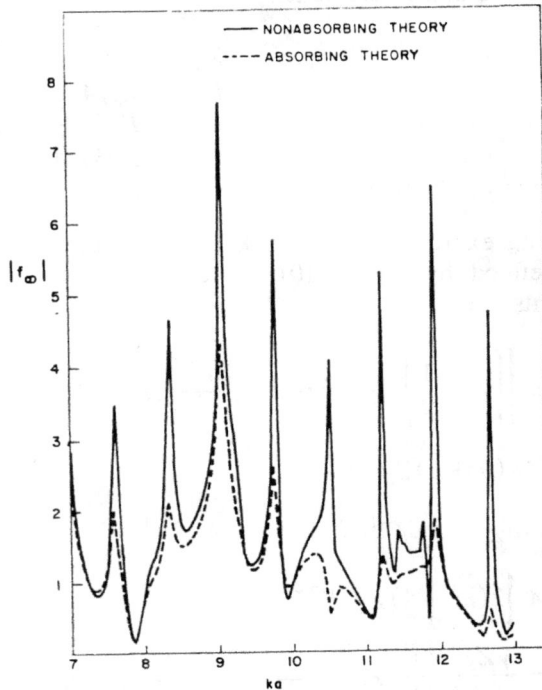

Fig 5 — Comparison between calculations of the reflection from a lucite sphere with and without the inclusion of absorption.

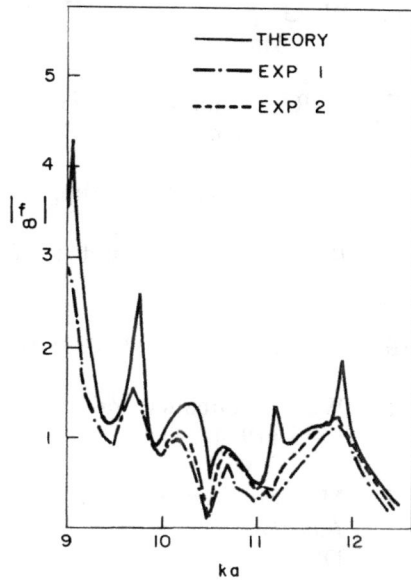

Fig. 6 — The reflection function calculated from the theory for an absorbing Lucite sphere is compared to reflection functions obtained from two independent experiments.

CONCLUSION

The monostatic reflection from a lucite sphere in water cannot be described by a plane-wave expansion technique unless the effects due to absorption are included in the analysis. The need for modification of the existing theory was demonstrated by a transient measurement technique and theoretical verification carried out. The results indicate that the measurement technique is capable of producing quantitative results with reasonable confidence on other elastic shapes not amenable to exact calculation.

REFERENCES

1. J. J. Faran, Jr., J. Acoust. Soc. Am. **23**, 405 (1951).

2. R. Hickling, J. Acoust. Soc. Am. **34**, 1582 (1962).

3. A. J. Rudgers, NRL Rep. 6551 (June 1967).

4. 4. T. Hasegawa, Y. Kitagawa, and Y. Watanabe, J. Acoust. Soc. Am. **62**, 1298 (1977).

5. L. D. Hampton and C. M. McKinney, J. Acoust. Soc. Am. **33**, 694 (1961).

6. W. G. Neubauer, R. H. Vogt, and L. R. Dragonette, J. Acoust. Soc. Am. **55**, 1123 (1974).

7. L. R. Dragonette, R. H. Vogt, L. Flax, and W. G. Neubauer, J. Acoust. Soc. Am. **55**, 1130 (1974).

8. W. E. Moore, NRL Rep. 7658 (1974).

9. R. R. Goodman and R. Stern, J. Acoust. Soc. Am. **34**, 338 (1962).

10. B. Hartmann and J. Jarzynski, J. Appl. Phys. **43**, 4304 (1972).

11. W. G. Neubauer and L. R. Dragonette, J. Acoust. Soc. Am. **36**, 1685 (1964).

12. M. Abramowitz and I. A. Stegun, Eds., *Handbook of Mathematical Functions* (Natl. Bur. Stand. Appl. Math. Ser. 55, Washington, DC, 1970), p. 362, Eq. 9.1.69 and p. 437, Eq. 10.1.1.

Chapter 11

LONGITUDINAL WAVES INCIDENT
ON AN ELASTIC SPHERICAL OBSTACLE
IN ANOTHER ELASTIC MATERIAL

INTRODUCTION

Interaction effects of elastic waves and bodies are the basic source of information used to attempt identification of flaws and inclusions in materials nondestructively. Even though analytical descriptions of some systems are possible, useful interpretation has often been severely limited by the lack of special function evaluations over sufficiently large ranges of parameters. Even when such calculations are available, interpretation is limited because basic wave-interface interactions are not completely understood. The spherical inclusion of one material in another is a system that is encumbered with all of these limitations. The problem of the spherical shape is interesting because it is a closed (finite) surface with a separable boundary value solution and, in some cases, is a reasonable approximation to an isotropic hole in a solid, or an inclusion or scatterer of one material in a matrix of a different material (often called the host material). Advances in computational capability [1] have permitted computation of a variety of combinations of hosts and inclusions.

Other solutions of spherical problems [2] indicate that erratic or unexpected behavior can occur in the scattering by a solid object in an elastic medium. A number of results, described by their scattering cross sections (SCS) as a function of frequency for spherically shaped inclusions in a different host material, are collected in Ref. 3. Some of the same combinations of materials assumed in the references, are assumed to be host and inclusion here. Earlier conclusions were derived from calculations that did not exceed a ka of 10 and in some cases only went as high as $ka = 3$. Certain conclusions about higher frequency behavior were borne out in some cases by higher ka computations. In other cases, the limited computations indicated incorrect conclusions. Direct experimental observations of total scattering cross sections are difficult, or perhaps impossible, to achieve. In spite of the fact that total scattering cross section is an experimentally unobtainable quantity, it has proven to be useful in allowing generalizations about classes of scattering problems. Unfortunately, the quantity called SCS does not have a universal definition. Here previously published curves are compared with alternate descriptive quantities for the cases treated elsewhere.

SCATTERING DESCRIPTIONS

Three different quantities are commonly used to quantitatively describe scattering from bodies as a function of frequency: total scattering cross section (TSCS), differential scattering cross section (DSCS) and form function $|f_\infty|$. Perhaps the most common of these is TSCS, which is usually defined [3] as the ratio of total energy scattered per unit time to the incident wave energy per unit area normal to the propagation direction per unit time. This is the quantity used to describe solutions in the area of nondestructive investigation or evaluation. However, even when TSCS is discussed in theoretical treatments it is not measured in the associated experiments [4]. The popularity of TSCS is probably a result of the spacial averaging inherent in the definition: i.e., for a body, it represents the total energy around the entire body. This is difficult to measure. In most cases measurements in a single direction, or a few directions, are all that can be made, and even those, often with great difficulty, especially if the medium containing the scattering body is a solid.

The definition of TSCS for spherical geometries leads to the expression

$$\text{TSCS} = 4\pi a^2 \left[\sum_{n=0}^{\infty} \frac{2n+1}{k_{L1}^2 a^2} |A_n|^2 + \frac{n(n+1)}{k_{L1} k_{S1} a^2} |B_n|^2 \right], \qquad (1)$$

where a is the scattering sphere radius, k is the wave number for longitudinal waves when subscripted by L and for shear waves when subscripted with S. The additional subscripts 1 and 2 refer to the outer (matrix or host) and inner (sphere or inclusion) medium respectively. The coefficients A_n and B_n are derived from the boundary value problem in the spherical geometry in which the wave equation is separable. The formulation of the problem is treated elsewhere [3,5,6], and in Chapter 14, so only the solutions and definitions of pertinent quantities necessary for computation will be given here. The matrices defining A_n and B_n, the elements in those matrices and the definitions of the quantities contained in them are given in the appendix.

In addition to the above definition of TSCS another quantity called the normalized (total) scattering cross section ($TSCS_N$) is defined as the ratio of the TSCS to the geometrical cross section (πa^2) of the sphere [3]. It is therefore the ratio of total power in the scattered wave to the power of the incident wave through the cross-sectional area of the scatterer (sphere).

The use of $TSCS_N$ seems to have been helpful in the evaluation of scatterers distributed in a volume since it takes into account all types of scattering, including mode conversions. It is a measure of the total scattered field, without taking into account any directional properties. At the same time, and for the same reasons, there exist limitations in the physical interpretation of scattering by evaluating $TSCS_N$. To compensate in part for this shortcoming, definitions have been made of $TSCS_N$ for an incident longitudinal wave or shear wave alone and the resultant longitudinal or shear wave reflected from the scatterer [4].

The DSCS is similar to the TSCS but includes the angular dependence of the scattering. It can be defined as the scattered power (energy per unit time) per steradian divided by the incident (longitudinal wave) intensity (energy flux). Solving the boundary value problem results in the expression

$$\text{DSCS} = a^2 \left[\left| \sum_{n=0}^{\infty} \frac{2n+1}{k_{L1} a} A_n P_n (\cos\theta) \right|^2 + \left| \sum_{n=0}^{\infty} \frac{2n+1}{k_{S1} a} B_n P_n' (\cos\theta) \right|^2 \right], \quad (2)$$

which again combines the effects of longitudinal and shear waves. Such a combined quantity cannot be evaluated experimentally with present day sensors.

For the case of backscattering ($\theta = \pi$), since

$$P_n' (\cos\pi) = 0 \text{ and } P_n (\cos\pi) = (-1)^n,$$

$$\text{DSCS} = a^2 \left| \sum_{n=0}^{\infty} \frac{(-1)^2 2n+1}{k_{L1} a} A_n \right|^2, \quad (3)$$

in which case no shear waves are involved in the scattering.

Another quantity used to describe the scattered field is the form function $|f_\infty|$ which results from the analyses of reflection from spheres and cylinders after the manner of Hickling [2] or Neubauer et al. [7] and others [8,9]. For a longitudinal wave incident on a sphere,

$$|f_\infty (ka, \theta)| = \frac{2}{k_{L1} a} \left| \sum_{n=0}^{\infty} (-1)^n (2n+1) A_n P_n (\cos\theta) \right|. \quad (4)$$

For backscattering $(\theta = \pi)$ this becomes

$$|f_\infty(ka, \pi)| = \frac{2}{k_{L1}a} \left| \sum_{n=0}^{\infty} (-1)^n (2n + 1) A_n \right|, \qquad (5)$$

so the relationship between DSCS and form function is

$$DSCS (\pi) = \frac{a^2}{4} |f_\infty(ka, \pi)|^2. \qquad (6)$$

Elasticity problems have most often been described by $TSCS_N$ and only occasionally by DSCS. In acoustical problems $|f_\infty|$ is probably the most common description and the one that has been compared quantitatively with experiments [4].

COMPUTATIONS

In this work backscattering calculations were carried out for combinations of seven different materials. Therefore, 42 cases of a spherical inclusion of one material in another were computed, as well as the seven materials with a spherical void or hole in each. The materials, their densities and longitudinal and shear wave speeds are listed in Table 1. The capability for computation of spherical Bessel functions achievable for large argument. That capability has since been achieved by Flax and Mason, [1] whose results were used for these computations. Here only an incident longitudinal wave will be considered in the host medium but both longitudinal and shear waves are generated on reflection. Absorption of bulk waves are included in the computations by using a complex wave number.

Table 1

Material	Density (gm/cc)	Longitudinal Wave Speed (m/sec)	Shear Wave Speed (m/sec)	v_L/v_S
Polyethylene	0.90	1950	540	3.61
Polymethyl methacrylate	1.18	2680	1380	1.94
Magnesium	1.74	5770	3050	1.89
Beryllium	1.87	12890	8800	1.46
Aluminum	2.70	6568	3149	2.09
Germanium	5.36	5285	3376	1.57
Iron	7.90	5790	3100	1.87

The cases of a spherical cavity in each material computed by Truell et al. [3] were taken only to a value of $k_1a = 7$. Calculations, which were carried out up to a k_1a of 40, support their conclusion that as k_1a increases, values of $TSCS_N$ of very nearly 2 are approached by all curves. For k_1a of 40, $TSCS_N$ was in all cases between 1% and 2% of the value 2. Since our curves show only a gradual smooth approach to those values with increasing k_1a they will not be reproduced here.

The case of a germanium (Ge) sphere in an aluminum (Al) matrix was computed (see Fig. 1) to permit a direct comparison with the curves of Truell et al. [3]. Agreement was obtained up to a k_1a of 10, which is the largest value calculated by Truell. Truell et al. speculated correctly that the curve was perhaps a "very gradual oscillation, with a maximum in the vicinity of $k_1a = 8$, upon which a fine structure of rapid oscillations is superimposed." That is exactly the behavior of this curve even up to a k_1a of 40. The only other features of any interest are the bursts of sharp, but small, spikes around values of k_1a of 19, 25, and 33. The case for these small oscillations is undoubtedly interference between the incident reflected wave and waves reemerging in the direction of incidence after refraction and reflection inside the sphere.

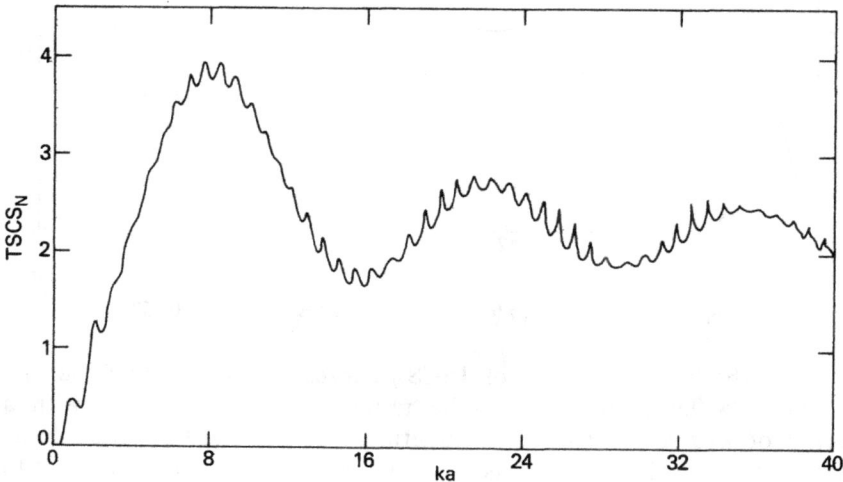

Fig. 1 — $TSCS_N$ vs k_1a for a Ge sphere in Al. $(v_1/v_2) = 1.243$.

On the basis of their limited calculations Truell et al. concluded that $TSCS_N$ curves are smoothest for cases in which there are large differences in longitudinal wave speeds of matrix and sphere material, while similar wave speeds result in highly oscillatory $TSCS_N$ curves.

This conclusion is not borne out by the calculations shown here. It was shown to be true in some cases and not in others, as evidenced by the results in Figs. 2-7. For a beryllium (Be) sphere in a polyethylene (PE) matrix, for which the longitudinal wave speed ratio v_1/v_2 is 0.152, the $TSCS_N$ curve is shown in Fig. 2 to be relatively smooth up to a k_1a of 30. Lest that be considered an isolated case, the example of a Be sphere in Ge, for which $v_1/v_2 = 0.410$, is given in Fig. 3. Whereas both of these cases fit the hypothesis of Truell et al., for the inverse case of PE in Be, for which $v_1/v_2 = 6.569$, the computations result in the oscillatory $TSCS_N$ curve shown in Fig. 4. This is in contradiction to the hypothesis of Truell et al. Again to eliminate the possibility of an isolated instance, the $TSCS_N$ of a Ge sphere in Be, $(v_1/v_2 = 2.348)$, is shown in Fig. 5 to be oscillatory also. The case of a magnesium (Mg) sphere in an iron (Fe) matrix, shown in Fig. 6, is one for which v_1 is very close to v_2, i.e ., $v_1/v_2 = 1.004$, and does have a $TSCS_N$ curve oscillatory in nature, which is consistent with the hypothesis. On the other hand the case of a Ge sphere in Fe shown in Fig. 7 which has a ratio of v_1/v_2 of 1.096 results in a smooth curve.

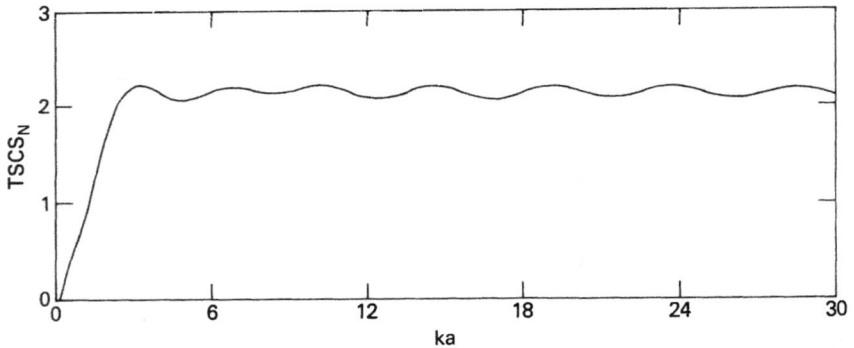

Fig. 2 — $TSCS_N$ vs k_1a for a Be sphere in PE. $(v_1/v_2) = 0.152$.

On the basis of this set of $TSCS_N$ curves one might conclude that for $v_1/v_2 > 0.9$ there seem to be rather regular oscillations with a period of $k_1a \leqslant 1$. These are superimposed on another larger basic period which itself decreases as v_1/v_2 increases and becomes difficult to distinguish as the small period superimposed oscillations increase. By viewing all of the 42 computed curves at one time, an attempt was made to distinguish between smooth- and oscillatory — (or perhaps, more correctly, jagged) — curve behavior and we concluded that for $v_1/v_2 < 1.3$ curves tend to be smooth and for $v_1/v_2 > 1.3$ curves tend to oscillate rapidly.

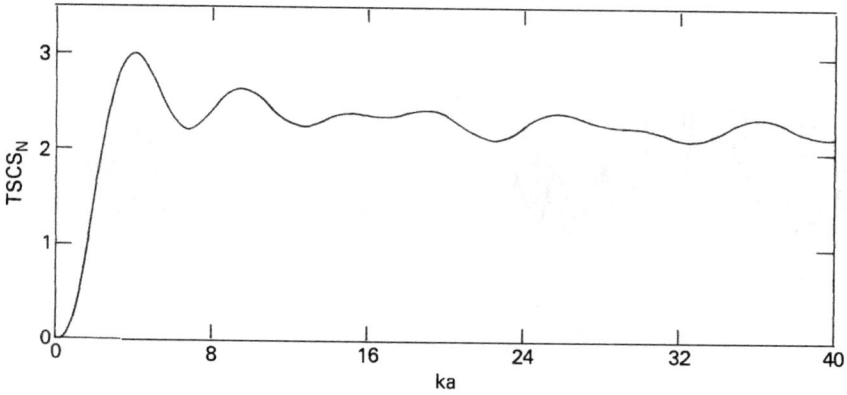

Fig. 3 — $TSCS_N$ vs k_1a for a Be sphere in Ge. $(v_1/v_2) = 0.410$.

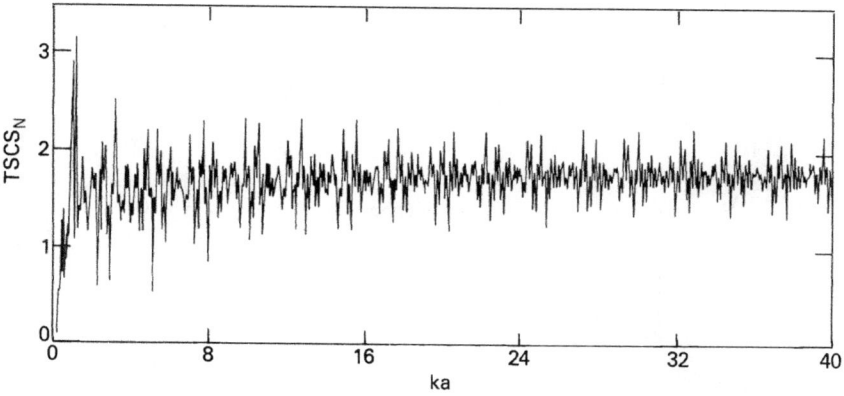

Fig. 4 — $TSCS_N$ vs k_1a for a PE sphere in Be. $(v_1/v_2) = 6.595$.

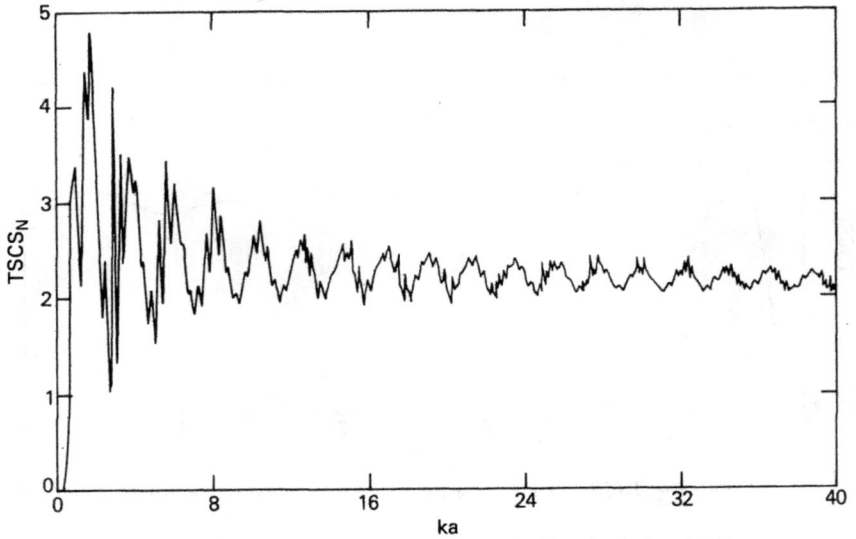

Fig. 5 — $TSCS_N$ vs k_1a for a Ge sphere in Be. $(v_1/v_2) = 2.438$.

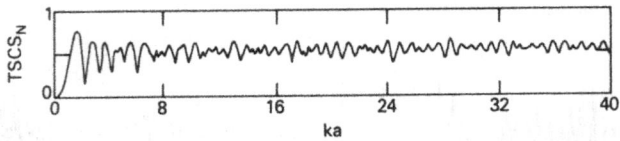

Fig. 6 — $TSCS_N$ vs k_1a for a Mg sphere in Fe. $(v_1/v_2) = 1.004$.

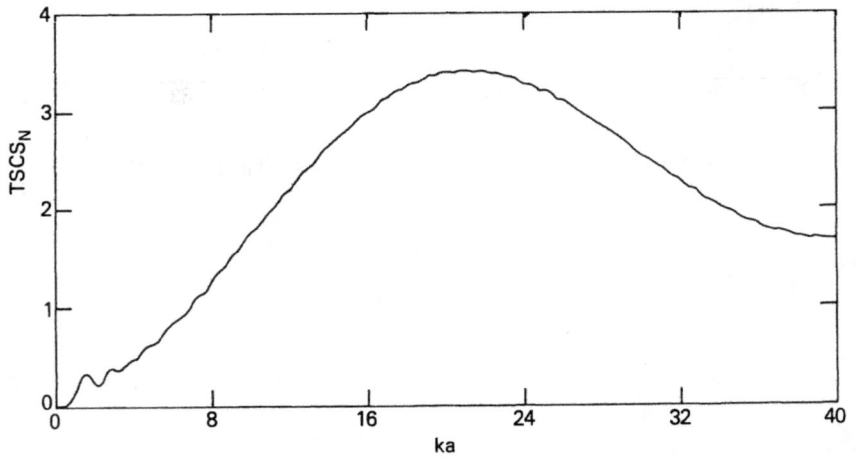

Fig. 7 — $TSCS_N$ vs k_1a for a Ge sphere in Fe. $(v_1/v_2) = 1.096$.

Similarities were noted among the curves for which the scattering sphere was assumed to be beryllium. An example of a typical one of these is given in Fig. 8. Similarly, for the cases for which beryllium was assumed to be the matrix material curves similar to Fig. 9 were obtained. These similarities are possibly related to the fact that beryllium has unusually large longitudinal and shear wave speeds (see Table 1).

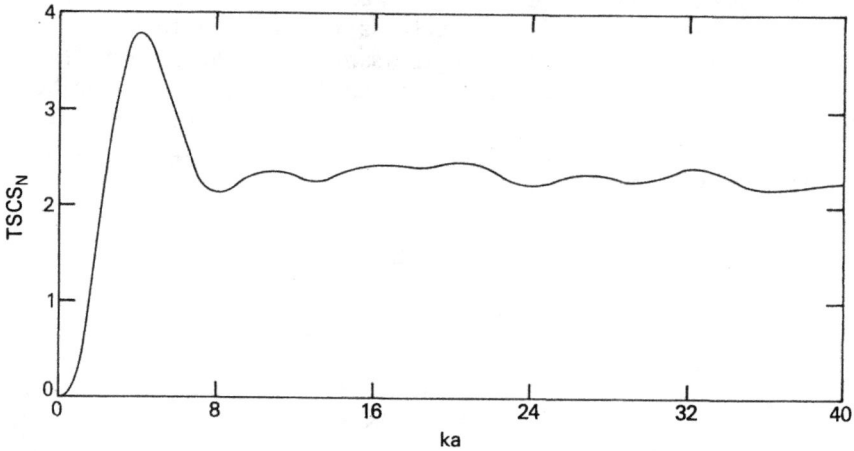

Fig. 8 — $TSCS_N$ vs k_1a for a Be sphere in Mg. $(v_1/v_2) = 0.448$.

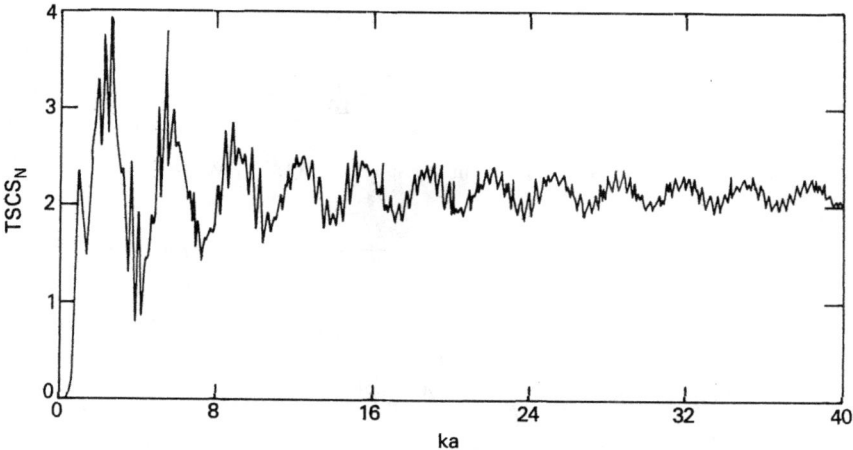

Fig. 9 — $TSCS_N$ vs k_1a for a Al sphere in Be. $(v_1/v_2) = 1.958$.

Many $TSCS_N$ curves were particularly devoid of behavior that one might attempt to interpret physically in terms of resonant or circumferential wave behavior. (See Figs. 7 and 10.) However, curves of DSCS or $|f_\infty|$ for the same cases have more features that may ultimately be subject to such interpretation. The two such cases given in Figs. 11 and 12 correspond respectively to Figs. 7 and 10. Oscillations of similar, but not the same, periods are found in the plots for $TSCS_N$ and $|f_\infty|$ for a Ge sphere in Al (Figs. 1 and 13 respectively). The opposite observation can be made on examination of Fig. 9, for an Al sphere in Be for which the $TSCS_N$ is regular and relatively well behaved, while the $|f_\infty|$ curve for the same combination (Fig. 14) is irregular. These cases indicate that it may be helpful for applications where characteristic behavior is sought, to calculate both types of functions until distinct physical causes can be attributed to specific features of curve behavior.

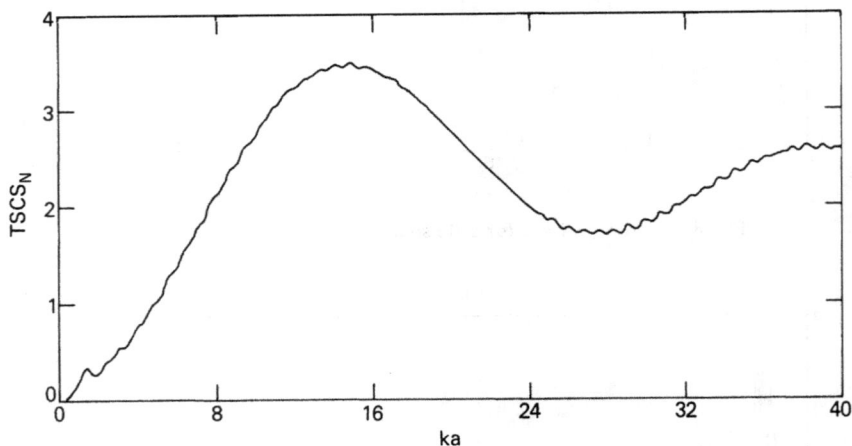

Fig. 10 — $TSCS_N$ vs k_1a for an Mg sphere in Al. $(v_1/v_2) = 1.138$.

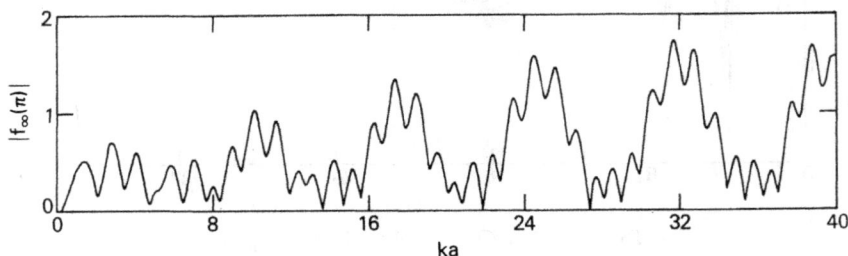

Fig. 11 — Form function $|f_\infty|$ vs k_1a for a Ge sphere in Fe. $(v_1/v_2) = 1.096$.

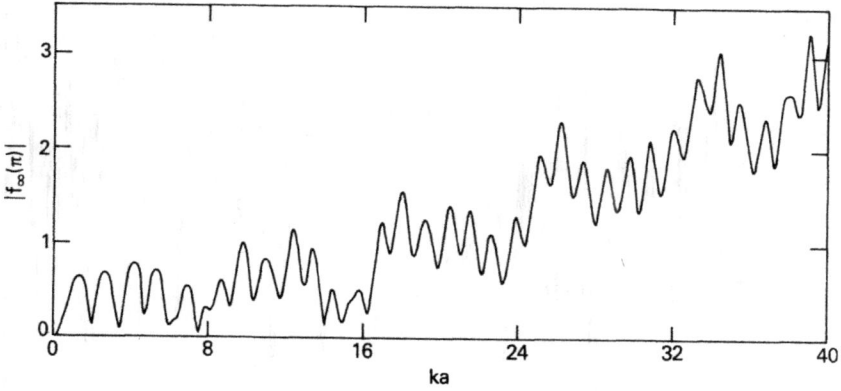

Fig. 12 — Form function $|f_\infty|$ vs $k_1 a$ for a Mg sphere in A1. $(v_1/v_2) = 1.138$.

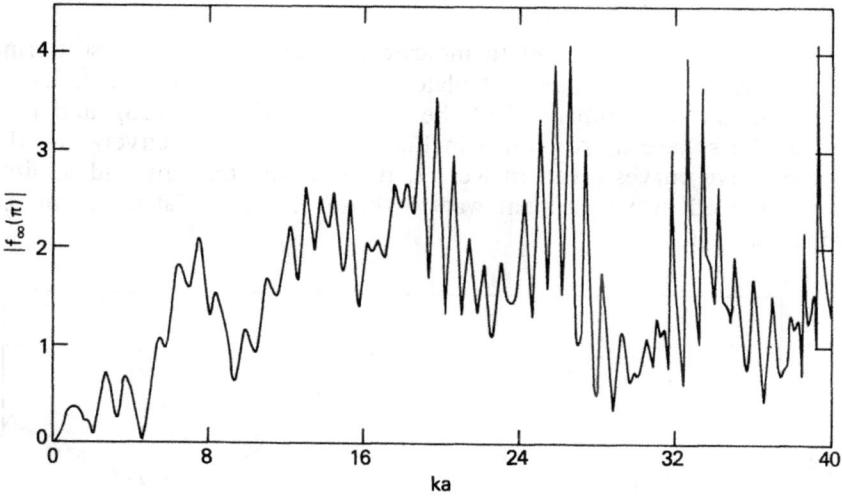

Fig. 13 — Form function $|f_\infty|$ vs $k_1 a$ for a Ge sphere in A1. $(v_1/v_2) = 1.243$.

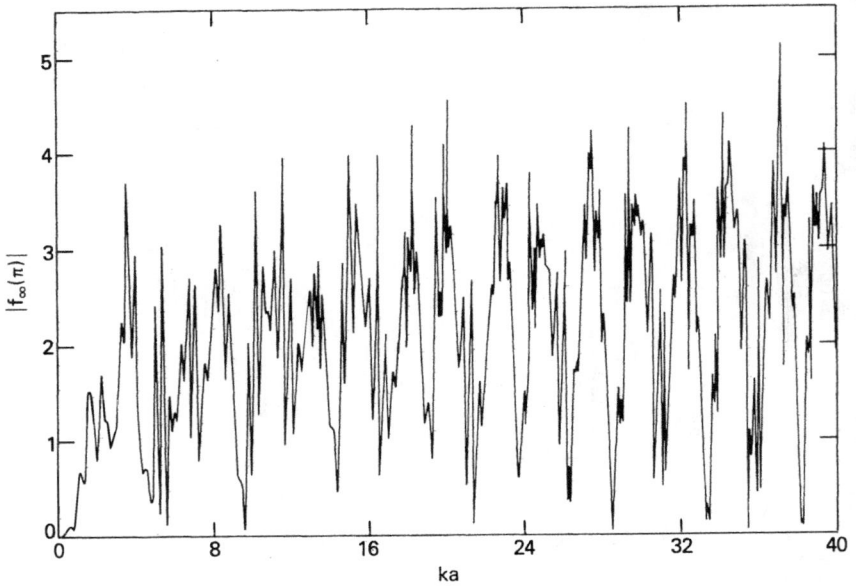

Fig. 14 — Form function $|f_\infty|$ vs $k_1 a$ for an Al sphere in Be. $(v_1/v_2) = 1.958$.

Some results claimed to indicate resonant behavior in scattering by an inclusion have been calculated by Flax and Überall [10]. However, on careful comparison of the curves describing $TSCS_N$ and $|f_\infty|$ for an Fe sphere in Al shown in Figs. 15 and 16 respectively and the partial wave curves given in Ref. 7, none of the maxima and minima corresponded in a consistent way to those of the partial waves in the spherical region.

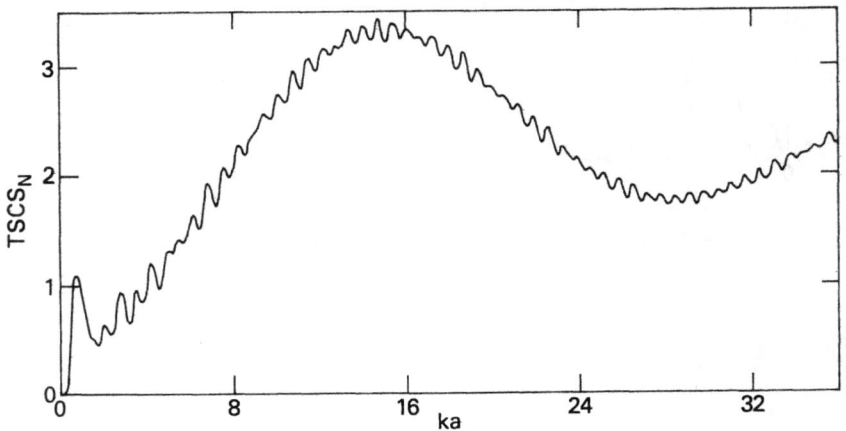

Fig. 15 — $TSCS_N$ vs $k_1 a$ for an Fe sphere in Al. $(v_1/v_2) = 1.134$.

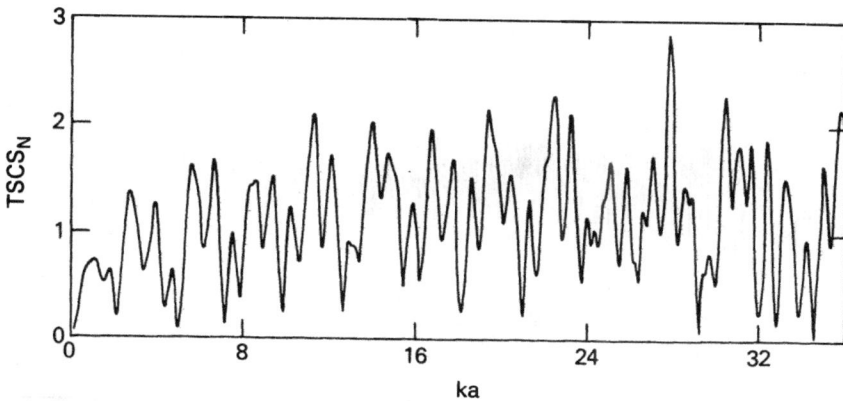

Fig. 16 — Form function $|f_\infty|$ vs $k_1 a$ for an Fe sphere in Al. $(v_1/v_2) = 1.134$.

The $TSCS_N$ curve for a spherical cavity in a solid has been noted as having a single maximum roughly corresponding to radial resonant behavior [3]. It is also anticipated that radial resonant behavior should be observable in elastic spherical inclusions [3]. However, experience with the reflection from solid spheres in water [7] leads one to conclude that the radial mode resonant peak may not be expected to be significant in the curves describing reflection from elastic spheres imbedded in another solid. Resonant behavior is an excitation of a natural mode of vibration of a body, a sphere in this case, and whether a peak is detected depends on whether that vibrational mode has significant radiation in the observation direction. For solid spheres and shells in fluids at low frequencies the radial mode is difficult to excite by an incident wave [6-12]. It is, therefore, also difficult to detect. There have been indications that resonant behavior cause minima in $|f_\infty|$ for backscattering because of circumferential modes in spheres and cylinders occurring at values of ka below those required for their radial or breathing mode [11]. The predominant modal indications are likely to result from the reinforcement of interface modes, as was the case for cylindrical shells in water [11]. It would, therefore, seem that a reasonable degree of understanding of these scattering curves for spherical bodies in a different material matrix must await the more detailed examination and understanding of the properties of waves that exist on the interface of two different materials.

By means of complex wave numbers, absorption was included in the solutions for spherical scattering and reflection. (See the appendix.) Results of computations assuming no absorption are shown in Fig. 17 for $TSCS_N$ for a PE sphere in Al. The same case computed assuming a longitudinal and shear wave absorption β_L and β_S of 0.0070 Np and

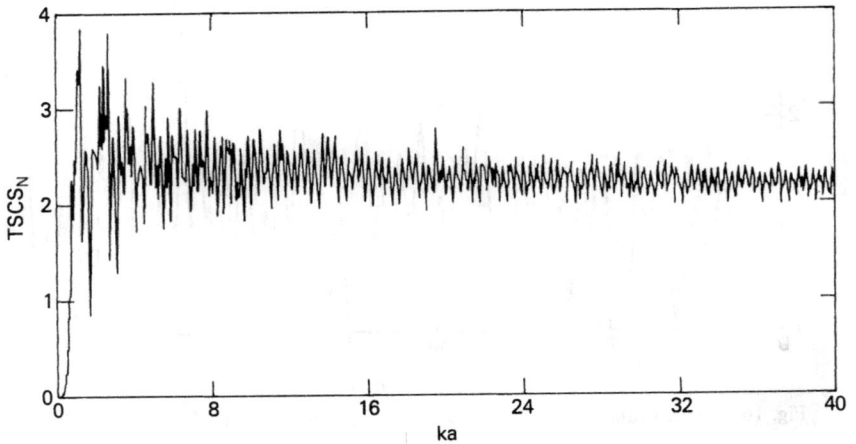

Fig. 17 — TSCS_N vs k_1a for a nonabsorbing PE sphere in Al. $(v_1/v_2) = 3.368$.

0.0212 Np respectively for the PE sphere is shown in Fig. 18. A curve of $|f_\infty|$ for the same case considered in Fig. 17, of a PE sphere in Al, again without absorption, is shown in Fig. 19. The form function $|f_\infty|$ for the same case of a PE sphere in Al but with absorption in the sphere of PE is given in Fig. 20 showing a more dramatic effect in the comparisons of the absorbing and nonabsorbing cases. Comparison of Figs. 17 and 18 shows a moderate difference in the TSCS_N curves with and without absorption. However, for the same case the differences shown in the curves of $|f_\infty|$ in Figs. 19 and 20 indicate large magnitude differences. The two descriptors, TSCS_N and $|f_\infty|$ (or DSCS) have different sensitivities to the damping effects of absorption.

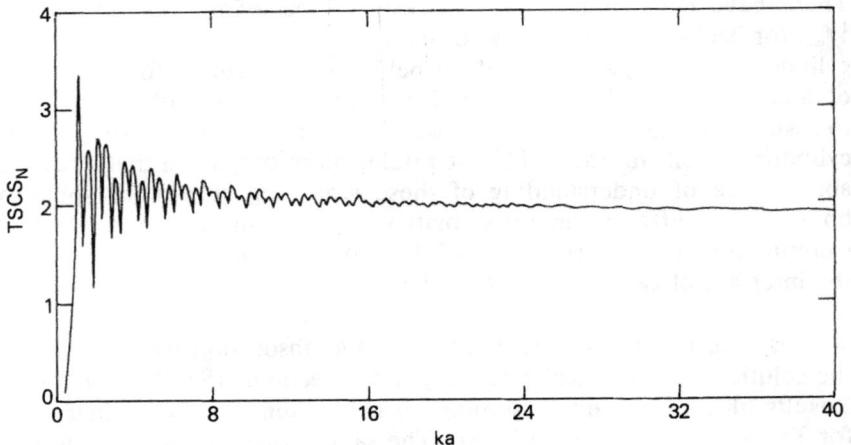

Fig. 18 — TSCS_N vs k_1a for an absorbing PE sphere in Al. $(v_1/v_2) = 3.368$.

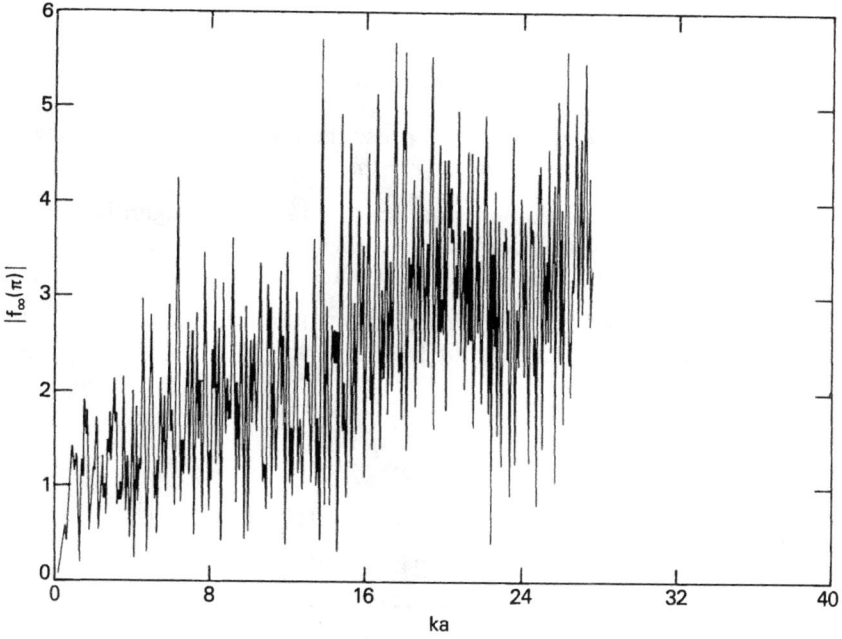

Fig. 19 — Form function $|f_\infty|$ vs k_1a for a nonabsorbing PE sphere in Al.
$(v_1/v_2) = 3.368$.

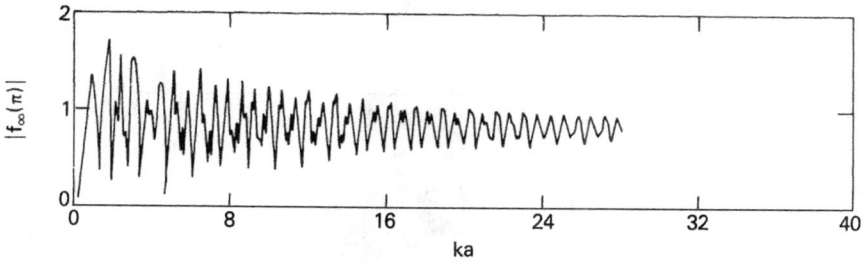

Fig. 20 — Form function $|f_\infty|$ vs k_1a for an absorbing PE sphere in Al. $(v_1/v_2) = 3.368$.

APPENDIX

Four boundary conditions are prescribed at the interface between the host medium and the sphere at $r = a$:

(i) radial displacements are continuous, (ii) tangential displacements are continuous, (iii) radial stresses are equal, and (iv) tangential stresses are equal.

Since only waves in the outer medium will be considered here, the required coefficients A_n and B_n for Eqs. (1-6) are

$$A_n = \frac{\begin{vmatrix} a_{11} & a_{12} & a_{14} & a_{15} \\ a_{21} & a_{22} & a_{24} & a_{25} \\ a_{31} & a_{32} & a_{34} & a_{35} \\ a_{41} & a_{42} & a_{44} & a_{45} \end{vmatrix}}{\Delta},$$

$$B_n = \frac{\begin{vmatrix} a_{11} & a_{12} & a_{13} & a_{15} \\ a_{21} & a_{22} & a_{23} & a_{25} \\ a_{31} & a_{32} & a_{33} & a_{35} \\ a_{41} & a_{42} & a_{43} & a_{45} \end{vmatrix}}{\Delta},$$

where $\Delta = \begin{vmatrix} a_{11} & a_{12} & a_{13} & a_{14} \\ a_{21} & a_{22} & a_{23} & a_{24} \\ a_{31} & a_{32} & a_{33} & a_{34} \\ a_{41} & a_{42} & a_{43} & a_{44} \end{vmatrix}$

The elements are given below.

$$a_{11} = \rho_2/\rho_1 \, [j_n(Z_{L2}) - 2G_1 j_n''(Z_{L2})]/(1 + 2G_2)$$

$$a_{12} = -2n(n+1) \, (\rho_2/\rho_1) \, [Z_{S2}j_n'(Z_{S2}) - j_n(Z_{S2})]/Z_{S2}^2$$

$$a_{13} = [h_n^{(2)}(Z_{L1}) - 2G_1 h_n^{(2)}(Z_{L1})]/(1 + 2G_1)$$

$$a_{14} = -2n(n+1) \, [Z_{S1}h_n^{(2)'}(Z_{S1}) - h_n^{(2)''}(Z_{S1})]/Z_{S1}^2$$

$$a_{15} = [j_n(Z_{L1}) - 2G_1 j_n''(Z_{L1})]/(1 + 2G_1)$$

$$a_{21} = Z_{L2}j_n'(Z_{L2})$$

$$a_{22} = n(n+1)j_n(Z_{S2})$$

$$a_{23} = Z_{L1}h_n^{(2)'}(Z_{L1})$$

$$a_{24} = n(n+1)h_n^{(2)}(Z_{S1})$$

$$a_{25} = Z_{L1}j_n'(Z_{L1})$$

$$a_{31} = 2[Z_{L2}j_n'(Z_{L2}) - j_n(Z_{L2})] \, [\mu_2 Z_{S1}^2/\mu_1 Z_{S2}^2]$$

$$a_{32} = [Z_{S2}^2 j_n''(Z_{S2}) + (n+2)(n-1)j_n(Z_{S2})] \, [\mu_2 Z_{S1}^2/\mu_1 Z_{S2}^2]$$

$$a_{33} = 2[Z_{L1}h_n^{(2)'}(Z_{L1}) - h_n^{(2)}(Z_{L1})]$$

$$a_{34} = Z_{S1}^2 h_n^{(2)''}(Z_{S1}) + (n+2)(n-1)h_n^{(2)}(Z_{S1})$$

$$a_{35} = 2[Z_{L1}j_n'(Z_{L1}) - j_n(Z_{L1})]$$

$$a_{41} = j_n(Z_{L2})$$

$$a_{42} = j_n(Z_{S2}) + Z_{S2}j_n'(Z_{S2})$$

$$a_{43} = h_n^{(2)}(Z_{L1})$$

$$a_{44} = h_n^{(2)'}(Z_{S1}) + Z_{S1}h_n^{(2)}(Z_{S1})$$

$$a_{45} = j_n(Z_{S1})$$

Where ρ_1 and ρ_2 are the densities of medium 1 and 2 respectively, μ_1 and μ_2 are the Lamé rigidity moduli and j_n, h_n are the spherical Bessel and Hankel functions, respectively.

The following definitions are used to simplify the notation somewhat.

$$Z_{L1} \equiv k_{L1}a$$

$$Z_{S1} \equiv k_{S1}a$$

$$Z_{L2} \equiv k_{L2}a$$

$$Z_{S2} \equiv k_{S2}a$$

$$G_1 \equiv Z_{L1}^2 Z_{S1}^2 - Z_{L1}^2$$

$$G_2 \equiv Z_{L2}^2 Z_{S2}^2 - 2Z_{L2}^2.$$

In the expression for the elements, the primes indicate differentiation with respect to the argument of the spherical Bessel and Hankel functions. The effects of material properties that cause attenuation of longitudinal and shear waves are accounted for in the theory by introducing complex longitudinal and shear wave numbers defined as

$$\kappa_{L1} \equiv k_{L1}(1 - i\beta_{L1})$$
$$\kappa_{S1} \equiv k_{S1}(1 - \beta_{S1})$$
$$\kappa_{L2} \equiv k_{L2}(1 - i\beta_{L2})$$
$$\kappa_{S2} \equiv k_{S2}(1 - i\beta_{S2})$$

where $\beta's$ are absorption factors with dimensions of Np. The relation to absorption coefficients $\alpha_L \lambda_L$ and $\alpha_S \lambda_S$ (λ = wavelength) in dB is

$$\beta_{L,S} = \alpha_{L,S}/(40\pi \ e).$$

Formulations apply to lossless cases by letting $\beta_{L,S} = 0$. As a result of the use of complex wave numbers, the elements of the determinants defining A_n and B_n contain spherical Bessel functions of complex arguments. Techniques developed by Flax and Mason [1] for calculating these functions were used. Far field approximations are assumed for the scattered field.

REFERENCES

1. L. Flax and J.P. Mason, "FORTRAN Subroutines to Evaluate, in Single or Double Precision, Bessel Functions of the First and Second Kinds for Complex Arguments," NRL Report No. 7997, Washington, DC (1976).

2. R. Hickling, J. Acoust. Soc. Am. **34**, 1582-1592 (1962).

3. R. Truell, C. Elbaum, and B.B. Chick, (Academic Press, Inc., New York, 1969).

4. R.M. White, J. Acoust. Soc. Am. **30**, 771-785 (1958).

5. C.F. Ying and R. Truell, J. Appl. Phys. **27**, 1086-1097 (1956).

6. L. Flax and H. Überall, J. Acoust. Soc. Am. **67**, 1432-1442 (1980).

7. W.G. Neubauer, R.H. Vogt, and L.R. Dragonette, J. Acoust. Soc. Am. **55**, 1123-1129 (1974).

8. L.R. Dragonette, R.H. Vogt, L. Flax, and W.G. Neubauer, J. Acoust. Soc. Am. **55**, 1130-1137 (1974).

9. R.H. Vogt and W.G. Neubauer, J. Acoust. Soc. Am. **60**, 15-22 (1976).

10. L. Flax and H. Überall, J. Acoust. Soc. Am. **64**, S130 (1978).

11. L.R. Dragonette, "Evaluation of The Relative Importance of Circumferential or Creeping Waves in The Acoustic Scattering From Rigid and Elastic Solid Cylinders and From Cylindrical Shells," NRL Report No. 8216, Washington, DC (1978).

12. L. Flax, L.R. Dragonette, and H. Überall, J. Acoust. Soc. Am. **63**, 723-731 (1978).

Chapter 12

RUBBER CYLINDERS AND SPHERES*

INTRODUCTION

The scattering of an infinite plane acoustic wave by isotropic cylinders and spheres fabricated of lossless materials has been considered by numerous authors both theoretically [1-3] and experimentally [4-6]. The harmonic series analysis modified to include the effect of absorption in the elastic solid [7-9] were found to agree with experiment [7,10] to within the uncertainty in the elastic constants of the material used.

Here a parametric study will be described of the reflection and scattering of sound by silicone rubber cylinders and spheres in water. As will be shown, the use of silicone rubber allows a significant simplification of the theory. These materials are of special interest due to the fact that their longitudinal sound speed [11] c_L is significantly lower than that of water. Furthermore, the product of the density ρ_s and c_L can be made to approach that of water. Finally, at ultrasonic frequencies the longitudinal transmission loss is low. From a practical standpoint materials such as these have proven useful in the construction of acoustic-slow waveguides [12] and lenses [13].

FORM FUNCTION CALCULATIONS

The ratio of the absolute values of the scattered, $|p_s|$, to the incident, p_0, pressures can be expressed as

$$|p_s|/|p_0| = (a/2r)|f_\infty|_s \qquad (1)$$

for spheres and

$$|p_s|/|p_0| = \left(\frac{a}{2r}\right)^{1/2} |f_\infty|_c \qquad (2)$$

---

*This work first appeared in: C. M. Davis, L. R. Dragonette and L. Flax, J. Acoust. Soc. Am. 63, 1694-1698 (1978).

for infinite cylinders, where a is the radius of the sphere or cylinder, r is the distance from the center of the scatterer to the receiver and $|f_\infty|$ is defined as the amplitude of the form function. Derivation of the far-field expressions for $|f_\infty|$ are given in Chapter 10 and Chapter 4 for spheres and cylinders, respectively.

The variation of $|f_\infty|$ with ka ($k = 2\pi/\lambda$, where λ is the wavelength of sound in water) for the case of monostatic scattering of sound from silicone rubber spheres and cylinders in water is calculated. Values of ρ_s, elastic moduli, and sound-absorption coefficients charac-teristic of commercially available material [11] are considered. Follow-ing Mason, [14] the value of shear speed, c_s, is taken to be maximum of $c_L/10$ and the shear absorption coefficient a minimum of 25 times the corresponding longitudinal absorption coefficient α. As is generally the case for rubber, the value of α for silicone rubber has been shown [11] to be proportional to frequency.

Curves of $|f_\infty|_c$ obtained with and without shear rigidity and absorption were indistinguishable from each other. This is a result of the abnormally low shear speed which in turn leads to negligible shear conversion at the water-rubber interface. Thus, all subsequent calcula-tions have omitted the effect of shear and so correspond to a liquid cylinder. In Fig. 1 a comparison is made with and without absorption for values of c_L and ρ_s typical of silicone rubber. The curves with $\alpha \neq 0$ were obtained using a value approximately equal to the average of those reported by Folds [11]: specifically $\alpha/k = 5 \times 10^{-3}$ where α is expressed in Np/cm. As can be seen, the effect of absorption in these materials is simply to attenuate the amplitude of the oscillations. As will be shown, these oscillations are due to the interaction between various lens waves which suffer attenuation in the material and, in some cases, specular reflections which do not. In order to simplify both the calculations and the interpretation, the investigation will be restricted for the most part to the lossless case.

In Fig. 2, $|f_\infty|_c$ is represented for $c_L = 1.0$ km/s and variable ρ_s such that $\rho_s c_L = \rho_s$. Silicone rubber with $\rho_s > 1.5$ g/cm^3 can be achieved in practice by loading with metal oxide. Curves from Fig. 1 corresponding to $\rho_s = 1.0$ and 1.5 g/cm^3 are repeated. For ρ_s (and therefore $\rho_s c_L$) = 1.5 the value of $|f_\infty|_c$ increases linearly with ka up to a ka value of approximately 5, at which point the value of $|f_\infty|_c$ begins to oscillate about a value of approximately 0.5. For values of $\rho_s c_L$ different than 1.5, an oscillation appears in the low ka region whose amplitude increases with increasing deviation of $\rho_s c_L$ from 1.5.

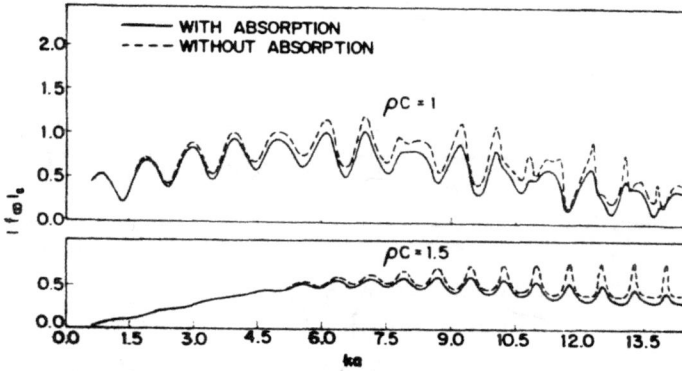

Fig. 1 — A comparison of computed curves of $|f_\infty|$ vs ka for silicone rubber cylinders without and with longitudinal absorption included.

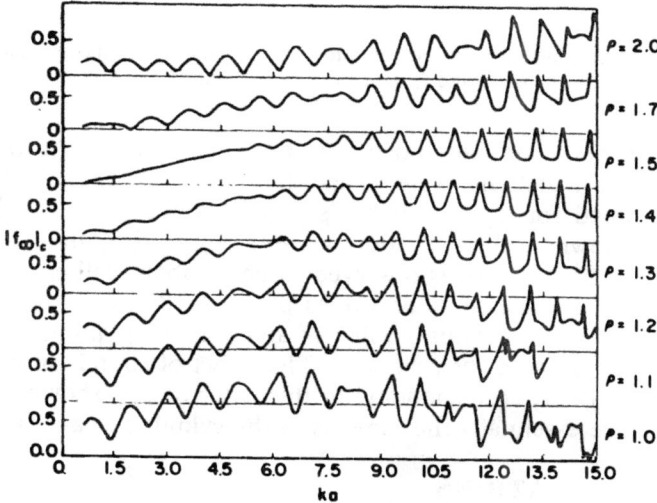

Fig. 2 — Form-function vs ka curves computed for cylinders with $c_L = 1.0$ km/s and $1.0 \leqslant \rho_s \leqslant 2.0$ g/cm^3.

In Fig 3 various combinations of ρ_s and c_L were chosen such that the value of $\rho_s c_L$ is maintained at a value of 1.5. In each case the form function increases linearly with ka, with the linear region extending to high values of ka as c_L increases (or ρ_s decreases). For the upper curve ($c_L = 1.3$) the value of $|f_\infty|_c$ deviates only slightly from zero over the entire range of ka shown. This is to be expected since when both c_L and ρ_s equal the corresponding values in water, the cylinder becomes essentially a water cylinder in water and as such produces no scattering.

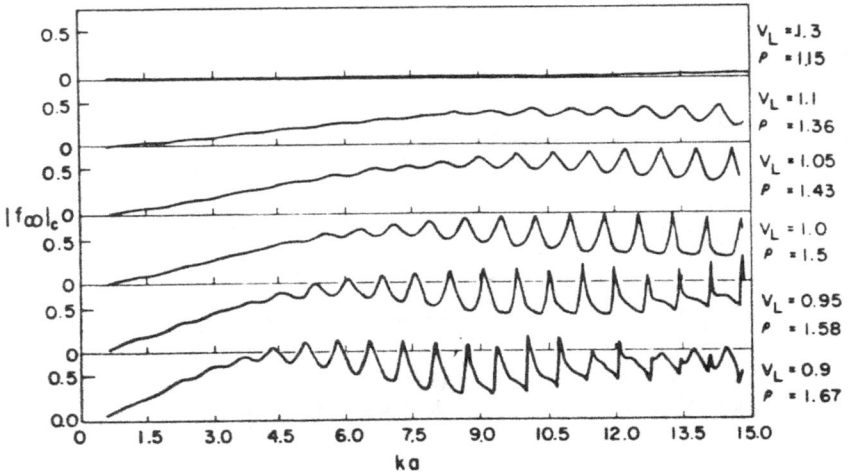

Fig. 3 — Form-function vs ka curves computed for cylinders with combinations
of ρ_s and c_s giving $\rho_s c_s = \rho_c$ (water).

Similar behavior should occur for other geometries. To check this assumption, calculations were made for spherical geometries. The results obtained for $c_L = 1.0$ km/s and ρ_s values 1.3, 1.5, and 1.7 g/cm^3 are shown in Fig. 4. For $\rho_s = \rho_s c_L = 1.5$, the monotonic increase in $|f_\infty|_s$ with ka is observed to be parabolic rather than linear as in the cylindrical case. For values of $ka > 8$, the value of $|f_\infty|_s$ exhibits an oscillation about a value of approximately 4.0, compared to the 0.5 value observed for cylinders. The effect of sound absorption on the curve for $|f_\infty|_s$ corresponding to this latter case is shown in Fig. 5. The results are essentially the same as in the cylindrical case (Fig. 1).

PULSE CALCULATIONS

The form function curves seen in Figs. 1-5 describe the steady state, backscattered response of the cylinders to an incident, infinite plane wave, and give no obvious clues to the individual mechanisms which account for the return. Isolation of the individual backscattered echoes, which contribute to the steady-state response, is achieved by calculating the response of the cylinders to an incident acoustic pulse short enough in duration that the backscattered return is a series of separate echoes whose magnitude and time of arrival can be observed and analyzed. If $g(ka)$ is the wave-number spectrum of the incident pulse, the backscattered response at a distance r much greater than the radius a is given by [2,3,6]

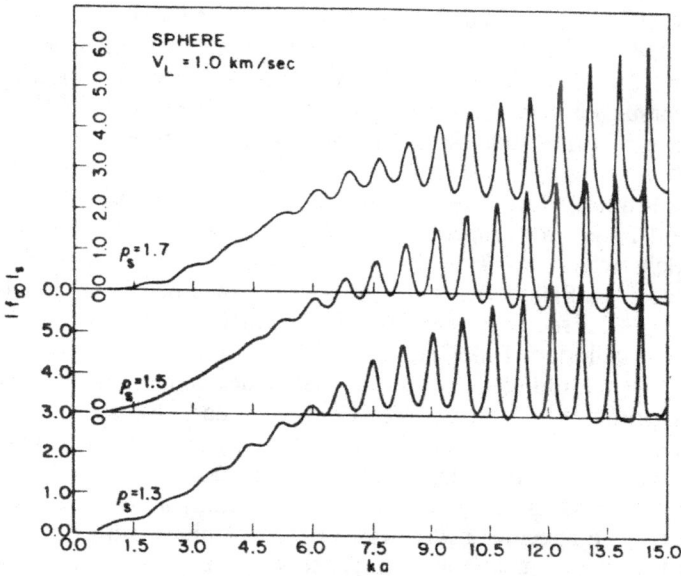

Fig. 4 — Form-function vs ka curves computed for spherical targets with $c_L = 1.0$ km/s and $\rho_s = 1.3$, 1.5, and 1.7 g/cm³.

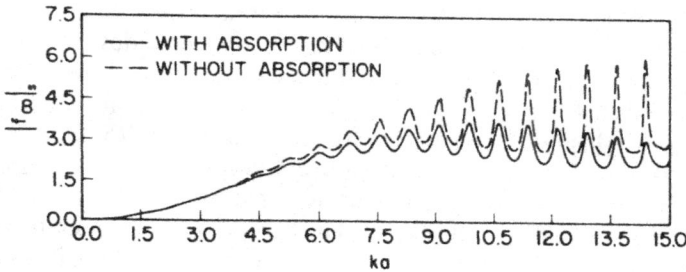

Fig. 5 — A comparison of computed form-function vs ka curves for silicone rubber spheres without and with absorption included.

$$p_r = p_0 \left(\frac{a}{2r}\right)^{1/2} \mathrm{Re} \int_0^\infty g(ka) |f_\infty(ka)| e^{ika\tau} d(ka), \qquad (3)$$

where $|f_\infty|_c$ has been described previously in Eq. (2), and τ is a dimensionless time parameter

$$\tau = (c_w t - r)/a \qquad (4)$$

normalized to be zero at the time the incident wave would reach the position occupied by the center of the cylinder in the absence of the cylinder, and c_w is the speed of sound in water.

Calculations were made of the response of rubber cylinders to an incident gated sine-wave pulse. The densities and sound speed of the cylinders were chosen to be within the values considered in the form function versus ka curves shown in Figs. 1-3. Figures 6(b)-6(d) give the response of three targets to the incident pulse shown in Fig. 6(a). In all three cases the targets are cylinders with a negligible shear speed and a longitudinal sound speed of 1.0 km/s. The densities are ρ_s = 1.0, 1.5, and 20 g/cm^3, respectively. The incident pulse is centered at a dimensionless frequency $k_0 a$ = 12, where k_0 = $2\pi f_0/c_w$. The echo responses seen in Figs. 6(b) and 6(d) show a specular reflection beginning at $\tau = -2$ (recall $\tau = 0$ when the incident wave reaches the center of the cylinder) followed by a train of pulses which have taken internal paths beginning at $\tau \simeq 4$. The pulse train of equally spaced echoes beginning at $\tau \simeq 4$ exists in all three cases. No specular reflection exists for the case ρ_s = 1.5 g/cm^3 where $\rho_s c_L$ equals the specific acoustic impedance of the surrounding water. Even in this case significant energy returns in the backscattered direction, as can be observed in Fig. 6(c). The specular return is 180° out of phase with the incident pulse when ρ = 1.0 g/cm^3 and in phase when ρ_s = 2.0 g/cm^3 (Figs. 6(b) and 6(d), respectively). The form function is a steady-state parameter, which reflects the long pulse interference of the individual returns isolated in Fig. 6. Most consideration is given to the cases when $\rho_s c_L$ = 1.5, since specular backscattered reflection is zero, and for the uses discussed in the introduction (waveguides and lenses) specular backscatter is not desirable. In addition to the calculation seen in Fig. 6(c) at $k_0 a$ = 12, backscattered echoes were computed at $k_0 a$ values of 2.0, 3.0, 6.0, 7.5, and 9.0 for the case ρ_s = 1.5 g/cm^3, c_L = 1.0 km/s. These calculations showed a buildup of the pulse amplitude at $\tau = 4$ at the rate of 0.08 $p_0/k_0 a$ up to a maximum of approximately 0.6 p_0. This accounts for the magnitude of $|f_\infty|_c$ in the region from 0 $\leqslant k_0 a < 7$. The oscillation in $|f_\infty|_c$ are caused by the interference between this initial return and the rest of train of echoes. The first echo in Fig. 6(c) had an amplitude of 0.5 p_0. The echoes which follow in the train have reached amplitudes of 0.16 p_0, 0.09 p_0, and 0.05 p_0 and occur at τ = 12.1, 20.2, and 28.3, respectively.

If the sequence of equally spaced echoes seen in Fig. 6(c) is assumed to be the result of the continuous repetition of the same travel path, within the cylinder a circumferential wave speed may be ascribed to the return. The ratio of the circumferential speed, c^*, to c_w is given by

$$c^*/c_w = 2\pi/\Delta\tau, \tag{5}$$

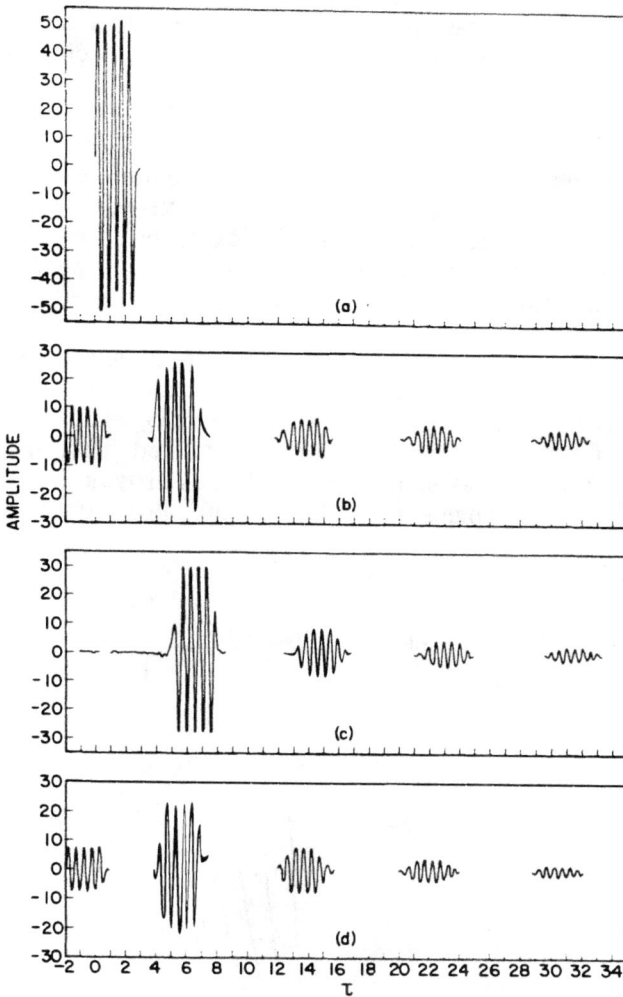

Fig. 6 — A computation of the echoes backscattered when a five cycle pulse (a) is incident on cylindrical targets with $C_l = 1.0$ km/s, (b) $\rho_s = 1.0$ g/cm^3, (c) $\rho_s = 1.5$ g/cm^3, and (d) $\rho_s = 2.0$ g/cm^3.

where $\Delta\tau$ is the time between echoes as discussed earlier. Peaks in the form function occur when the individual echoes add in phase as the incident pulse is lengthened. This occurs when the circumference of the cylinder is an integer number of wavelengths. Leading to a second relation [15]:

$$c^*/c_w = (\Delta ka)_{peak}; \tag{6}$$

therefore, if one starts with $|f_\infty|_c$ vs ka, the circumferential speed may be determined from the spacing of the successive equally spaced peaks of $|f_\infty|_c$. For the target described in Fig. 6(b), the form function gives $(\Delta ka)_{peak} = 0.7$, and Fig. 6(c) gives $\Delta\tau = 8.1$. The speed ratio c^*/c_w obtained from Eqs. (5) and (6) both give $c^*/c_w = 0.7$.

The relationship between surface (or circumferential) wave characteristics [17] and guided wave propagation has been discussed previously [16,17]. If the case $\rho_s = 1.5$ g/cm^3, $c_L = 1.0$ km/s is taken as typical of the speeds and densities considered here, the refracted internal ray paths are as shown in Fig. 7. Seven typical paths are followed with incident angles θ of 10°, 20°, 30°, 40°, 50°, 80°, and 90°.

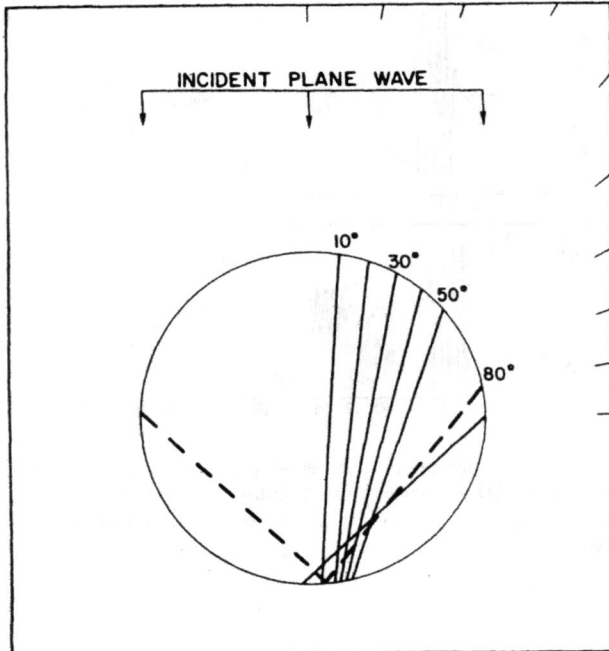

Fig. 7 — A cross section of a cylindrical target with $\rho_s = 1.5$ g/cm^3, $c_L = 1.0$ km/s, showing the internally refracted ray paths.

The converging lens effect is seen in the bending of the incident wave towards the normal, and all internal rays generated from normal to tangential incidence exit the cylinder at angles within 14° of the forward-scattering direction. At incidence angles up to approximately 55°, the reflection coefficient curves [18] predict from 90% to 100% of the energy is transmitted into and through the cylinder back into the water continuing approximately in the forward direction. Thus, energy in the ray paths beginning at 10° through 50° incidence shown in the figure is almost entirely transmitted back into the water. For rays beginning at 65° to 85° incidence angles, reflection-coefficient calculations predict both significant transmission and significant internal reflection. Two legs of the ray path beginning at 80° are seen in Fig. 7, and multiple reflections and transmission from rays generated at angles greater than 65° incidence angle give the backscattered echoes seen in Fig. 6(b). The path of the limiting internal ray, generated at tangential incidence ($\theta = 90°$) is also seen in Fig. 7. Repetition of the same circumferential path gives the apparent circumferential behavior. A plot of $|f_\infty|_c$ vs angle at a ka of 12.0 for the material considered above is given in Fig. 8. This shows the convergence of energy in the forward scattering direction ($\theta_s = 0°$) with the steady-state pressure amplitude scattered in the forward direction 20 dB greater than that backscattered ($\theta_s = 180°$) and 14 dB greater than that incident.

CONCLUSIONS

The scattering of sound from silicone rubber cylinders and spheres was found to be equal to that of an identically shaped liquid exhibiting the same density and longitudinal velocity. The resultant cw scattering can be considered in two regions: (1) a low ka region defined as that over which the first (or primary) lens wave is growing to its equilibrium value and (2) the region of higher ka. In the low ka region the scattering can be explained almost entirely in terms of interference between this primary lens wave and specular reflection. In the case where $\rho_s c_L = \rho_w c_w$, no specular reflection occurs but there is nevertheless a return which at low ka is due essentially to the primary lens wave. The corresponding low ka region of $|f_\infty|$ reveals the manner in which the primary lens wave builds up. In the cylindrical case this buildup is observed to be linear with ka, reaching an amplitude approximately one-half that of the incident wave for lossless materials. For spherical configurations the build up is quadratic in ka, and in the lossless case the amplitude approaches four times that of the incident wave. In the higher ka region scattering is due to the interference between various lens waves and (when $\rho_s c_L \neq \rho_w c_w$) the specular return.

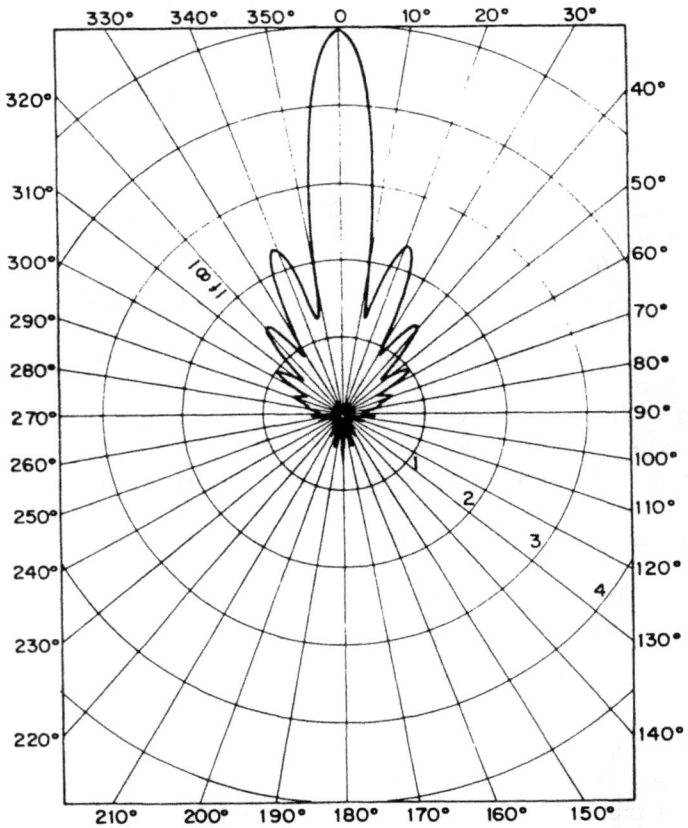

Fig. 8 — The bistatic-refraction-function curve ($|f_\infty|$ vs θ_s) for a cylindrical target with $\rho_s = 1.5$ g/cm^3, $c_L = 1.0$ km/s. Backscattering is 180°, forward scattering 0°.

REFERENCES

1. J. Faran, Jr., J. Acoust. Soc. Am. **23**, 405-418 (1951).

2. R. Hickling, J. Acoust. Soc. Am. **34**, 1582-1592 (1962).

3. A. Rudgers, NRL Report 6551 (1967).

4. L. Hampton and C. McKinney, J. Acoust. Soc. Am. **33**, 694 (1961).

5. W. Neubauer, R. Vogt, and L. Dragonette, J. Acoust. Soc. Am. 55, 1123-1129 (1974).

6. L. Dragonette, R. Vogt, L. Flax, and W. Neubauer, J. Acoust. Soc. Am. 55, 1130-1137 (1974).

7. R. Vogt, L. Flax, L. Dragonette, and W. Neubauer, J. Acoust. Soc. Am. 57, 558-561 (1975).

8. R. Vogt, L. Flax, L. Dragonette, and W. Neubauer, J. Acoust. Soc. Am. 62, 1315 (1977).

9. L. Flax and W. Neubauer, J. Acoust. Soc. Am. 61, 307-312 (1977).

10. L. Scheutz and W. Neubauer, J. Acoust. Soc. Am. 61, 513-517 (1977).

11. D. Folds, J. Acoust. Soc. Am. 56, 1295-1296 (1974).

12. P. Rogers and W. Trott, J. Acoust. Soc. Am. 56, 1111-1117 (1974).

13. D. L. Folds and J. Hanlin, J. Acoust. Soc. Am. 58, 72-77 (1975).

14. W. Mason, "Dispersion and Absorption of Sound in High Polymers," in *Handbuch der Physik, Vol. XI, Acoustics I,* edited by S. Flugge (Springer-Verlag, Berlin, 1961).

15. H. Überall, L. Dragonette, and L. Flax, J. Acoust. Soc. Am. 61, 711-715 (1977).

16. M. Harrison, U. S. David Taylor Model Basin Report 872 (January 1954).

17. W. Neubauer and L. Dragonette, J. Acoust. Soc. Am. 48, 1135-1149 (1970).

18. W. Mayer, J. Appl. Phys. 34, 909-911 (1963).

19. W. Mayer, J. Appl. Phys. 34, 3286-3290 (1963).

Chapter 13

ELASTIC CYLINDERS AND SPHERES
AT HIGH ka*

The scattering of sound by elastic cylinders and spheres immersed in fluids has been discussed by a number of authors [1-4]. In these papers, and in others mentioned in their references, it is assumed that the elastic body has incident on it a plane sound wave of pressure amplitude p_0 and circular frequency $\omega = kc$, where k is the wave number and c is the speed of sound in the fluid medium of density ρ. The density and compressional and shear-wave speeds in the solid material of the cylinder or sphere are assumed to be ρ_1, c_L, and c_s, respectively. The radii of these bodies are denoted by a.

In general, one obtains the exact normalized scattered steady-state pressure amplitude per unit incident pressure (form function) by means of a partial wave expansion or normal mode solution. The solution contains all creeping, as well as through, waves [5]. The form function is usually expressed in terms of a nondimensional frequency parameter ka. Agreement between partial wave theory and experiment is presented in Refs. 2-4. To obtain much higher frequency values usually required elaborate expansions of asymptotic series [6] or a creeping-wave formalism [7].

The present investigation shows that even at a ka of 950.0, no limiting value of the form function is approached. Calculated results are obtained for cylinders and spheres made of aluminum, for which the material parameters are given in Table 1.

Table 1 — Material Properties

	$\rho \times 10^3$ (kg/m)	$c_L \times 10^3$ (m/sec)	$c_s \times 10^3$ (m/sec)
Aluminum	2.7118	6.370	3.136
Water	0.998	1.482	. . .

*These results first appeared in: Lawrence Flax, J. Acoust. Soc. Am. **62**, 1502-1503 (1977)

The mathematical solution for the scattered sound field is given in Refs. 1-4. Essentially, the scattered monostatic pressure at large distance R from the sphere can be represented by the expression

$$\frac{p_0 a}{2R} \left| f_\infty \left(ka, \ \frac{\rho_1}{\rho}, \ \frac{c_L}{c}, \ \frac{c_s}{c} \right) \right|, \tag{1}$$

where f_∞ is a complex function of the nondimensional frequency ka and other nondimensional parameters related to the elastic constants and geometric properties of spheres. Similarly, the farfield monostatic reflection for cylinders is

$$p_0 (a/2R)^{1/2} \left| f_\infty \left(ka, \ \frac{\rho_1}{\rho}, \ \frac{c_L}{c}, \ \frac{c_s}{c} \right) \right|. \tag{2}$$

The evaluation of f_∞ entails computations of sums of Bessel functions of the first and second kind and their derivatives. Integer-order Bessel functions are required for a cylindrical geometry and half-integer order for a spherical geometry. These Bessel functions must be evaluated for high orders and large arguments. The ability to compute the scattered pressure depends upon two factors: the range of the Bessel functions and the convergence properties of the form-function summations. The calculations have been checked by single- as well as double-precision techniques, experimental verification has been obtained to a ka of 50 [1-4], and series convergence has been checked by numerous methods. For a detailed discussion of validation procedures, see Refs. 8 and 9.

The form function for a sphere and a cylinder are shown in Figs. 1 and 2, respectively. Values of $|f_\infty|$ were calculated for a range of ka

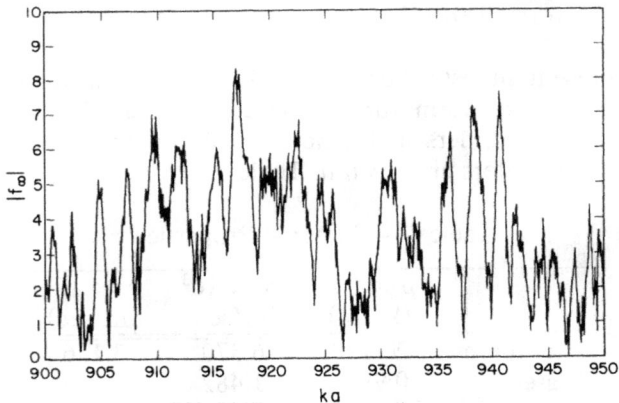

Fig. 1 — Theoretical calculations of the monostatic
reflection ($|f_\infty|$) for an aluminum sphere

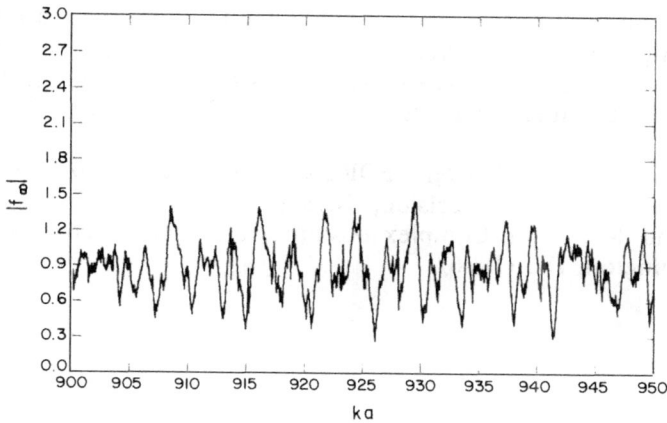

Fig. 2 — Theoretical calculations of the monostatic
reflection ($|f_\infty|$) for an aluminum cylinder

from 900.0 to 950.0 at intervals of 0.05. These calculations demonstrate that the form function does not settle down to a limiting value.
It is also apparent that the magnitude of $|f_\infty|$ for a sphere in some
ranges has an amplitude over three times that of a cylinder. In these
calculations, no absorption was included.

REFERENCES

1. R. Hickling, J. Acoust. Soc. Am. **34**, 1582-1592 (1962).

2. R. H. Vogt, L. Flax, L. R. Dragonette, and W. G. Neubauer, J.
   Acoust. Soc. Am. **57**, 558-561 (1975).

3. L. Flax and W. G. Neubauer, J. Acoust. Soc. Am. **61**, 307-312
   (1977).

4. W. W. Ryan, Jr., "Acoustic Backscattering from Cylinders," ARL
   Report No. TR-73-20, University of Texas, Austin, TX (1973).

5. D. Brill and H. Überall, J. Acoust. Soc. Am. **50**, 921-939 (1971).

6. T. T. Wu, Phys. Rev. **104**, 1201-1212 (1956).

7. R. D. Doolittle, H. Überall, and P. Uginčius, J. Acoust. Soc. Am.
   43, 1-14 (1968).

8. J. P. Mason and R. V. Baier, "FORTRAN Subroutines to Evaluate, in Single or Double Precision, Bessel Functions of the First and Second Kinds for Integer or Half Odd-Integer Orders and Positive Real Arguments," NRL Memo Report 2493 (July 1972).

9. L. Flax and J. P. Mason, "FORTRAN Subroutines to Evaluate, in Single or Double Precision, Bessel Functions of the First and Second Kinds for Complex Arguments," NRL Report No 7997, Washington, DC (1976).

Chapter 14

REFLECTION AND VIBRATIONAL MODES
OF ELASTIC SPHERES*

INTRODUCTION

Effects of resonances on the scattered sound by solid elastic spheres immersed in water have been observed by several investigators. [1,2] The normalized scattered steady-state pressure amplitude per unit incident pressure regarded as a function of frequency, i.e., the form function, shows features attributable to the free vibrations of the elastic sphere. For the case of backscattering by lossless metal spheres considered here, the low-order characteristic frequencies of the free sphere are separated enough for the effects of its resonant behavior to be seen in the form function. However, when the frequencies of the natural modes of oscillation of the vibrating sphere lie close to one another, the form function does not necessarily show the separated effect of each resonance.

Faran [1] first considered the origin of the near nulls observed in the backscattered pressure amplitude, but for cylinders in water instead of spheres. He observed that they occurred close to the frequencies of natural vibration of the free elastic cylinder. Effects of resonances on the scattering of sound by a fluid-filled cylindrical cavity in an elastic solid have also been observed. [3] In Faran's work, the distribution in angle of the scattering was observed and computed for relatively few selected values of the frequency. In the present work, only backscattering and free vibration of spheres was considered and the monostatic form function was computed for a continuous range of frequency parameter (ka, where k is the wave number and a is the sphere radius) from 0 to as high as 28, thereby showing the relationship between several resonant frequencies of the free sphere and the shape of the form function.

---

*These results first appeared in: R. H. Vogt and W. G. Neubauer, J. Acoust. Soc. Am. **60**, 15 (1976).

241

Expressions which describe the scattering of sound by lossless elastic spheres in water are reviewed here briefly for purposes of definition. [1,4] Let a steady-state plane wave traveling in the direction of the positive z-axis and incident on a sphere of radius a centered at the origin be given by

$$p_0 \exp[i(\omega t - kz)], \tag{1}$$

where ω and k are the angular frequency and wave number. The sound scattered into the far field at a distance of r from the origin may be expressed in spherical polar coordinates as

$$p = (p_0 a/2r)\exp[i(\omega t - kr)]f_\infty, \tag{2}$$

where f_∞ is the far field form function,

$$f_\infty = -(2/ka) \sum_{n=0}^{\infty} (2n + 1) P_n(\cos\theta)\sin\eta_n \exp(i\eta_n) \tag{3}$$

and

$$\tan\eta_n = -(a_n/b_n), \tag{4}$$

with

$$a_n = \zeta_n j_n(ka) - (ka) j_n'(ka), \tag{5}$$

$$b_n = \zeta_n y_n(ka) - (ka) y_n'(ka). \tag{6}$$

The quantity ζ_n, defined in the Appendix, is a function of the angular frequency ω and the material properties of the sphere. The quantities j_n and y_n are spherical Bessel functions of the first and second kind and the prime denotes the derivative with respect to the argument. When ζ_n is set equal to zero in Eqs. (5) and (6), the expressions describe the steady-state scattering by a rigid sphere. In the case of monostatic scattering, the value of the Lengendre polynomial $P_n(\cos\theta)$ is $(-1)^n$.

The quantity ζ_n becomes infinite at the characteristic frequency of each normal mode of vibration of the free elastic sphere. Torsional modes are disregarded since they are not excited in the case under consideration. The equation determining the characteristic frequencies is given in the Appendix. Associated with each n in Eq. (3) there is a sequence of modes of vibration of the free elastic sphere. It is of interest to determine physically how descriptive a single term in Eq. (3) is of an isolated resonance effect observed in the form function.

DISCUSSION

Frequency Domain

The nth term in the expression for the scattered pressure [Eqs (2) - (6)] can be separated into a part which would be present if the sphere were rigid and a part describing elastic reradiation [5]. For the reradiated part, the elastic sphere is regarded as being driven by the nth components of the incident pressure and the "rigidly scattered" pressure. When the terms are separated as just described, the expression for the scattered pressure can be written

$$p = -\frac{ip_0}{kr}\exp[i(\omega t - kr)] \sum_{n=0}^{\infty} (2n + 1)\frac{P_n(\cos\theta)}{h_n(ka)}$$

$$\times \left[j_n(ka) + \frac{(\rho c)}{(ka)^2 h_n(ka)(Z_n + z_n)} \right], \tag{7}$$

where

$$z_n = i\rho c h_n(ka)/h'_n(ka)$$

is the radiation impedance and

$$Z_n = -i\rho cka/\zeta_n$$

is the modal impedance of the elastic sphere, $h_n(ka)$ is the spherical Hankel function, $j_n(ka) - iy_n(ka)$, and ρc is the specific acoustic impedance of the fluid medium surrounding the sphere. The modal impedance Z_n of the sphere becomes zero at the resonant frequencies of the free sphere and becomes infinite for the rigid sphere. The terms remaining for the case when all Z_n are set equal to infinity in Eq. (7) give the part of the pressure which would be scattered by the rigid sphere. A single term of the radiated part is referred to as a resonant term in the following and its amplitude, normalized in the same way as f_∞ in Eqs. (2) and (3), is designated $|p_r|$ where the term number n is not shown explicitly.

The amplitudes $|p_r|$ of the resonant terms of the immersed sphere labelled by the corresponding modes of vibration of the free elastic sphere are shown plotted in Fig. 1 in a range of ka from 0 to 15 for three materials. Sphere modal notation is described in the appendix. The size, spacing and overlap of the resonance terms can be seen in the figure. The width of the resonance is the ka interval between the point on either side of the maximum where the amplitude is $1/\sqrt{2}$ times the peak value. The width of the (2, 1) resonance of tungsten carbide is 0.15. The figure shows an increase in the width of the $(n, 1)$ resonances of aluminum and brass with increasing n. The various overtones have smaller widths and amplitudes. It appears that the fundamental resonance terms of a given material have nearly the same peak amplitudes. In the regions between the peaks of a given resonance, the nth resonant term varies smoothly in amplitude, as, for example, for brass in the region between the (4, 1) and (4, 2) resonances. The (1,

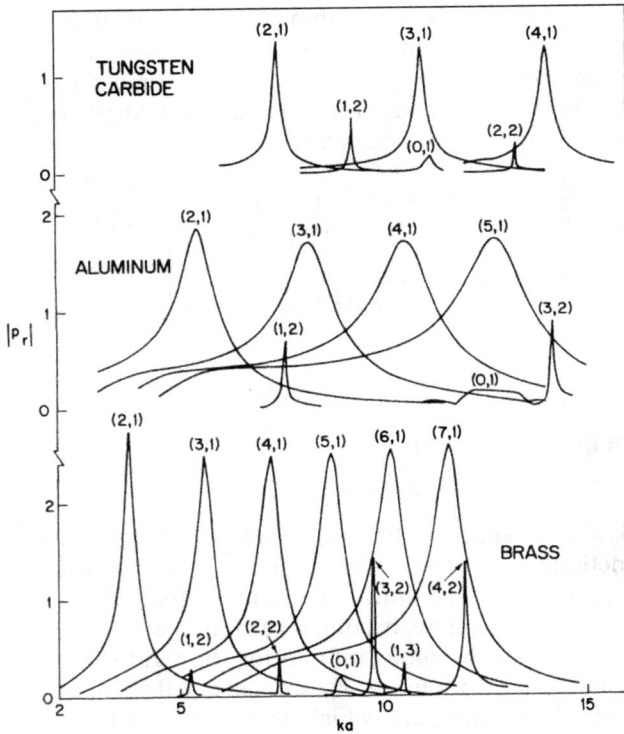

Fig. 1 — Modulus $|p_r|$ of resonant terms as a function
of ka for tungsten carbide, aluminum, and brass.

1) mode corresponds to a linear motion of the sphere as a whole and
does not appear so that the lowest $(1, 1)$ mode is $(1, 2)$. The $(2, 2)$
resonance in aluminum which is near ka of 10.8 has a negligibly small
amplitude and does not appear in the figure. The $(0, 1)$ and $(0, 2)$
resonances correspond to pure radial oscillations and are small and
wide.

For small values of the integer n, except 0 and 1, the fundamental
frequencies of the modes expressed as k_2a, vary slowly as a function of
poisson's ratio [1,6]. The wave number k_2 corresponds to the speed of
shear waves in the sphere material. In the particular case of the $(2, 1)$
mode, k_2a lies between 2.6 to 2.7 for a range of poisson's ratio between
0.05 and 0.45. The corresponding frequency in the scattering form
function is the product of k_2a and the ratio of the shear speed in the
material to the speed of sound in water. The effect in the form function is an increase in the value of ka corresponding to the lower resonances in proportion to the shear speed. The $(2, 1)$ modes of free

vibration have ka values of 7.43, 5.57, and 3.78 for tungsten carbide, aluminum and brass, respectively, which are very nearly equal to the ka values of the respective (2, 1) resonance peaks in Fig. 1. The corresponding shear speeds are 4185, 3120, and 2110 m/sec which are very nearly in the same ratios as the ka values of the resonance peaks.

Figure 2 shows the monostatic form function of a tungsten carbide sphere. A mode of vibration of the free elastic sphere is indicated in the figure by a short vertical line above or below the curve with the integer corresponding to n on the left and to l on the right. The ka value of the characteristic frequency of the mode is given by the abscissa of the short vertical line separating n and l on the scale of ka immediately below it. It can be seen in Fig. 2 that values of ka corresponding to frequencies of vibration of the free sphere, which are close to the peaks of the resonant terms of Eq. (7) plotted in Fig. 1, are also close to minima or other rapid changes in the modulus of the form function in tungsten carbide. The form function approaches that of the rigid sphere in the regions of ka lying between the first few widely spaced resonances. Figure 3 shows the form function for an armco iron sphere, similar in appearance to tungsten carbide for ka less than 15, where the resonances are widely separated and appear near minima. In Fig. 4 the curve shows the form function for an aluminum sphere. The small open and closed circles near the curve are experimental measurements discussed elsewhere [2]. Figure 1 shows the broad width of the lowest modal peaks (the fundamental set for which l is unity) in aluminum, i.e., at resonances (2, 1), (3, 1), (4, 1), and (5, 1). Broad dips are also observed in the form function of Fig. 4 at fundamental modal values. However, narrow peaks and dips are found in Fig. 1 and Fig. 4, respectively, for the overtones (1, 2) and (3, 2).

The form function for a brass sphere in Fig. 5 appears to have maxima near the resonant frequencies with the minima following on the side of higher ka, apparently the result of the addition of the phase of the nth elastic term to the total amplitude calculated for rigid scattering near the resonances. The narrow resonances of the overtones again appear in the form function as sharp breaks, in contrast to the more slowly varying features which are related to the large broad resonances at the fundamentals.

The effect of a single mode on the overall form function in the region of interest can be seen qualitatively by modifying the corresponding term in the expressions for the form function of a rigid sphere or of the elastic sphere. If the form function of the rigid sphere is computed using only the nth term given by Eq. (7) which characterizes the elastic sphere and all other terms appropriate for the rigid

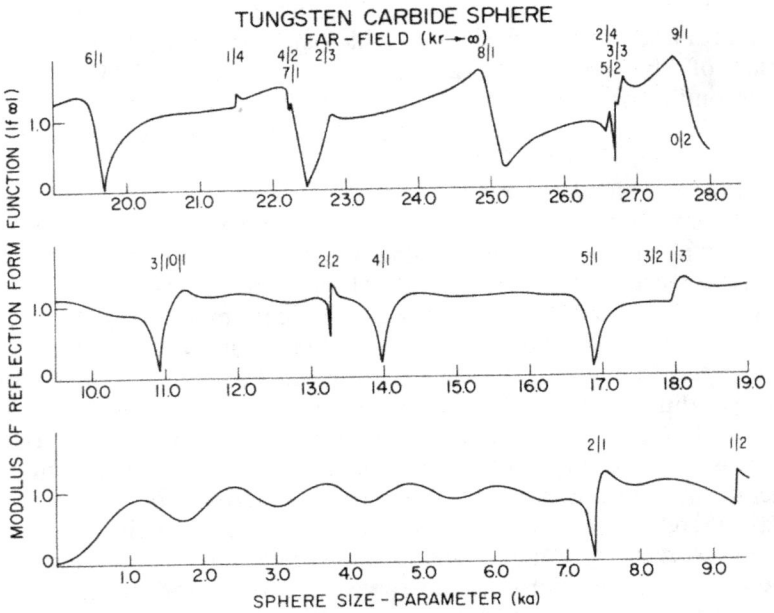

Fig. 2 — Computed form function and resonance positions
for a tungsten carbide sphere.

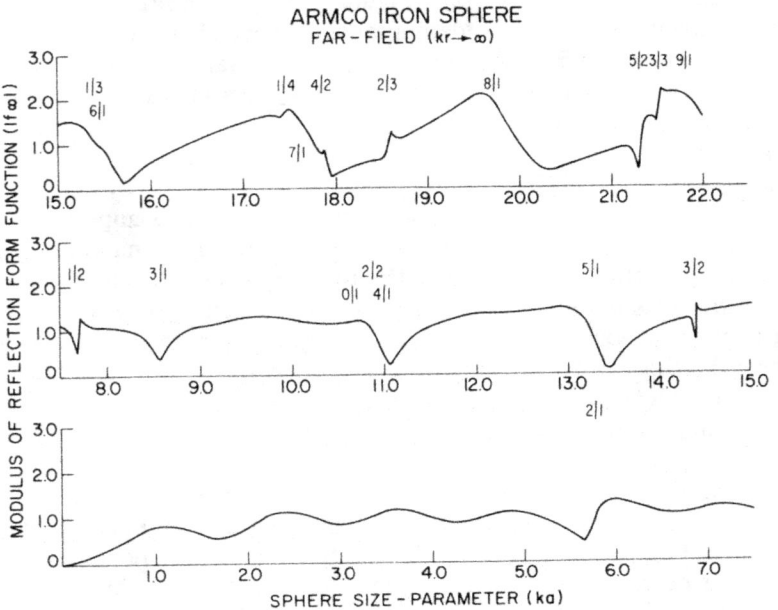

Fig. 3 — Computed from function and resonance positions
for an Armco iron sphere.

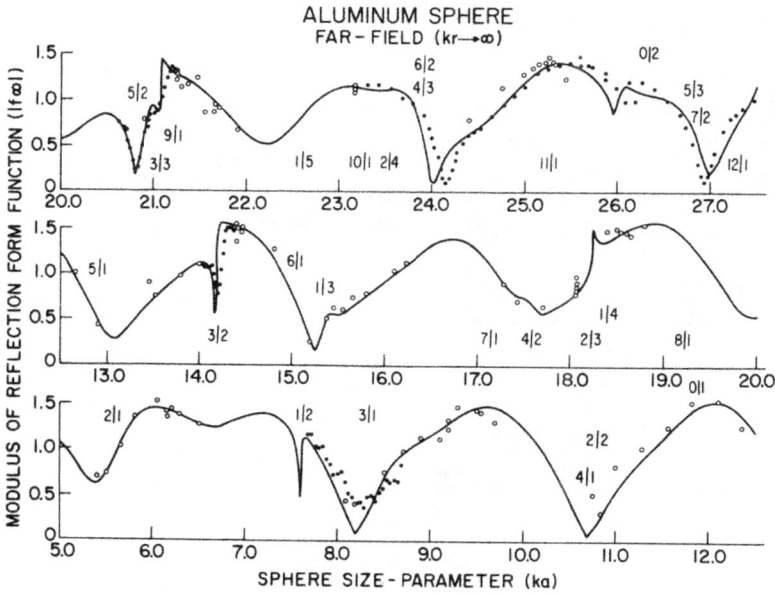

Fig. 4 — Computed form function and resonance positions
for an aluminum sphere.

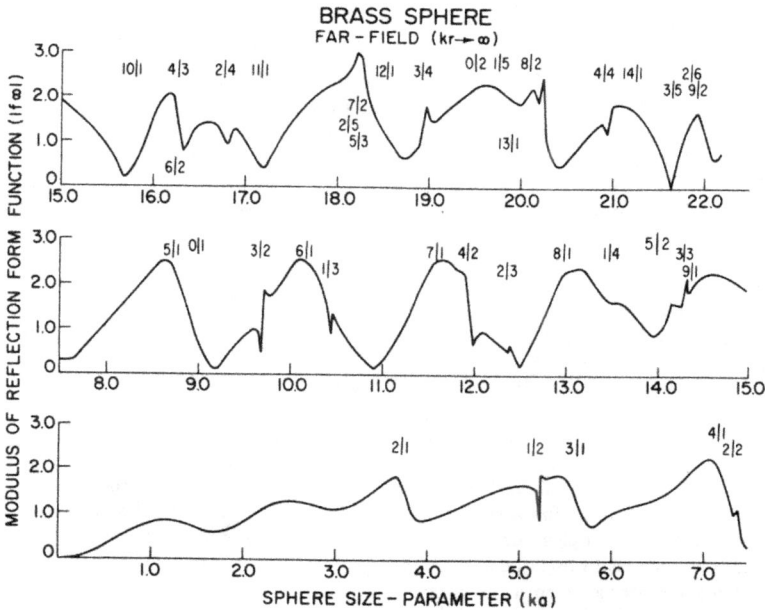

Fig. 5 — Computed form function and resonance positions
for a brass sphere.

sphere, the result is a rigid sphere form function showing the effect of a given fundamental and its overtones. Or if the form function of a given elastic sphere is computed with Z_n infinite in the nth term, thereby in effect substituting the nth term appropriate for a rigid sphere, the resulting form function is that of the elastic sphere without the nth resonances. For example, the solid curve in Fig. 6 is the form function for a tungsten carbide sphere. The solid circles which are connected by a curved line show points of the rigid sphere form function calculated in the range of ka from 5 to 15 with the term for $n = 2$ replaced by that for tungsten carbide. The circles appear close to the form function of the rigid sphere (within 2% for ka between 10 and 13, for example) except near the resonances corresponding to $n = 2$, namely, $(2, 1)$ and $(2, 2)$, where they approach the form function of tungsten carbide. Or if the form function of tungsten carbide is computed with the second term modified by setting Z_2 equal to infinity, the resonances corresponding to the second mode disappear. The results of

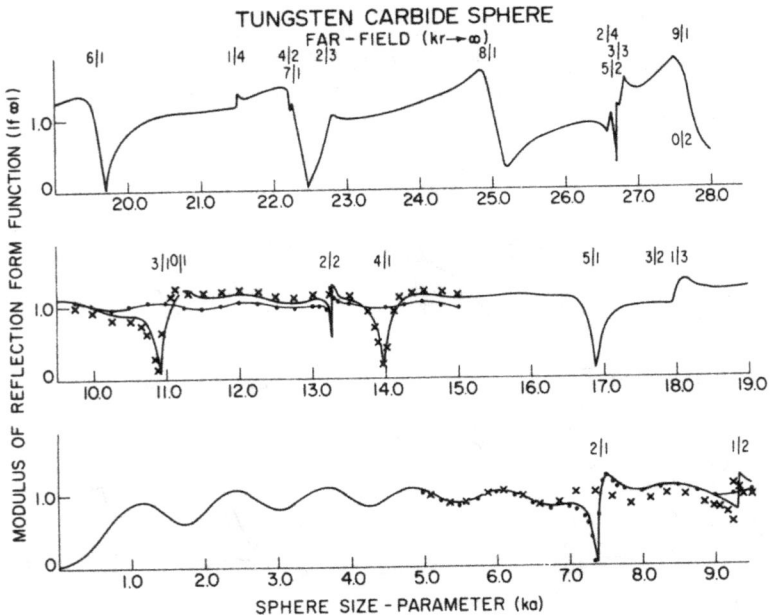

Fig. 6 — Computed form function for a tungsten carbide sphere (solid line). Points on the form function computed from terms appropriate for a rigid sphere except for $n = 2$ for which case the appropriate "elastic" term is used (solid circles connected by a curved line) and computed from terms appropriate for an elastic sphere except for $n = 2$ for which case the appropriate "rigid" term is used (x's).

such a calculation are shown by the x's in the figure which fall near the tungsten carbide form function except in the regions of the $(2, l)$ resonances. Figure 7 shows a similar calculation for the $(2, l)$ resonances in brass. If the points corresponding to the modified rigid sphere, shown by solid circles connected by a line, are followed starting at $ka = 3$, it can be seen that they approach the form function of brass at the resonance $(2, 1)$ and they show a dip corresponding to minima at resonances $(2, 2)$ and $(2, 3)$. A steplike increase can be seen at resonances $(2, 4)$ and $(2, 5)$. At values of ka sufficiently removed from the $(2, l)$ resonances, the circles approach the form function of the rigid sphere (within 2% for ka between 8 and 13, for example). Also, it can be seen that the x's are near the form function of brass except at the resonances $(2, 1)$, $(2, 2)$, $(2, 3)$, $(2, 4)$, and $(2, 5)$.

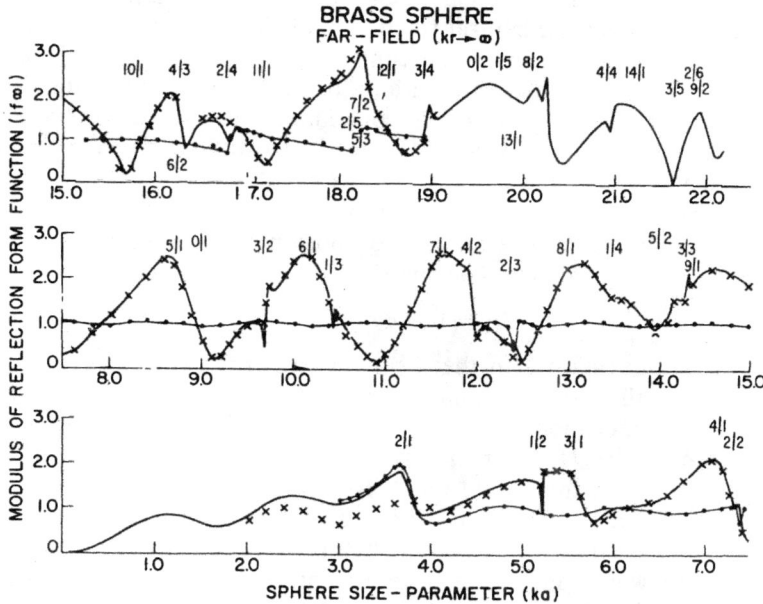

Fig. 7 — Computed form function for a brass sphere (solid line). Points on the form function computed from terms appropriate for a rigid sphere except for $n = 2$ for which case the appropriate "elastic" term is used (solid circles connected by a curved line) and computed from terms appropriate for an elastic sphere except for $n = 2$ for which case the appropriate "rigid" term is used (x's).

The Time Domain

In the case of an incident plane wave, the pulse scattered at a distance r from the center of the sphere of radius a is given by

$$p(\tau) = (p_0 a/2r)(1/2\pi) \text{Re} \int_0^\infty g(ka)f(ka)\exp(ika\tau)d(ka), \quad (8)$$

where $\tau = (ct - r)/a$ is a propagation parameter appropriate to an outgoing spherical wave and Re denotes the real part [7, 8]. The normalized complex spectrum of the incident plane wave is given by $g(ka)$, and $f(ka)$ is the form function of the elastic sphere immersed in a fluid. The shape of the form function $f(ka)$ within the frequency bandwidth of the incident pulse determines the behavior of the scattered pulse in time. Hickling [8] pointed out differences in the computed scattered pulse forms for a long truncated sinusoidal incident pulse. He showed in particular cases that the pulse form of the echo depends on whether the central frequency of the incident pulse is at the position of a maximum or an adjacent minimum of the form function. A simpler situation results when a long truncated sinusoidal pulse has a central frequency at a value of ka coinciding with the peak of a single isolated resonance in the form function. For a sufficiently long incident pulse, the vibrating sphere approaches a steady-state oscillation and at the termination of the pulse, the amplitude of the oscillations of the immersed sphere decays with a time constant related to the width of the resonance. By analogy with the simple oscillator

$$\Delta(ka)/(ka)_0 = 1/Q,$$

where $(ka)_0$ corresponds to the resonant frequency and as before, Δka is the full width of the resonance peak at $1/\sqrt{2}$ times the maximum amplitude. Experimentally Q was derived from the expression $p_1/p_0 = \exp(-\pi N/Q)$, where p_0 is the amplitude near the beginning of the decay and p_1 is the amplitude N cycles later. The $(2, 1)$ resonance of tungsten carbide at $(ka)_0$ of 7.37 and width $\Delta(ka)$ of 0.15, as previously mentioned, is a case for which a comparison of calculation and measurement was made. The scattered pulse form was computed by Eq. (8) for an incident truncated sinusoidal pulse of 60-cycles duration. the envelope of the computed pulse is shown in Fig 8. The abscissa is given in terms of τ, which in this case is such that each sinusoidal cycle corresponds to 0.85 units of τ and the pulse first appears at $\tau = -2$. It can be seen from the figure that the amplitude of the pulse envelope approaches steady-state after the initial transient and that after a duration in τ of about 51 corresponding to nearly 60 cycles, the duration of the incident pulse has passed and is followed by the transient oscillations of the radiating sphere. From the decay of the radiated amplitude, the Q is estimated from a semilogarithmic plot to be 49 which

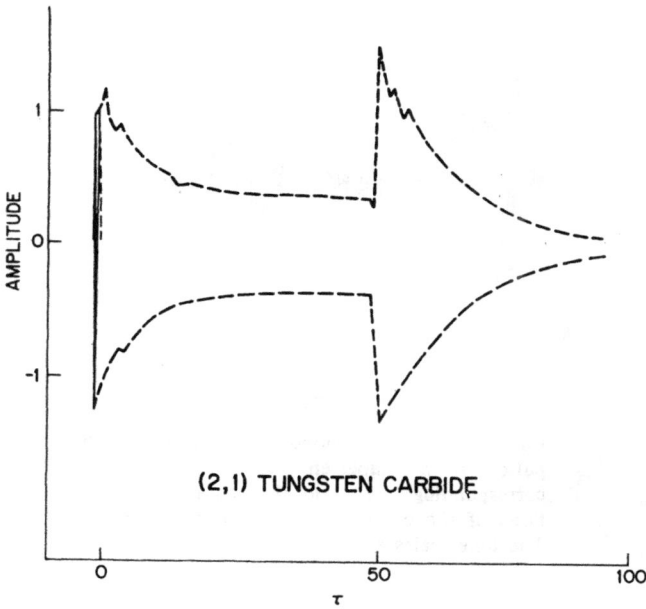

Fig. 8 — Computed echo envelope of a pulse monostatically reflected from a tungsten carbide sphere. The pulse spectrum is centered at the (2, 1) resonance dip in the form function for tungsten carbide. The first cycle within the envelope is shown.

agrees with the value obtained from the width of the computed resonance by the use of Eq. (9), $7.37/0.15 = 49$. Figure 9 shows the record of a scattered pulse observed when a truncated sinusoidal pulse of 300 μsec duration, very nearly 60 cycles, is incident on a tungsten carbide sphere at ka of 7.36 corresponding to the (2, 1) resonance. from the figure, the decay of the second transient gives a value of Q estimated to be 40. In the same way for the (3, 1), (4, 1), (5, 1), and (6, 1) resonances of the tungsten carbide sphere, the values for Q were found from the widths all to be 60 with corresponding experimentally measured values of 60, 50, 50, and 60.

A narrow resonance with small $\Delta(ka)$ has a large Q when compared with a nearby wide resonance with large $\Delta(ka)$. The resulting difference in decay times sometimes makes it possible to detect the presence of a small narrow resonance in the vicinity of a large one by observation of the "long" decay of the echo after the forced excitation of the sphere is turned off. As an example, the (1, 2) resonance for an aluminum sphere is narrow and small compared to the nearby (3, 1) resonance shown in Figs. 1 and 4. The echo return related to the (1, 2) resonance of aluminum sphere is shown in Fig. 10. The part of the

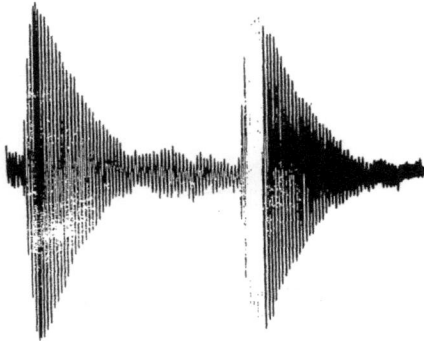

(2,1) TUNGSTEN CARBIDE

Fig. 9 — Observed monostatically reflected pulse from tungsten carbide sphere corresponding closely to theoretical conditions of the computation shown in Fig. 8. The time scales are different for Figs. 8 and 9.

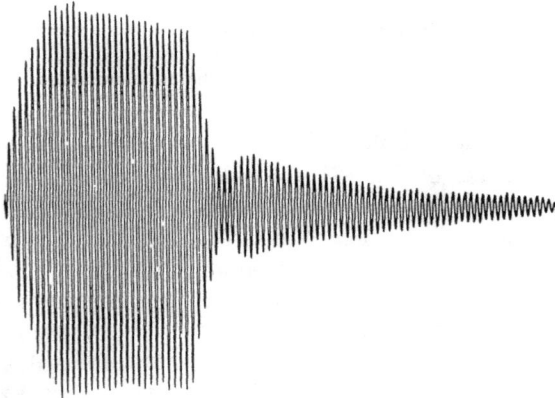

(1,2) ALUMINUM

Fig. 10 — Observed monostatically reflected pulse whose spectrum is centered at a k at which the (1, 2) resonance is computed. The reflecting sphere is of aluminum.

pulse with large amplitude corresponds to the steady-state forced oscillation of the sphere and the "tail" following shows the amplitude decay of the vibrating sphere and gives a Q value of 90 in comparison with the value 90 as obtained from the width of the (1, 2) peak in Fig. 1. The echo return from the (3, 1) resonance in Fig. 11 shows first the transient peak when the sphere is excited by the incident pulse and following, the transient pulse which appears when the incident pulse is turned off. Comparison with Fig. 10 illustrates the relatively rapid decay of the wide resonance. A measured Q of 7 was estimated in agreement with a value of 8 obtained from the width of the resonance peak. In the same way, the (4, 1) resonance of aluminum was found experimentally to have a Q of 10 and a computed Q of 9. It appears then that the experimentally determined decay of the undriven sphere is in approximate agreement with the value obtained from the computed resonance width in the case of observable isolated resonances or of narrow resonances situated near broad ones. When the computed resonance term indicates a sharp peak, the computed results are more affected by uncertainties in the values of the maximum amplitude and in the widths of the resonances. When the peak is broad and the decay rapid, the experimental measurements are difficult. The experimental and calculated results are, therefore, given with no indication of their accuracy.

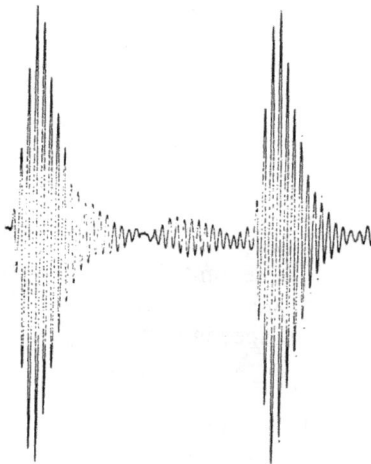

(3,1) ALUMINUM

Fig. 11 — Observed monostatically reflected pulse whose spectrum is centered at a ka at which the (3, 1) resonance is computed. The reflecting sphere is of aluminum.

APPENDIX

The monostatic far field form function f_∞ is given in the form [4]

$$f_\infty = - \frac{2}{ka} \sum_{n=0}^{\infty} (-1)^n (2n + 1) \sin\eta_n \, e^{i\eta_n},$$

with

$$\tan\eta_n = - \zeta_n j_n(ka) - (ka) j'_n(ka) \zeta_n y_n(ka) - (ka) y'_n(ka) \quad \text{(A1)}$$

in which $j_n(ka)$ and $y_n(ka)$ are nth order spherical bessel functions of the first and second kind and the primes denote derivatives with respect to the argument.

The quantity ζ_n in Eq. (A1) is defined by

$$\zeta_n = \frac{\rho}{\rho_1} \frac{[(k_2 a)^2/2][(A_n/B_n) - (C_n/D_N)]}{(E_n/B_n) - (F_n/D_n)} \quad \text{(A2)}$$

where

$$A_n = (k_1 a) j'_n(k_1 a),$$

$$B_n = (k_1 a) j'_n(k_1 a) - j_n(k_1 a),$$

$$C_n = 2n(n + 1) j_n(k_2 a),$$

$$D_n = (k_2 a)^2 j''_n(k_2 a) + (n - 1)(n + 2) j_n(k_2 a),$$

$$E_n = (k_1 a)^2 \left[\left(\frac{\sigma}{1 - 2\sigma} \right) j_n(k_1 a) - j''_n(k_1 a) \right],$$

$$F_n = 2n(n + 1) [j_n(k_2 a) - (k_2 a) j'_n(k_2 a)],$$

with double primes denoting second derivatives. In the expression for E_n, Poisson's ratio of the sphere material is σ given in terms of the longitudinal and shear sound speeds c_1 and c_2, respectively, by

$$\sigma = (c_1^2 - 2c_2^2)/2(c_1^2 - c_2^2).$$

The wave numbers k_1 and k_2 corresponding, respectively, to the longitudinal and shear waves in the sphere material are related to the wave number k in the surrounding fluid by

$$k_1 a = (c/c_1) ka$$

and

$$k_2 a = (c/c_2) ka.$$

The characteristic equation for the resonant frequencies of the spheroidal modes of the free vibrating sphere results from setting the denominator in Eq. (A2) set equal to zero,

$$D_n E_n - B_n F_n = 0.$$

The modes are denoted by integers (n, l). For each value of n there is a fundamental frequency and overtones which are given by l with $l = 1$ representing the fundamental [9]. (The fundamental is sometimes denoted by $l = 0$.) The $(0, l)$ modes are purely radial oscillations. The $(1, 1)$ mode corresponds to rigid body motion of the sphere as a whole and does not appear as a resonant frequency of free vibration. When the sphere is immersed in water, however, the rigid body motion appears in the overall shape of the form function below about $ka = 5$ as shown by Hickling and Wang [10]. The spheroidal mode $(2, 1)$ has the lowest frequency for a free sphere and corresponds to a motion with the shape of the sphere alternating between a prolate and an oblate spheroid.

REFERENCES

1. J.J. Faran, Jr., J. Acoust. Soc. Am. **23**, 405-418 (1951).

2. W.G. Neubauer, R.H. Vogt, and L.R. Dragonette, J. Acoust. Soc. Am. **55**, 1123-1129 (1974).

3. Y. Pao and W. Sachse, J. Acoust. Soc. Am. **56**, 1478-1486 (1974).

4. A.J. Rudgers, NRL Report 6551 (1967) (unpublished).

5. M.C. Junger and D. Feit, *Sound, Structures and Their Interaction* (MIT Press, Cambridge, MA), pp. 333.

6. D.B. Fraser and R.C. Lecraw, Rev. Sci. Instrum. *B35*, 1113 (1964).

7. A.J. Rudgers, J. Acoust. Soc. Am. **45**, 900-910 (1969).

8. R. Hickling, J. Acoust. Soc. Am. **34**, 1582-1592 (1962).

9. Y. Sato and T. Usami, Geophys. Mag. **31**, 15 (1962-1963).

10. R. Hickling and N.M. Wang, J. Acoust. Soc. Am. **39**, 276-279 (1966).

Chapter 15

RESONANCE EXCITATION AND SOUND SCATTERING*

INTRODUCTION

The formalism of the classical resonance theory of nuclear reactions [1, 2] is applied to the problem of acoustic scattering from submerged elastic circular cylinders and spheres. This approach demonstrates in a direct fashion that the strongly fluctuating behavior of the cross section for sound scattering from elastic bodies, as shown, e.g., by numerical normal-mode calculations of the backscattering amplitude due to Hickling [3], is caused by a superposition of (generally narrow) resonances in the individual normal-mode amplitudes (or partial waves). This was inferred [4] from considerations of the total backscattering amplitude, where the connection of the scattering resonances with creeping waves ("Regge poles") on the surface of the scatterer, and with the eigenvibrations of the elastic body had been established together with a physical condition for excitation of these resonances.

Here the excitation of elastic-body resonances by incident-plane acoustic waves are discussed by considering the individual partial waves in the Rayleigh series. The resonance scattering formalism is developed for solid elastic cylinders and spheres, employing the "method of linear approximation" in frequency as used in nuclear resonance reaction theory. The scattering amplitudes are obtained as a sum of resonance terms and of a geometric background amplitude, while the elastic waves in the interior are found to be given as pure resonance terms. Numerical values of the partial-wave scattering amplitudes are presented, which identify the background as the scattering amplitude of a rigid body, and demonstrate the Regge pole behavior of the elastic resonances appearing in the amplitudes. Various types of interferences between resonances and background are analyzed, and the time decay ("ringing") of an excited resonance is shown to be inversely related to the width of the resonance.

---

*These results were first reported in: L. Flax, L. R. Dragonette and H. Überall, J. Acoust. Soc. Am. **63**, 723 (1978).

The phases of the partial waves, demonstrate that phase jumps of π take place when one passes through a resonance. An attempt is made to identify the lowest "Franz-type" creeping-wave resonance in this way. Angular distributions and resonance effects appearing therein are also considered. A connection is established between the total rigid-body scattering amplitude and certain resonance nulls appearing in the angular distributions.

RESONANCE THEORY OF ACOUSTIC SCATTERING

Scattering From Elastic Cylinders

An infinite plane acoustic wave $\exp i(kx - \omega t)$ with propagation constant $k = \omega/c$, incident along the x axis on a solid elastic cylinder of radius a and density ρ_c, produces the following scattered field p_{sc} at a point $P(r, \phi)$ located in the fluid of density ρ_w surrounding the cylinder [5, 6]:

$$p_{sc} = - \sum_{n=0}^{\infty} \epsilon_n \, i^n \frac{L_n J_n(X) - X J_n'(X)}{L_n H_n^{(1)}(X) - X H_n^{(1)'}(X)}$$

$$\times H_n^{(1)}(kr) \cos n\phi; \tag{1a}$$

here

$$\epsilon_n = 1 \ (n = 0), \quad 2 \ (n > 0),$$

and

$$L_n = (\rho_w/\rho_c)(D_n^{(1)}/D_n^{(2)}), \tag{1b}$$

where $D_n^{(1)}$ and $D_n^{(2)}$ are 2×2 determinants:

$$D_n^{(1)} = \begin{vmatrix} a_{21} & a_{23} \\ a_{31} & a_{33} \end{vmatrix}, \quad D_n^{(2)} = \begin{vmatrix} a_{11} & a_{13} \\ a_{31} & a_{33} \end{vmatrix} \tag{1c}$$

whose elements are given in Chapter 5. The argument X of the Bessel and Hankel functions in Eq. (1a) is $X = ka = \omega a/c_w$ with c_w the speed of sound in water. The primes denote differentiation with respect to the arguments. The matrix elements a_{lm} of Eq. (1c) contain Bessel functions with arguments $X_L = k_L a = \omega a/c_L$ and $X_T = k_T a = \omega a/c_T$, where c_L and c_T are, respectively, the speeds of the compressional and transverse waves in the cylindrical material.

For simplicity in future expressions we define

$$R_n(X) = - \frac{L_n J_n(X) - X J_n'(X)}{L_n H_n^{(1)}(X) - X H_n^{(1)'}(X)}. \tag{1d}$$

The scattered pressure given in Eq. (1a) may be rewritten as

$$p_{sc} = \frac{1}{2} \sum_{n=0}^{\infty} \epsilon_n i^n (S_n - 1) H_n^{(1)}(kr) \cos n\phi, \tag{2a}$$

where we introduced the scattering function $S_n \equiv \exp(2i\delta_n)$ familiar from nuclear scattering theory [7], with δ_n the so-called scattering phase shifts. In the present case, one has

$$S_n = 2R_n + 1. \tag{2b}$$

The far field value of p_{sc} can be obtained by employing the Hankel asymptotic form for $kr \gg n$,

$$H_n^{(1)}(kr) \sim (2/\pi ikr)^{1/2} i^{-n} e^{ikr}, \tag{3}$$

and we may introduce the far field scattering "form function"

$$f(\phi) = \sum_{n=0}^{\infty} f_n(\phi), \tag{4}$$

consisting of "partial-wave" contributions

$$f_n(\phi) = (\pi ika)^{-1/2} \epsilon_n (S_n - 1) \cos n\phi. \tag{5}$$

In terms of $f(\phi)$, p_{sc} becomes asymptotically

$$p_{sc} \sim (a/2r)^{1/2} e^{ikr} f(\phi). \tag{6}$$

For a rigid cylinder, one has asymptotically [8] for the case of back-scattering ($\phi = \pi$)

$$|f^{(r)}(\pi)|^2 \sim 1. \tag{7a}$$

This can be obtained by considering the limit of $R_n(X)$ as $\rho_c \to \infty$:

$$[R_n(X)]_{\text{rigid}} = - J_n'(X)/H_n^{(1)'}(X). \tag{7b}$$

(The corresponding soft-cylinder expression is, incidentally, given by dropping the primes.)

Following Ref. 6, the expression for S_n, Eq. (2b), may be represented in either of the two forms

$$S_n = S_n^{(s)}(L_n - z_2)/(L_n - z_1) \tag{8a}$$

$$= S_n^{(r)}(L_n^{-1} - z_2^{-1})/(L_n^{-1} - z_1^{-1}), \tag{8b}$$

where

$$S_n^{(s)} = - H_n^{(2)}(X)/H_n^{(1)}(X) \equiv e^{2i\xi_n^{(s)}} \tag{9a}$$

and

$$S_n^{(r)} = - H_n^{(2)'}(X)/H_n^{(1)'}(X) \equiv e^{2i\xi_n^{(r)}} \tag{9b}$$

represent, respectively, the S functions for scattering from a soft or a rigid cylinder, with $\xi_n^{(s)}$, $\xi_n^{(r)}$ the corresponding phase shifts. The latter can be shown to be given by the real quantities:

$$\tan \xi_n^{(s)} = J_n(X)/Y_n(X), \quad \tan \xi_n^{(r)} = J_n'(X)/Y_n'(X). \tag{10}$$

In Eqs. (8), use was made of

$$z_i = X H_n^{(i)'}(X)/H_n^{(i)}(X), \quad i = 1, 2. \tag{11}$$

Equation (1) used with Eqs. (8) shows that the limits of a soft ($\rho_c \ll \rho_w$) or rigid ($\rho_c \gg \rho_w$) cylinder correctly reproduce the corresponding soft- or rigid-cylinder scattering functions $S_n^{(s)}$, $S_n^{(r)}$, respectively.

The quantities z_i of Eq. (11) may be separated into real and imaginary parts as follows:

$$z_{1,2} = \Delta_n^{(s)} \pm i s_n^{(s)}, \tag{12a}$$

or

$$z_{1,2}^{-1} = \Delta_n^{(r)} \pm i s_n^{(r)}, \tag{12b}$$

with

$$\Delta_n^{(s)} = X \frac{J_n(X) J_n'(X) + Y_n(X) Y_n'(X)}{J_n^2(X) + Y_n^2(X)}, \tag{13a}$$

$$S_n^{(s)} = \frac{2}{\pi} \frac{1}{J_n^{(2)}(X) + Y_n^2(X)} \quad (>0), \tag{13b}$$

$$\Delta_n^{(r)} = \frac{1}{X} \frac{J_n(X) J_n'(X) + Y_n(X) Y_n'(X)}{[J_n'(X)]^2 + [Y_n'(X)]^2}, \tag{14a}$$

$$S_n^{(r)} = - \frac{2}{\pi X^2} \frac{1}{[J_n'(X)]^2 + [Y_n'(X)]^2} \quad (<0). \tag{14b}$$

We now employ the linear-approximation method of nuclear resonance theory, in which we define "resonance frequencies" $X_n^{(s)}$ or $X_n^{(r)}$ by the conditions:

$$L_n(X_n^{(s)}) = \Delta_n^{(s)}, \tag{15a}$$

$$L_n^{-1}(X_n^{(r)}) = \Delta_n^{(r)}, \tag{15b}$$

which may lead to a multiplicity of solutions for a given n. The quantities $(L_n - \Delta_n^{(s)})$ or $(L_n^{-1} - \Delta_n^{(r)})$ are assumed to be linearly varying with frequency, and expanded in a Taylor series in X, in the vicinity of any one of the resonance frequencies

$$L_n(X) \cong \Delta_n^{(s)} + \beta_n^{(s)}(X - X_n^{(s)}), \tag{16a}$$

$$L_n^{-1}(X) \cong \Delta_n^{(r)} + \beta_n^{(r)}(X - X_n^{(r)}); \tag{16b}$$

here, we did not distinguish the multiplicity of resonance frequencies explicitly by a further index. One may also introduce a "resonance width" Γ_n by the definitions

$$\Gamma_n^{(s)} = -2S_n^{(s)}/\beta_n^{(s)},$$ (17a)

$$\Gamma_n^{(r)} = -2S_n^{(r)}/\beta_n^{(r)}.$$ (17b)

The S function may then be rewritten in either of the two resonance forms:

$$S_n \equiv e^{2i\delta_n} = S_n^{(s)} \frac{X - X_n^{(s)} - \frac{1}{2} i\Gamma_n^{(s)}}{X - X_n^{(s)} + \frac{1}{2} i\Gamma_n^{(s)}}$$ (18a)

$$= S_n^{(r)} \frac{X - X_n^{(r)} - \frac{1}{2} i\Gamma_n^{(r)}}{X - X_n^{(r)} + \frac{1}{2} i\Gamma_n^{(r)}}.$$ (18b)

Note that because of the reality of the quantities involved one has the "unitarity relations"

$$|S_n| = |S_n^{(s)}| = |S_n^{(r)}| = 1,$$ (19)

expressing the fact that no energy is lost in the scattering process (for the elastic scatterer, this is due to our implied assumption of no absorption, i.e., real values of X_L, X_T). Furthermore, S_n is seen to possess not only a resonance pole at the complex frequency

$$X = X_p^{(n)} \equiv X_n - \frac{1}{2} i\Gamma_n,$$ (20a)

but also a resonance zero at

$$X = X_z^{(n)} \equiv X_n + \frac{1}{2} i\Gamma_n,$$ (20b)

necessary in order to satisfy the unitarity condition. Since the resonance widths Γ_n as defined in Eqs. (17a) and (17b) will be seen below to be positive quantities, the pole is thus located in the lower half of the complex X plane, $\frac{1}{2}\Gamma_n$ determining its distance from the real axis, and the zero in the upper half plane at the same distance above the axis.

The quantity $(S_n - 1)$ which appears in the scattering amplitude or in the partial waves, Eqs. (2a) or (5), and which has the forms

$$S_n - 1 \equiv 2ie^{i\delta_n} \sin \delta_n \tag{21a}$$

$$= 2ie^{2i\xi_n^{(s)}} \left[\frac{S_n^{(s)}}{L_n - \Delta_n^{(s)} - iS_n^{(s)}} + e^{-i\xi_n^{(s)}} \sin \xi_n^{(s)} \right] \tag{21b}$$

$$= 2ie^{2i\xi_n^{(r)}} \left[\frac{S_n^{(r)}}{L_n^{-1} + \Delta_n^{(r)} - iS_n^{(r)}} + e^{-i\xi_n^{(r)}} \sin \xi_n^{(r)} \right], \tag{21c}$$

may be represented by resonance expressions

$$(S_n - 1)/2i \equiv e^{i\delta_n} \sin \delta_n \tag{22a}$$

$$= e^{2i\xi_n^{(s)}} \left[\frac{\frac{1}{2}\Gamma_n^{(s)}}{X_n^{(s)} - X - \frac{1}{2}i\Gamma_n^{(s)}} + e^{-i\xi_n^{(s)}} \sin \xi_n^{(s)} \right] \tag{22b}$$

$$= e^{2i\xi_n^{(r)}} \left[\frac{\frac{1}{2}\Gamma_n^{(r)}}{X_n^{(r)} - X - \frac{1}{2}i\Gamma_n^{(r)}} + e^{-i\xi_n^{(r)}} \sin \xi_n^{(r)} \right]. \tag{22c}$$

With Eq. (5), one then has the partial-wave form functions

$$f_n(\phi) = 2i\epsilon_n (\pi ika)^{-1/2} e^{i\delta_n} \sin \delta_n \cos n\phi. \tag{23a}$$

From Eqs. (22), these consist of a (soft or rigid) background or "potential-scattering" contribution:

$$f_n^{(s)}\phi = 2i\epsilon_n (\pi ika)^{-1/2} e^{i\xi_n^{(s)}} \sin \xi_n^{(s)} \cos n\phi, \tag{23b}$$

$$f_n^{(r)}\phi = 2i\epsilon_n (\pi ika)^{-1/2} e^{i\xi_n^{(r)}} \sin \xi_n^{(r)} \cos n\phi \tag{23c}$$

(it will be seen below that for the example of an aluminum cylinder in water, the background corresponds to rigid-cylinder scattering) with, in general, a large number of resonances superimposed.

In the vicinity of one of these, the partial wave is given by (e.g., for a rigid background)

$$f_n(\phi) = 2i\epsilon_n (\pi ika)^{-1/2} e^{2i\xi_n^{(r)}}$$

$$\left[\frac{\frac{1}{2}\Gamma_n^{(r)}}{X_n^{(r)} - X - \frac{1}{2}i\Gamma_n^{(r)}} + e^{-i\xi_n^{(r)}} \sin \xi_n^{(r)} \right] \cos n\phi. \tag{23d}$$

When calculating the form function as Eq. (4), interferences between the resonances and the background will appear, which may sometimes impart striking shapes to the resonance curves, as will be seen below.

Several authors have observed [4, 9, 10] that the location in X of the resonances obtained both theoretically and experimentally in $f(\pi)$ and $f_n(\pi)$ coincide with the eigenfrequencies of elastic vibrations of the scattering body. In fact, these eigenfrequencies are obtained from the condition of a vanishing denominator in Eqs. (8), which leads to real eigenfrequencies for a free body ($\rho_w \ll \rho_c$), but to complex eigenfrequencies for the fluid-loaded body. (For the example of an aluminum cylinder in water as discussed below, the real part of the latter agrees exceedingly closely with the real eigenfrequencies.) From our resonance representation, it is now clear that the complex eigenfrequencies of the scatterer are precisely the locations of the resonance poles [Eq. (20a)] in the complex frequency plane, whose real parts determine the resonance frequencies in the scattering amplitudes. It should be noted, however, that the scattering body may possess more modes of eigenvibrations than those appearing in the amplitude of acoustic scattering, if the latter cannot excite the corresponding mode. For example, the $n = 0$ "breathing modes" have been found to be excited with a very small amplitude only, both in cylinder and sphere scattering [4, 10].

It will be instructive to consider, in addition to the external scattering amplitude, also the solution of the scattering problem in the interior of the scattering body. There, the displacement field u of the cylinder is represented in terms of a scalar (Ψ) and a vector potential (A):

$$u = - \nabla\Psi + \nabla \times A, \qquad (24)$$

with the solutions [8]

$$\Psi = \sum_{n=0}^{\infty} \epsilon_n i^n C_n J_n(k_L r) \cos n\phi, \qquad (25a)$$

$$A_z = \sum_{n=0}^{\infty} \epsilon_n i^n E_n J_n(k_T r) \sin n\phi. \qquad (25b)$$

The coefficients in these partial-wave series are given in terms of preceding notation as

$$C_n = \frac{2i}{\pi \rho_w \omega^2} \frac{(X_T^2 - 2n^2) J_n(X_T) + 2X_T J_n'(X_T)}{X H_n^{(1)'}(X) D_n^{(1)}(z_1^{-1} - L_n^{-1})}, \qquad (26a)$$

$$E_n = \frac{2i}{\pi \rho_w \omega^2} \frac{2n [J_n(X_L) - 2X_L J_n'(X_L)]}{X H_n^{(1)'}(X) D_n^{(1)}(z_1^{-1} - L_n^{-1})}, \qquad (6b)$$

where only the version leading to the rigid background has been written. The expansion of Eq. (16b) then leads immediately to the resonance expressions (in the vicinity of each resonance):

$$\Psi = \frac{2}{\pi i \rho_w \omega^2} \sum_{n=0}^{\infty} \frac{\epsilon_n i^n}{\beta_n^{(r)}} \frac{(X_T^2 - 2n^2) J_n(X_T) + 2 X_T J_n'(X_T)}{X H_n^{(1)'}(X) D_n^{(1)}}$$

$$\times \frac{J_n(k_L r) \cos n\phi}{X - X_n^{(r)} + \frac{1}{2} i \Gamma_n^{(r)}}, \qquad (27a)$$

$$A_z = \frac{2}{\pi i \rho_w \omega^2} \sum_{n=0}^{\infty} \frac{\epsilon_n i^n}{\beta_n^{(r)}} \frac{2n [J_n(X_L) - X_L J_n'(X_L)]}{X H_n^{(1)'}(X) D_n^{(1)}} \times \frac{J_n(k_T r) \sin n\phi}{X - X_n^{(r)} + \frac{1}{2} i \Gamma_n^{(r)}}.$$

$$(27b)$$

These results are of a pure resonance form, without (rigid or soft) background terms, which is logical since rigid or soft objects are impenetrable and admit no fields in their interior.

The above results, together with the scattering amplitudes of Eqs. (22), indicate the remarkable fact that for scattering at a frequency in between two eigenfrequencies of vibration of the elastic body, the scatterer appears as an impenetrable object and scatters sound accordingly ("potential scattering"), while at and near an eigenfrequency, a scattering resonance is excited by the incident wave and the field penetrates into the body. In the scattering amplitude, the resonance scattering interferes with the potential scattering, and this interference causes the strongly oscillating structure observed previously in the total scattering amplitudes [3, 4].

In some literature, the resonance features appearing in sound scattering from elastic bodies have been discussed in a qualitative way, both for steady-state and for transient scattering [11], showing that resonances should appear where the mechanical plus the radiation impedance go to zero. The present treatment goes beyond this earlier work by (a) indicating that the background scattering does not necessarily correspond to that of a rigid body [soft-body (for air bubbles) or even intermediate-background types (for thin shells) [12] may and do appear also] and (b) providing mathematically explicit forms for the resonances (as well as for the background), as the main features of the present quantitative approach. This together with the explicit eigenvalue equations for the real resonance frequencies (as distinct from the complex eigenfrequency of vibration of the fluid-loaded scatterer), and also the explicit expressions of the resonance widths, serves readily for interpreting and classifying in a quantitative manner the resonances that appear in the complicated patterns of calculated or measured total scattering amplitudes. After the elastic sphere is considered a study of resonances in the separate partial-wave amplitudes will be done and will lead to an interpretation of certain families of resonances in terms of circumferential waves (or Regge poles).

Scattering From Elastic Spheres

The case of plane-wave acoustic scattering from an elastic sphere of radius a can be treated quite analogously to that of a cylinder. With plane incident wave

$$p_{\text{inc}} \equiv e^{ikr\cos\theta} = \sum_{n=0}^{\infty} (2n + 1) i^n j_n(kr) P_n(\cos\theta), \tag{28a}$$

one obtains [3,9] the scattered amplitude

$$p_{sc} = \frac{1}{2} \sum_{n=0}^{\infty} (2n + 1) i^n (S_n - 1) h_n^{(1)}(kr) P_n(\cos\theta), \tag{28b}$$

where we shall again write $S_n \equiv \exp(2i\delta_n)$. With the asymptotic form

$$h_n^{(1)}(kr) \sim (1/kr) i^{-n-1} e^{ikr}, \tag{29}$$

and introducing the form function

$$f(\theta) = \sum_{n=0}^{\infty} f_n(\theta) \tag{30}$$

consisting of partial-wave form functions

$$f_n(\theta) = (2/ka)(2n + 1) e^{i\delta_n} \sin \delta_n P_n(\cos\theta), \tag{31}$$

we obtain the asymptotic scattering amplitude

$$p_{sc} \sim (a/2r) e^{ikr} f(\theta). \tag{32}$$

For a rigid sphere, one has asymptotically [3] for the case of back-scattering $(\phi = \pi)$

$$|f^{(r)}(\pi)|^2 \sim 1. \tag{33}$$

The S function for the sphere is obtained as

$$S_n = S_n^{(s)} (L_n - z_2)/(L_n - z_1) \tag{34a}$$

$$= S_n^{(r)} (L_n^{-1} - z_2^{-1})/(L_n^{-1} - z_1^{-1}), \tag{34b}$$

where

$$S_n^{(s)} = -\frac{h_n^{(2)}(X)}{h_n^{(1)}(X)} \equiv e^{2i\xi_n^{(s)}} \tag{35a}$$

and

$$S_n^{(r)} = -\frac{h_n^{(2)'}(X)}{h_n^{(1)'}(X)} \equiv e^{2i\xi_n^{(r)}}; \tag{35b}$$

Eqs. (35) again represent the S functions for soft- or rigid-sphere scattering, respectively, with corresponding phase shifts

$$\tan \xi_n^{(s)} = j_n(X)/y_n(X), \qquad \tan \xi_n^{(r)} = j_n'(X)/y_n'(X). \tag{35c}$$

In Eq. (34), use was made of

$$z_i = Xh_n^{(i)'}(X)/h_n^{(i)}(X), \qquad i = 1, 2. \tag{36}$$

For the sphere, the real quantity L_n is given by Hickling [3] (who calls it F_n), Vogt et al. [13] and in Chapter 10; it contains spherical Bessel functions with arguments X_L and X_T, and it is proportional to (ρ_w/ρ_s), where ρ is the density of the sphere. As for the cylinder, the correct soft $(\rho_s \ll \rho_w)$ and rigid $(\rho_s \gg \rho_w)$ limits for the S function are obtained from Eqs. (34).

Since the preceding Eqs. [(34) and (35)] for the sphere are formally equivalent to the corresponding Eqs. [(8) and (9)] for the cylinder, the subsequent development of the resonance formalism, Eqs. (12) and (15)-(22), can identically be taken over for the sphere, with only account taken of the different form of z_i, which leads to a replacement of Eqs. (13) and (14) by

$$\Delta_n^{(s)} = X \frac{j_n(X)j_n'(X) + y_n(X)y_n'(X)}{j_n^2(X) + y_n^2(X)}, \tag{37a}$$

$$s_n^{(s)} = \frac{1}{X} \frac{1}{j_n^2(X) + y_n^2(X)}, \tag{37b}$$

and

$$\Delta_n^{(r)} = \frac{1}{X} \frac{j_n(X)j_n'(X) + y_n(X)y_n'(X)}{[j_n'(X)]^2 + [y_n'(X)]^2}, \tag{38a}$$

$$s_n^{(r)} = -\frac{1}{X^3} \frac{1}{[j_n'(X)]^2 + [y_n'(X)]^2}, \tag{38b}$$

respectively. The resonance formalism then goes through in terms of these quantities, and with an expansion of the new function $L_n(X)$ as in Eqs. (16), one thus arrives at a partial-wave form function, e.g., for the case of a rigid background

$$f_n(\theta) = \frac{2}{ka} (2n + 1)e^{2i\xi_n^{(r)}}$$

$$\left[\frac{\frac{1}{2}\Gamma_n^{(r)}}{X_n^{(r)} - X - \frac{1}{2}i\Gamma_n^{(r)}} + e^{-i\xi_n^{(r)}} \sin \xi_n^{(r)} \right] P_n(\cos \theta), \tag{39}$$

which again consists of resonances superimposd on a background corresponding to scattering from an impenetrable sphere. All the conclusions drawn from this expression for the cylinder apply equally to the case of a sphere.

RESONANCES IN PARTIAL-WAVE AMPLITUDES

In this section, results of the preceding resonance theory will be illustrated by numerical calculations for aluminum cylinders and spheres immersed in water. It is pointed out, however, that these numerical results were not obtained from resonance formulas such as Eqs. (18), which are approximations, but from the corresponding exact equations such as Eqs. (8).

Figure 1(a) shows a plot versus $X \equiv ka$ of the modulus of the cylinder backscattering form function $|f(\pi))|$ (Eq. (4)) and Fig. 1(b) shows a similar plot for the sphere (Eq. (30)). Features of the rapid oscillations occurring in these plots have been analyzed previously [4], where it was shown that successive minima are caused by a coincidence of the speeds of various elastic-type creeping waves with the phase velocities of various normal-mode vibrations of the scattering body, which happens at the eigenfrequencies of the latter.

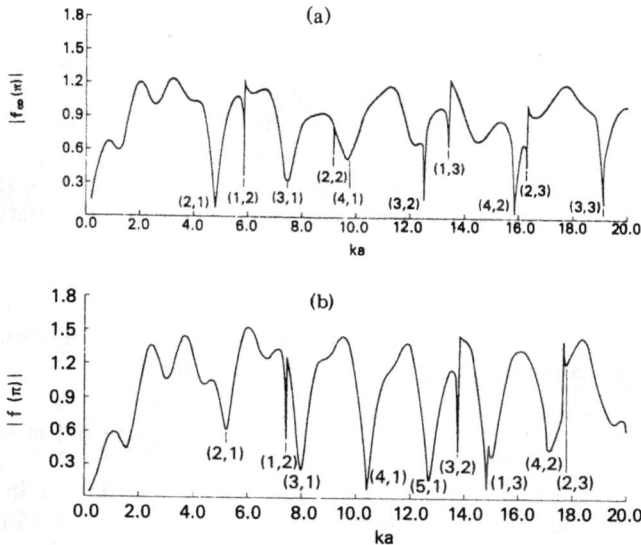

Fig. 1 — Form function modulus $|f(\pi)|$ plotted versus ka, for aluminum bodies in water [(a) cylinder, (b) sphere]. Features are labeled by (n, L). See text.

A more detailed picture of the situation is obtained if one plots the individual partial-wave absolute amplitudes $|f_n(\pi)|$ vs ka This is done in Fig. 2(a) for the cylinder and in Fig. 2(b) for the sphere, for $n = 0$ up to $n = 5$. In each amplitude curve, a number of narrow resonance features can be recognized superimposed on a smooth background, which is found to be given by the scattering amplitude of the corresponding rigid body. This is seen by a comparison with Figs. 3. Partial-wave amplitudes for the rigid, soft, and elastic cylinder are shown in Fig. 3(a) and similar plots for the sphere are given in Fig. 3(b). The figures clearly demonstrate that the elastic case follows the rigid background except in the regions where resonances occur. A further demonstration of the rigid-background nature for cylinder scattering is given in Fig. 4, where the modulus of $|f_2(\pi) - f_2^{\text{rigid}}(\pi)|$ from Eqs. (23) is plotted. This procedure leaves the pure resonance amplitudes only.

A correspondence can be recognized to exist between the partial-wave resonances of Fig. 2 and those of the total form function, Fig. 1. However, Fig. 2 reveals a more detailed picture. One may discern families of resonances labeled $L = 1, 2, 3, \ldots$ whose members appear in parentheses in all the partial waves, shifting to higher frequencies from one partial wave to the next. They may be identified with the elastic creeping-wave poles discussed in Refs. 4 and 5, $L = 1$ representing the Rayleigh wave and $L = 2, 3, \ldots$ the whispering gallery modes [14]. Trajectories of the positions ν_L of these modes in the complex ν plane (obtained by making n a complex variable ν) as functions of ka were obtained previously [15, 16]; they are all located in the first quadrant of the ν plane. Every time such a creeping-wave pole moving along its trajectory passes near an integer n, a resonance will appear at the corresponding value of ka in the nth partial wave, with a width determined by the distance of the pole from the real axis. The partial-wave resonance families are therefore successive manifestations of the various creeping-wave modes (known as "Regge poles" in the nuclear physics literature) as they move along their trajectories, and a picture such as Fig. 2 may serve to make these connections evident, without having to go into the complex plane [5, 15, 16] (see Chapter 3).

We note that for the breathing mode ($n = 0$), only the $(n, L) = (0, 3)$ We note that for the breathing mode ($n = 0$), resonance appears, while all members of all Regge pole families appear in the $n \geq 1$ amplitudes, indicating that these resonances will be excited in acoustic scattering. The numbers shown at each resonance in Fig. 2 are the eigenfrequencies (ka values) of the various free eigenvibrations of the elastic body (which were calculated by us for the body in a vacuum; this agrees closely with the values for the water-loaded body).

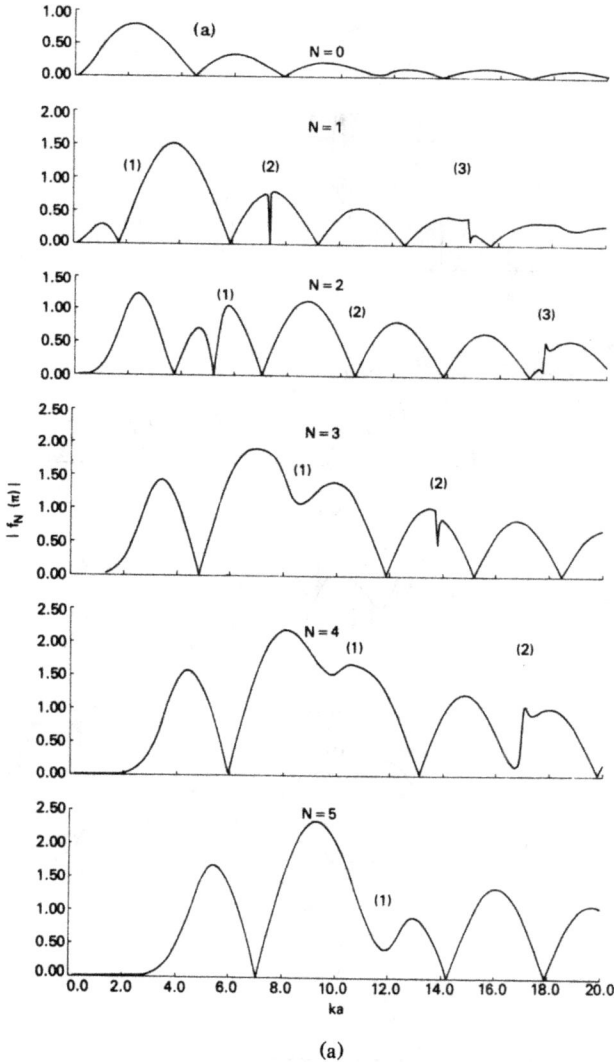

Fig. 2 — Partial-wave amplitude moduli $|f_n(\pi)|$ for the first six partial waves ($n = 0 - 5$), plotted versus ka for aluminum bodies in water [(a) cylinder, (b) sphere]. Positions of the eigenfrequencies are labeled by integer L in each partial wave.

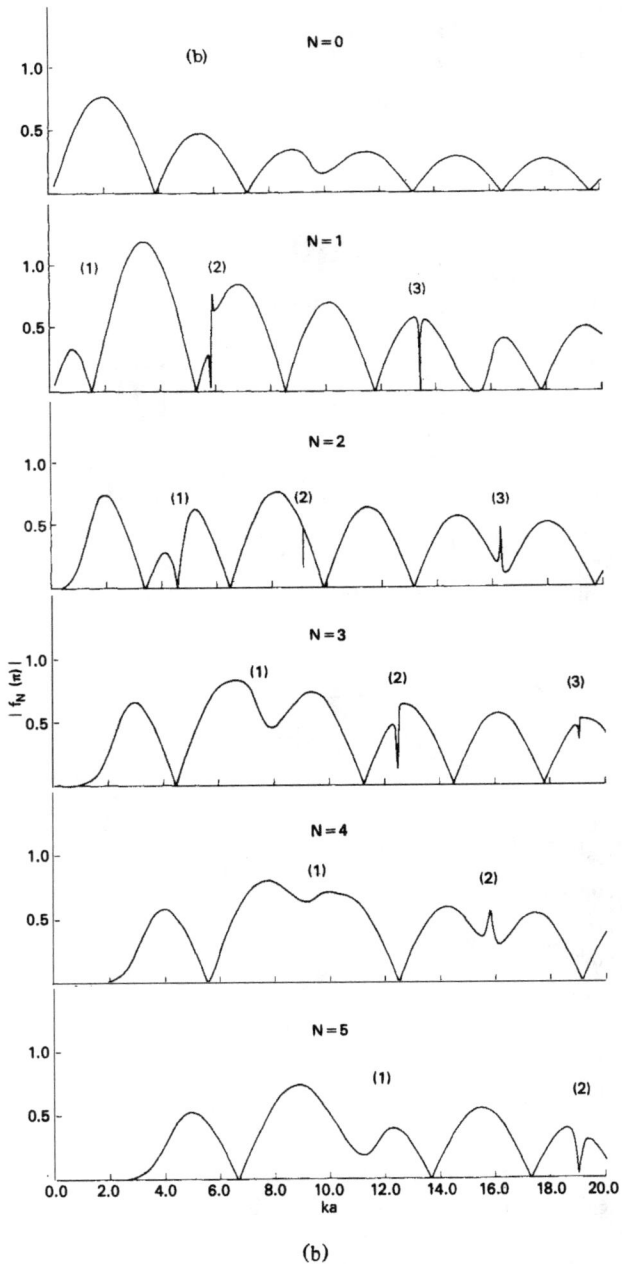

(b)

Fig. 2 — Partial-wave amplitude moduli $|f_n(\pi)|$ for the first six partial waves ($n = 0 - 5$), plotted versus ka for aluminum bodies in water [(a) cylinder, (b) sphere]. Positions of the eigenfrequencies are labeled by integer L in each partial wave.

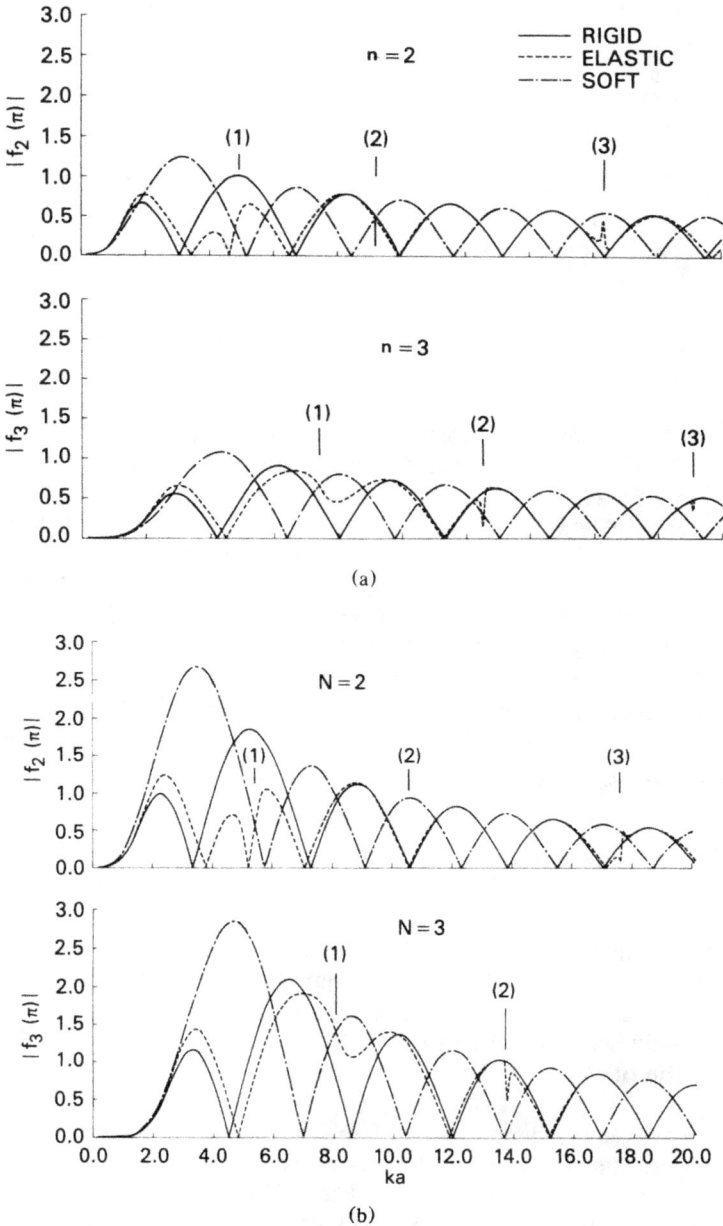

Fig. 3 — Comparison of partial-wave amplitude moduli $|f_n(\pi)|$ for the $n = 2$ and $n = 3$ partial waves of aluminum bodies in water [(a) cylinder, (b) sphere] with rigid-body and soft-body amplitudes as indicated.

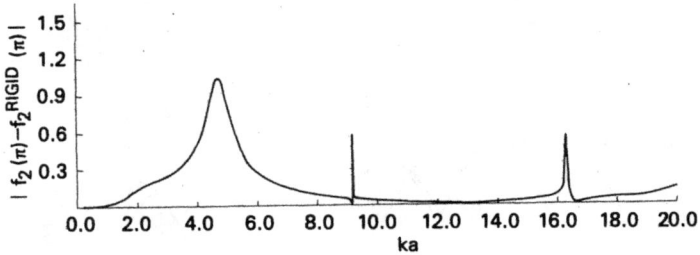

Fig. 4 — Modulus of $n = 2$ partial-wave scattering amplitude with rigid background subtracted, $|f_2(\pi) - f_2^{\text{rigid}}(\pi)|$, for aluminum cylinder in water.

Figure 2 reveals mostly narrow resonances with strikingly different shapes. However, these features may all be understood as follows. Classify as type A a "pure" resonance which shows a peak in $|f_n(\pi)|$; examples are the resonances $(n, L) = (2, 3)$ or $(4, 2)$ in Fig. 2(a), and for $n \geqslant 3$ the $(n, 1)$ Rayleigh resonances (which, incidentally, are seen to have considerably larger widths than the whispering gallery resonances). This type of resonance nearly coincides in frequency with a null of the rigid background, i.e., where $\xi_n^{(r)} \simeq 0$, so that only the resonance contribution in f_n [Eq. (23a)] is present here. Type B are resonances that nearly coincide with a peak of $|f_n^{(r)}(\pi)|$; examples are $(n, L) = (1,3)$, $(3, 3)$, $(5, 2)$, and the second one of the Rayleigh resonances $(2, 1)$. Here, $\xi_n^{(r)} = \frac{1}{2}\pi$ so that the background term in Eq. (23d) becomes $-i$, while the resonance term at $X = X_n^{(r)}$ equals $+i$, causing a cancellation and thus the strikingly narrow holes in $|f_n(\pi)|$ of Fig. 2(a). Finally, type C resonances are the cases, e.g., $(1, 2)$, in Fig. 2(a) which do not coincide with peaks nor nulls of the background, showing a dip on one side and a peak on the other side of the resonance. This is explained by the sign change of the resonance term in Eq. (23d) as X passes through the resonance; it will hence interfere constructively with the background on one side of $X_n^{(r)}$, and destructively on the other.

The frequency dependence of the resonance is given by the resonance expression which is the first term in Eqs. (23d) or (39), referred to in nuclear physics as having the "Breit-Wigner form" [2]. If we take the temporal Fourier transform $f_n(t)$ of the resonance amplitude, we find

$$|f_n^{\text{res}}(t)|^2 \propto e^{-(c/a)\Gamma_n t}. \qquad (40a)$$

This indicates that when excited by acoustic scattering or by other means, the resonance will decay with a lifetime

$$t = (a/\Gamma_n c)\, s. \tag{40b}$$

The quantity $1/\Gamma_n$ thus measures the lifetime of the resonance in units of the time a/c it takes the acoustic wave to traverse the length of one radius. For the $(n, 1)$ resonances, this time is fairly short $(1/\Gamma_n \leq 2)$ due to the large width of the Rayleigh poles, but it is considerably longer for the whispering gallery resonances. Corresponding long-lasting excitations ("ringing") of the resonances have been searched for by Faran [9], and have been found in some cases.

PHASE JUMP AT A RESONANCE

The phase of a resonance term is known to jump through a value of π as the frequency passes through the resonance frequency. To verify this the phase of $f_n(\pi)$ [Eq. (23a)] is examined which is seen to be given by

$$\phi_f = \delta_n + \frac{1}{2}\,\pi. \tag{41a}$$

The resonance form of Eq. (18b) leads to

$$\delta_n = \xi_n^{(r)} + \tan^{-1}\left[\frac{\Gamma_n^{(r)}}{2(X_n^{(r)} - X)}\right], \tag{41b}$$

which indicates that for $\Gamma_n^{(r)} > 0$, the phase $\delta_n - \xi_n^{(r)}$ jumps from 0 to π as X moves from $-\infty$ to $+\infty$, and that this phase jump will occur very abruptly just around $X = X_n^{(r)}$ if the width $\Gamma_n^{(r)}$ is small. Equation (41b) also shows that a corresponding phase jump of ϕ_f will appear superimposed on the progression of the background phase $\xi_n^{(r)}$, which is expected to vary smoothly with X.

To investigate this quantitatively for an aluminum cylinder, in Fig. 5 the phase ϕ_f (mod 2π) is plotted for $n = 2$ on top of the absolute amplitude $|f_n(\pi)|$. Apart from a bending over near $ka = 0$, the background phase $\xi_n^{(r)}$ (dashed curve) moves along a straight line. However, the total phase δ_n (solid curve) is seen to carry out a more or less abrupt jump *upward* (indicating that $\Gamma_n^{(r)} > 0$) by an amount π every time a resonance is passed, due to the second term in Eq. (41b); afterwards, it again parallels the background phase. We note that from a practical standpoint, if a resonance is so narrow that it may be overlooked in a computer plot of $|f_n|$ due to insufficient resolution the corresponding phase jump will still appear and will alert us to the presence of the resonance.

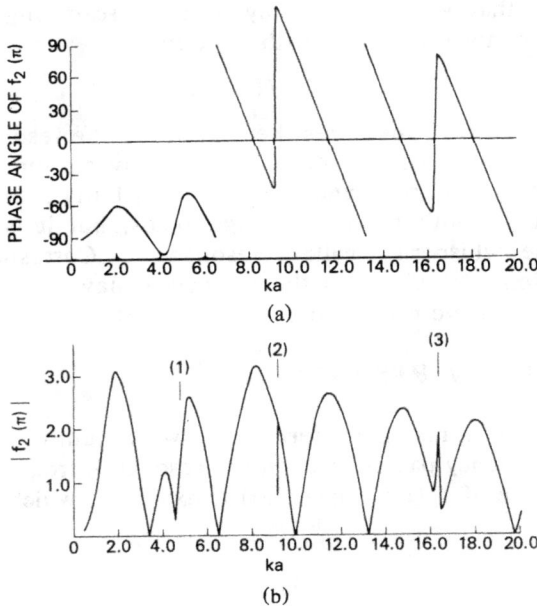

Fig. 5 — (a) Phase and (b) modulus of $n = 2$ partial-wave scattering amplitude for an aluminum cylinder in water.

The mentioned bending over near $X = 0$ (actually, between $X = 0.5$ and 2.5) of ϕ_f for $n = 2$ is present in all phases ϕ_f for $n \geq 1$, and its midpoint moves out approximately proportionally with X. This is exactly where one would expect, using, e.g., the tables of Ref. 5, the resonance of the lowest Franz pole F_1 to occur. The Franz waves are creeping-wave modes which propagate about the scatterer with a speed close to the sound speed in the ambient fluid, being concentrated in the fluid rather than inside the scatterer; hence they are present even around impenetrable bodies (although they are much less intense for soft than for rigid scatterers). No corresponding resonance is visible in our plots of $|f_n(\pi)|$; however, F_1 is located close enough to the real axis in order to cause the mentioned "pseudophase jump," while $F_{2,3}$ probably are not. Note that the total form function $f(\pi)$ for the rigid sphere shows a succession of such phase jumps (Ref. 3, Fig. 2) which, in connection with a corresponding structure of $|f(\pi)|$ given in the same reference, makes the connection of this feature of ϕ_f with F_1 rather suggestive.

ANGULAR DISTRIBUTIONS

The angular dependence of $|f_n(\phi)|$ (Eq. (23d)) is simply given by $|\cos n\phi|$ and hence not very informative. It has been noted, however [9], that the total form function $|f(\phi)|$, Eq. (4a) shows a deep near null at $\phi = 180°$ at the (2, 1) resonance [see also Fig. 1(a)], while similar near-nulls were seen at $\phi = \pm 120°$ at the (3, 1) resonance. In Figs. 6(b) and 7(b), $|f(\phi)|$ is plotted as a solid curve at $ka = 4.79$ [the (2, 1) resonance] and at $ka = 7.47$ [the (3, 1) resonance], respectively, while Figs. 6(a) and 7(a) show $|f(\phi)|$ at 3% below, and Figs. 6(c) and 7(c) at 3% above these values of ka. (The dashed curves show the corresponding angular distributions of the rigid background.) The mentioned holes in the elastic angular distributions at these resonances (not present for the rigid cylinder) are clearly visible. It is noted that for a type B resonance such as (2, 1), the nth partial wave is absent since here, the resonance amplitude has canceled the rigid background. Hence, from Eqs. (4a) and (23),

$$|f^{(r)}(\phi)| = |f(\phi) - 2\epsilon_n(\pi ika)^{-1/2} \cos n\phi|. \qquad (42)$$

Now, from Eq. (7a), one has $|f^{(r)}(\pi)| \sim 1$. However, at the (2, 1) resonance frequency $X_n^{(r)} = 4.79$, and with $\phi = \pi$, the magnitude of the second term on the right-hand side of Eq. (42) is 1.03, which is compatible with $f(\pi) \cong 0$. The half-width of this hole is expected to be that of $|\cos 2\phi|$, i.e., $\cong 35°$, which is compatible with Fig. 6(b). Similarly, for a type A resonance, one again has Eq. (42) except for a plus sign, and for the (3, 1) resonance at $X_n^{(r)} = 7.47$, a similar argument applies but now at $\phi = \dfrac{2}{3}\pi$.

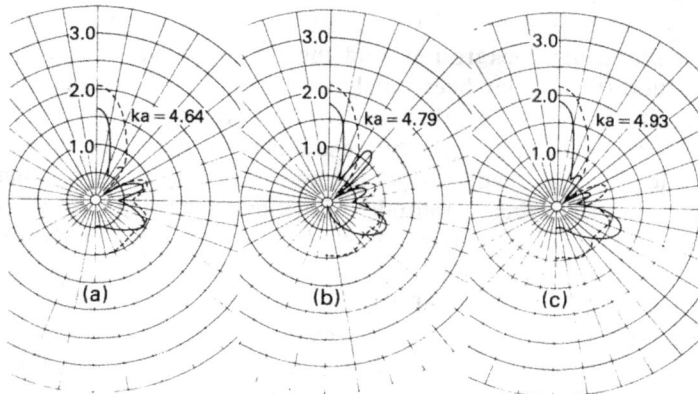

Fig. 6 — Bistatic-scattering amplitude $|f(\phi)|$ for an aluminum cylinder in water, (a) 3% below, (b) at, and (c) 3% above the (2, 1) resonance frequency $ka = 4.79$ (solid curve); comparison with rigid-cylinder amplitude (dashed curve).

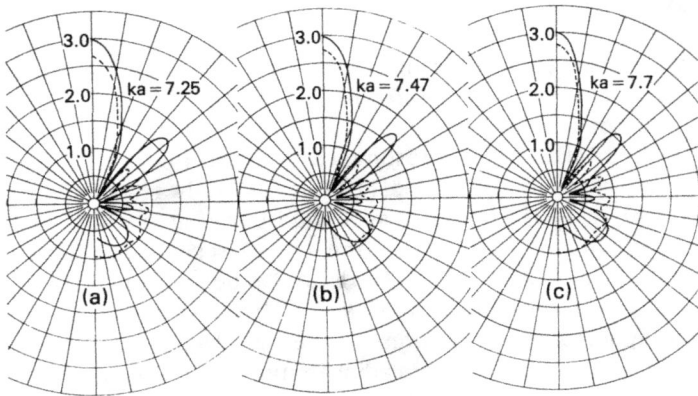

Fig. 7 — Bistatic-scattering amplitude $|f(\phi)|$ for an aluminum cylinder in water (a) 3% below, (b) at, and (c) 3% above the $(3, 1)$ resonance frequency $ka = 7.47$ (solid curve); comparison with rigid-cylinder amplitude (dashed curve).

CONCLUSIONS

The resonance scattering of acoustic waves from elastic cylinders and spheres, has been analyzed by resolving the scattering amplitudes into their individual partial waves, and by following the movement of families of resonances, each corresponding to a Regge pole of the scattering amplitude, through the successive partial waves. The adaptation of the resonance formalism of nuclear reaction theory to the acoustic-scattering problem, together with a numerical evaluation of the amplitudes, has demonstrated the resonances to be superimposed on a rigid-body scattering background, and the various types of interferences resulting therefrom have been analyzed. The merit of this approach lies primarily in the fact that in contrast to previous theory, the behavior of both background and resonances is made mathematically explicit, and this may readily serve for an interpretation of calculated or measured total scattering amplitude. Further, scattering problems with both rigid and soft background may be treated; and explicit expressions are available for both the resonance frequencies of the fluid-loaded scattering and of the width of the resonances. The resonance width determines the lifetime of a resonance, and the phases of the partial scattering amplitudes increase abruptly by π when passing through a resonance. Finally, some arguments have been advanced concerning the existence of near nulls in the total angular distributions at certain resonances.

REFERENCES

1. E.P. Wigner, Phys. Rev. **70**, 15 (1946); P.L. Kapur and R. Peierls, Proc. R. Soc. London Ser. A 166, 277 (1938); H.A. Bethe and G. Placzek, Phys. Rev. **51**, 450 (1937).

2. See also N. Bohr, Nature 137, 344 (1936); G. Breit and E.P. Wigner, Phys. Rev. **49**, 519 (1936); H. Feshbach, D.C. Peaslee, and V.F. Weisskopf, Phys. Rev. **71**, 145 (1947).

3. R. Hickling, J. Acoust. Soc. Am. **34**, 1582 (1962).

4. H. Überall, L.R. Dragonette, and L. Flax, J. Acoust. Soc. Am. **61**, 711 (1977).

5. R.D. Doolittle, H. Überall, and P. Uginčius, J. Acoust. Soc. Am. **50**, 921 (1971).

6. L. Flax and W.G. Neubauer, J. Acoust. Soc. Am. **61**, 307 (1977).

7. See, e.g., S. de Benedetti, *Nuclear Interactions* (Wiley, New York, 1964).

8. H. Überall, R.D. Doolittle, and J.V. McNicholas, J. Acoust. Soc. Am. **39**, 564 (1966); R.D. Doolittle and H. Überall, J. Acoust. Soc. Am. **39**, 272 (1966).

9. J.J. Faran, J. Acoust. Soc. Am. **23**, 405 (1951).

10. W.G. Neubauer, R.H. Vogt, and L.R. Dragonette, J. Acoust. Soc. Am. **55**, 1123 (1974); R.H. Vogt and W.G. Neubauer, J. Acoust. Soc. Am. **60**, 15 (1976).

11. M.C. Junger, J. Acoust. Soc. Am. **24**, 366-373 (1952); see also M.C. Junger and D. Feit, *Sound Structures and Their Interaction* (MIT, Cambridge, MA, 1972), p. 330; or H. Überall and H. Huang, "Acoustical Response of Submerged Elastic Structures Obtained through Integral Transforms," in *Physical Acoustics*, edited by W.P. Mason and R.N. Thurston (Academic, New York, 1976), Vol. 12, p. 217.

12. J.D. Murphy, E.D. Breitenbach, and H. Überall, J. Acoust. Soc. Am. **64**, 677 (1978).

13.   R.H. Vogt, L. Flax, L.R. Dragonette, and W.G. Neubauer, J. Acoust. Soc. Am. **57**, 558 (1975).

14.   Note that Fig. 2 uses the letter L to label these families of resonances, while in Ref. 4 the label l was used, which differs from L for $n = 0$ and 1.

15.   G.V. Frisk, J.W. Dickey, and H. Überall, J. Acoust. Soc. Am. **58**, 996 (1975).

16.   J.W. Dickey, G.V. Frisk, and H. Überall, J. Acoust. Soc. Am. **59**, 1339 (1976).

Chapter 16

CALIBRATION OF ACOUSTIC SCATTERING MEASUREMENTS USING SPHERE REFLECTIONS*

INTRODUCTION

Scattering measurements, in which the acoustic characteristics of a submerged target are investigated by comparing the response of the target with that of a standard scatterer, have been considered previously in connection with underwater measurements [1]. The sphere is a logical choice for a calibrating object because it is a finite, three-dimensional target whose response is independent of rotation, and because exact analytic scattering solutions can be obtained in the spherical geometry [2-9]. Considerations of the sphere as a standard, in comparative target strength (target strength) measurements depend on the invariance of the target strength of a rigid sphere with frequency. As given by Urick [10], the TS of a sphere is

$$(TS) = 20 \log (a/2), \tag{1}$$

where a is the radius expressed in yards. Equation (1) expresses the well-known fact that for $a/\lambda > 1$ the diffracted or creeping wave component of the scattering by a rigid sphere is negligible and the backscattered response is made up almost entirely of specular reflection (λ refers to the wavelength of the incident sound wave). The idea of a standard sphere in lake or ocean measurements was abandoned for such practical considerations as size, fabrication, and mounting; additionally, no sphere in water satisfies rigid boundary conditions, and in fact, the computed and measured responses for submerged solid and hollow metal spheres show steady-state variations from Eq. (1) of as much as 20 dB [4-9]. As will be demonstrated here, despite these large steady-state variations from the rigid body solution, spherical targets can be accurately and conveniently utilized as standard targets in laboratory measurements which utilize broadband techniques and a single source/receiver transducer.

---

*Description of this procedure first appeared in: L. R. Dragonette, S. K. Numrich, and L. J. Frank, J. Acoust. Soc. Am. **69**, 1186 (1981).

Differential measurement techniques, which enable absolute scattering measurements to be made without the necessity of a calibrated receiver, are standardly used in the scattering literature [7, 8, 11-16]. Absolute calibration can be avoided by using the same transducer to measure both the reflected and incident pressure amplitudes, and expressing both the analytic and measured results as a ratio of these amplitudes. A typical scattering experiment is seen schematically in Fig. 1. The target is centered at $P_0(x_0, y_0, z_0)$ and the backscattered return is measured by a receiver at $P_1(x_1, y_0, z_0)$. The incident pressure amplitude can be obtained with relative ease when the source and receiver are separate transducers [7, 8, 11-15]. For cases in which source and receiver are the same, a replica of the incident pulse may be obtained from the water surface or from a flat plate [16] interposed between source/receiver and target. In the latter procedure, the plate must be large enough to avoid illumination of its edges, thick enough to provide separation between front and rear surface echoes, and carefully positioned for normal incidence; in addition, amplitude corrections based on the reflection coefficient of the plate must be made. A more convenient and accurate alternate method is given below.

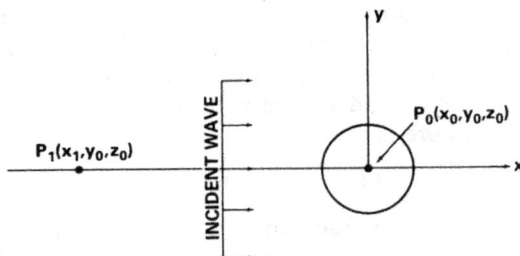

Fig. 1 — Schematic of a typical laboratory
scattering experiment.

MEASUREMENT METHODS

The steady-state, backscattered form function of the target seen schematically in Fig. 1 can be obtained empirically as

$$|f_\infty| = (2r/a)(p_r/p_0), \qquad (2)$$

where r is the range, a is the characteristic dimension of the target, and p_r and p_0 are, respectively, the reflected pressure amplitude measured at $P_1(x_1, y_0, z_0)$ and the incident pressure amplitude measured at $P_0(x_0, y_0, z_0)$. Steady-state measurements of $|f_\infty|$ when the target is an elastic sphere have been made previously and the results agreed, to within a few percent, with exact normal mode series computations given in Chapter 8. Measurements of the steady-state form function can also be made utilizing transient techniques, with the advantage that $|f_\infty|$ is obtained over a broad range of frequencies from a single measurement. The form function of a three-dimensional target illuminated by a short pulse, $p_i(t)$, is given by

$$|f_\infty(ka)| = (2r/a)[|g_r(ka)|/|g_i(ka)|], \qquad (3)$$

where $g_i(ka)$ is the transform of the incident pulse, $p_0(t)$, and $g_r(ka)$ is the transform of the reflected echo, $p_r(t)$. Comparisons between transient experiments and exact theory for elastic spheres in Chapter 9, absorbing spheres in Chapter 10, and infinite elastic cylinders given in Chapter 6, have demonstrated excellent agreement.

Solid tungsten carbide spheres are a convenient, accurate means of obtaining a quantitative measure of $g_i(ka)$, in scattering experiments in which a single transducer is used as both source and receiver. Tungsten carbide was chosen both for its high specific acoustic impedance and because precisely fabricated sizing balls made of tungsten carbide are available as on-the-shelf items.

CALIBRATION BY A RIGID SPHERE

For a rigid sphere of radius a_s positioned at a range r_s from a source/receiver transducer, the backscattered form function would be given as a function of frequency ν as

$$|f_\infty^s(\nu)| = (2r_s/a_s)[|g_s(\nu)|/|g_i(\nu)|], \qquad (4a)$$

where g_i and g_s are, respectively, the Fourier transforms of the incident pulse and of the pulse reflected from the rigid sphere. For frequencies at which the Franz wave does not contribute significantly to backscattering $|f_\infty^s(\nu)| = 1$ and

$$|g_i(\nu)| = (2r_s/a_s)|g_s(\nu)|. \qquad (4b)$$

The form function of any three-dimensional target at a range r_T from a source/receiver is given in terms of its characteristic dimension a_T by

$$|f_\infty^T(\nu)| = (2r_T/a_T)[|g_T(\nu)|/|g_i^T(\nu)|]. \qquad (4c)$$

Here g_T is the Fourier transform of the backscattered return from the target under investigation and g_i^T is the transform of the wave incident on the target. Assuming $1/r$ spreading, the transforms of the wave incident at the reference sphere and at the target are related by

$$g_i^T = g_i r_s/r_T, \qquad (4d)$$

and the far field form function of the target, f_∞^T can be obtained by comparing its reflection to that of the rigid sphere from Eqs. (4b) and (4c) as

$$|f_\infty^T(\nu)| = (r_T^2/r_s^2)(a_s/a_T)[|g_T(\nu)|/|g_s(\nu)|].\tag{5}$$

APPROXIMATION OF A RIGID SPHERE
BY TUNGSTEN CARBIDE

No sphere in water is rigid, even one made of a material with as high a specific acoustic impedance as tungsten carbide. This can be verified by comparing calculations of the form functions for tungsten carbide and for a rigid sphere. The exact normal mode series solutions, as formulated in Chapter 8, were used to analytically obtain the elastic and rigid form functions and these are plotted over the range $0 \leqslant ka \leqslant 75$ in Figs. 2(a) and 2(b). Significant differences between the steady-state solutions for the rigid and tungsten carbide curves exist over the range $ka > 8$, and these differences will continue until acoustic absorption in the material becomes significant (see Chapter 13). Acoustic absorption in tungsten carbide is insignificant over the ka ranges discussed in this paper.

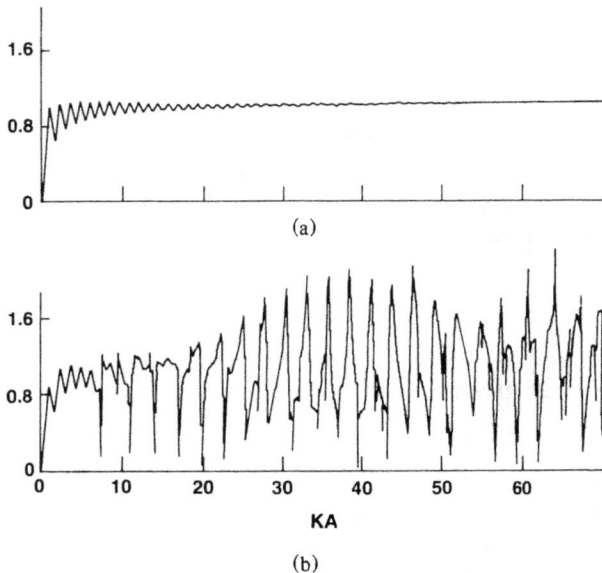

Fig. 2 — The form function for a rigid (a) and a tungsten carbide (b) sphere over the range $0 \leqslant ka \leqslant 75$.

The elastic responses of submerged spheres have been described in terms of free-body resonances as in Chapters 4, 14, and 15. Elastic effects dominate the form function curve for all ka values greater than $(ka)_{2,1}$ which is the ka position at which the (2, 1) free body resonance is excited [18-21]. (The initial number in the label is the normal mode number, the second number is the eigenfrequency.) This is true not only of tungsten carbide but of all materials with wave speeds and densities greater than the velocity of sound and density of water. A cursory look at the curves shown in Fig. 2 would seem to indicate that a tungsten carbide sphere could be used as a standard reflector only in the region $ka \leqslant 7.5$, since this is the region over which rigid and elastic responses are most nearly the same, but this is not the case. A significant finding in Chapters 4, 14, and 15 was the demonstration that the scattering from metal spheres and cylinders, with the material properties discussed above, could be described in terms of a background of rigid-body scattering onto which the resonance or elastic behavior is superimposed. The rigid body and elastic responses cannot be conveniently separated in a steady-state measurement, but they can be isolated in short pulse or broadband measurements. This isolation can be achieved even in the ka range over which the deep nulls and large peaks occur in the form function.

Figure 3 shows the response of a 2.54-cm-diam tungsten carbide sphere to a short incident pulse. The source/receiver transducer was driven by a square wave from a Panametrics Model 5055 Pulser/Receiver, and was located at a range of 73 cm from the sphere. The reflected echoes were obtained in digital form by a Biomation model 8100 analog-to-digital converter interfaced to a PDP 11/34 A computer. The center frequency of the echoes seen in Fig. 3 is 0.5 MHz, and the sampling rate used was 0.5 μs. The response seen in Fig. 3 is made up of several separated echoes. The first echo is labeled rigid reflection and as will be shown below, this echo is an excellent replica of the return that a truly rigid sphere would give. As discussed earlier, it has been demonstrated that the steady-state reflection from spheres and cylinders is made up of a rigid body return on to which the elastic response is superimposed. In Fig. 3 the rigid reflection is separated in time from the elastic echoes; moreover, the acoustic energy which generated the series of elastic echoes enters the sphere at critical angles off of normal incidence so that the backscattered specular reflection replicates a rigid sphere return both in amplitude and frequency content.

A simple demonstration that the return labeled rigid reflection in Fig. 3 is a replica of the return from a rigid sphere can be given by obtaining the form function for a tungsten carbide sphere from

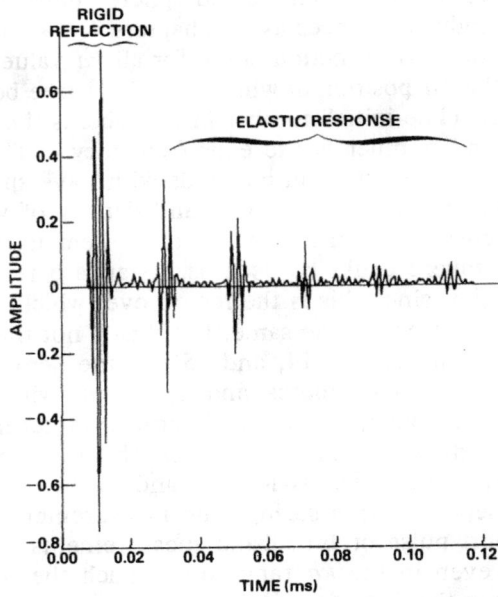

Fig. 3 — The backscattered response of a
tungsten carbide sphere.

$$|f_\infty(ka)| = (2r/a)[|g_r(ka)|/|g_1(ka)|].$$ (6)

In Eq. (6), g_1 is the transform of only the first or rigid echo seen in
Fig. 3, and g_r is the transform of the entire return given in Fig. 3. The
Fourier transforms of the initial or rigid return, and of the entire
reflected echo were computed, and the operations described in Eq. (6)
were carried out on the computer. A comparison of the form function
empirically obtained by this method and a computation made from
exact theory is given in Fig. 4, and quantitative agreement is obtained
over the range $14 \leqslant ka \leqslant 40$. A similar, higher frequency experiment
was performed on a 2.46-cm-diam sphere at a range of 110 cm. The
center frequency of the echoes in this latter experiment was 1.0 MHz,
and the sampling rate was 0.2 μs. A comparison of the analytic and
measured results over the range $40 \leqslant ka \leqslant 70$ is seen in Fig. 5. The
broad ka range covered by the two experiments, whose results are
given in Figs. 4 and 5, also evinces the efficiency and accuracy of using
the broadband, short pulse technique to obtain the steady-state
response of a target.

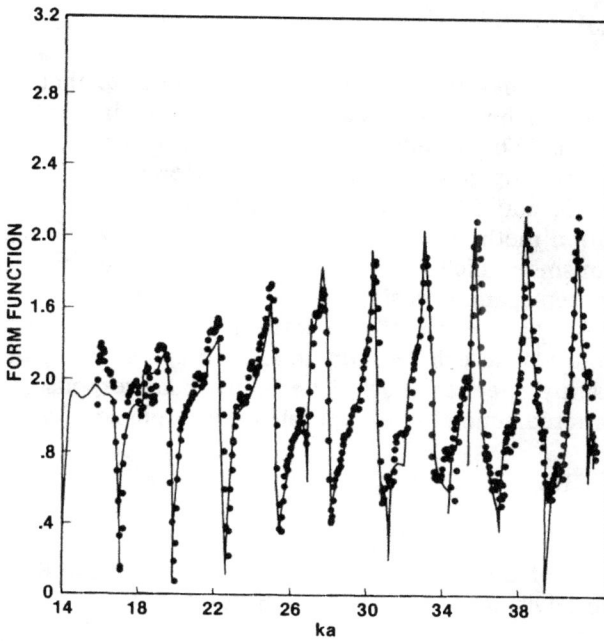

Fig. 4 — Experimentally obtained form function (the points) and analytic computation (the solid line) for a tungsten carbide sphere over the range $14 \leqslant ka \leqslant 40$.

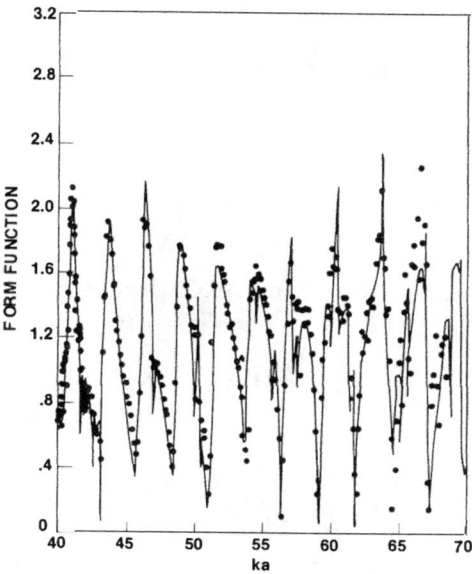

Fig. 5 — Comparison of experimentally obtained form function (the points) and analytic computation (the solid line) for a tungsten carbide sphere over the range $14 \leqslant ka \leqslant 70$.

SUMMARY

The results described in Figs. 3-5 demonstrate that tungsten carbide spheres can be used as standard targets in differential broadband measurements. The steady-state response, or form function of the sphere, is made up of the long pulse interference of echoes such as seen in Fig. 3, and differs greatly from the response of a rigid sphere. The calibration method works because the rigid body and elastic scattering mechanisms which make up the steady-state response of the spheres can be separated, using short pulses, over almost the entire frequency range. In the case of backscattering, the acoustic energy, which generates the elastic echoes, penetrates the sphere at angles which do not significantly effect the backscattered specular response, and this initial backscattered echo closely replicates the response of a rigid sphere.

REFERENCES

1. "Physics of Sound in the Sea," National Defense Research Committee Division 6, Summary Tech. Rep. 8 (1946).

2. J.J. Faran, ONR Tech. Memo. 22, 15 March (1951).

3. J.J. Faran, J. Acoust. Soc. Am. **23**, 405-418 (1951).

4. R. Hickling, J. Acoust. Soc. Am. **34**, 1582-1592 (1962).

5. R. Hickling, J. Acoust. Soc. Am. **36**, 1124-1137 (1964).

6. A.J. Rudgers, Nav. Res. Lab. Rep. 6551 (1967).

7. W.G. Neubauer, R.H. Vogt, and L.R. Dragonette, J. Acoust. Soc. Am. **55**, 1123-1129 (1974).

8. L.R. Dragonette, R.H. Vogt, L. Flax, and W.G. Neubauer, J. Acoust. Soc. Am. **5**, 1130-1137 (1974).

9. R.H. Vogt, L. Flax, L.R. Dragonette, and W.G. Neubauer, J. Acoust. Soc. Am. **57**, 558-561 (1975).

10. R.J. Urick, *Principles of Underwater Sound for Engineers*, (McGraw-Hill, New York, 1967), pp. 244-245.

11. L.D. Hampton and C. McKinney, J. Acoust. Soc. Am. **33**, 664-673 (1961).

12. G.R. Barnard and C.M. McKinney, J. Acoust. Soc. Am. **33**, 226-238 (1961).

13. K.J. Dierks, T.G. Goldsberry, and C.W. Horton, J. Acoust. Soc. Am. **35**, 59-64 (1963).

14. C.W. Horton and M.V. Mechler, J. Acoust. Soc. Am. **51**, 295-303 (1972).

15. R.E. Bunney, R.R. Goodman, and S.W. Marshall, J. Acoust. Soc. Am. **46**, 1223-1233 (1969).

16. H.D. Dardy, J.A. Bucaro, L.S. Scheutz, and L.R. Dragonette, J. Acoust. Soc. Am. **62**, 1373-1376 (1977).

17. L. Flax, J. Acoust. Soc. Am. **62**, 1502-1503 (1977).

18. R.H. Vogt and W.G. Neubauer, J. Acoust. Soc. Am. **60**, 15-22 (1976).

19. H. Überall, L. R. Dragonette, and L. Flax, J. Acoust. Soc. Am. **61**, 711-715 (1977).

20. L. Flax, L.R. Dragonette, and H. Überall, J. Acoust. Soc. Am. **63**, 723-731 (1978).

21. L.R. Dragonette, "Evaluation of the Relative Importance of Circumferential or Creeping Waves in the Acoustical Scattering from Rigid and Elastic Solid Cylinders and from Cylindrical Shells," NRL Report 8216, September (1978).

Chapter 17

EVALUATION OF A
REFLECTION-REDUCTION COATING*

INTRODUCTION

The ratio of reflected acoustic pressure p_r to incident acoustic pressure p_i is called the reflection coefficient R and is alternately expressed as $|R| \exp i\theta$, where θ is the phase shift caused at the reflecting interface. The reflection-coefficient amplitude $|R|$ affords an insufficient description of a lossy material intended to modify, usually reduce, the reflection from a body. In addition, the phase θ is desired and often measurements at various angles of incidence are desired as well. Usual methods of evaluating such reflection-reduction materials are the measurement of pressure in a wave reflected from a sample of the material attached to a flat backing plate either in a free field or at the end of a tube. That measurement is compared to the reflection from an uncoated flat plate or surface. Tube measurements have the disadvantage of relatively small sample sizes and are limited to normal wave incidence. Problems of limited sample size and diffraction, resulting from the edges of plates, are experienced in the reflected field. Diffraction problems, particularly troublesome for non-normal incidence determinations, limit conventional and proposed methods involving finite plates. In addition, such methods result in well-defined reflection coefficients only for certain relations among material constants such as their wave speeds (see Ref. 1). Those relations may not be known in advance. The procedure described here has no such limitations. However, quantities descriptive of reflection that require material parameter values that can be derived from the methods described here can be evaluated after the parameters have been determined. Such a quantity is the reflection coefficient given by Brekhovskikh [2] as

$$V = \frac{Z_1 \cos^2 2\gamma_1 + Z_t \sin^2 2\gamma_1 - Z}{Z_1 \cos^2 2\gamma_1 + Z_t \sin^2 2\gamma_1 + Z},$$

---

*This treatment was first presented in: W. G. Neubauer, J. Acoust. Soc. Am. **62**, 1024 (1977).

where the impedances of the incident wave in the fluid, the longitudinal and transverse waves in the solid are, respectively,

$$Z = \rho c/\cos\theta, \quad Z_1 = \rho_1 c_{L1}/\cos\theta_1$$

and

$$Z_t = \rho_1 c_{S1}/\cos\gamma_1.$$

The incident angle to the surface normal is θ, and the refracted angles of the longitudinal and shear waves are θ_1 and γ_1, respectively. The ambient fluid density is ρ and that of the solid is ρ_1. The wave speed in the ambient fluid is c, c_{L1} and c_{S1} are the longitudinal and shear wave speeds, respectively, in the solid.

A quantity such as the total impedance of the liquid-solid boundary given by Brekhovskikh [3] as

$$Z_{\text{tot}} = Z_1 \cos^2 2\gamma_1 + Z_t \sin^2 2\gamma_1$$

may be evaluated as well. Here, as in the expression for the reflection coefficient, the attenuation of longitudinal and transverse waves can be included by considering Z_1 and Z_t to be complex resulting from complex wave speeds.

BACKGROUND

As a point of departure in describing the new method that follows, descriptions of pertinent results will be briefly stated. The classical problem of the reflection from a two-layer shell has been reported [4, 5] (see Chapter 5). The monostatic or bistatic field can be calculated as a function of frequency. The solutions are not limited to a far field approximation. The medium inside the cylindrical shell can be any fluid or a vacuum and the outside fluid, in which the shell is immersed, is arbitrary. Both layers of the shell can be arbitrarily chosen solids that may cause wave attenuations that can be described in the theory by complex wave numbers (see Fig. 1).

Other results given in Chapter 6 show that accurate measurements can be made with nonspecialized acoustic sources, of reflections from long cylinders and the results are found to be in reasonably close agreement with field descriptions derived from exact solutions with one dimension (length) infinite. To demonstrate this, a least-squares fit of the exact theoretical calculations of the form function $|f_\infty(ka, \phi)|$ for a solid lucite cylinder is shown versus bistatic angle ϕ (180° is backscattering) in Fig. 2. The incident wave number is k and a is the

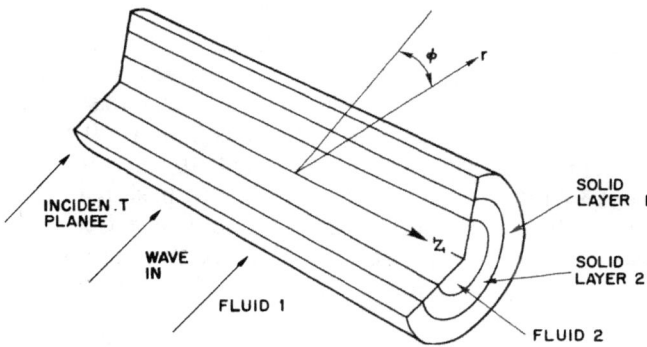

Fig. 1 — Geometry for the solution of the problem
of the reflection from a two-layer shell in a fluid

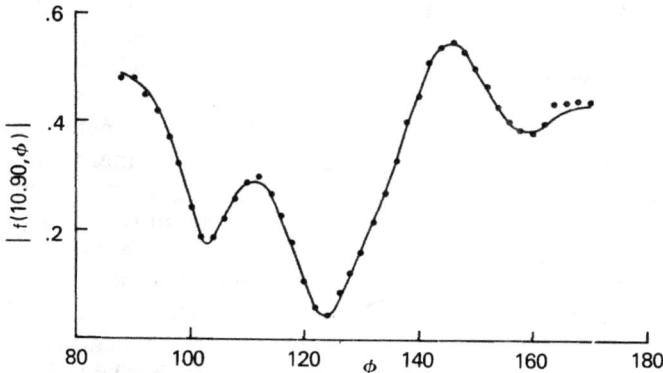

Fig. 2 — The reflected form functions $|f_\infty(ka, \phi)|$ for a lucite
cylinder for a ka of 10.90 versus ϕ, the angle subtended from
the cylinder axis by the source and receiver (180° is backscattering). Exact theory (——) least-squares fit to the experimental
data (●).

cylinder radius. The points are derived from a short-pulsed wave
experiment. Additional similar results for backscattering form function
versus ka appear in Chapter 6.

Experiments have demonstrated [6, 7] reasonable agreement
between cylindrical solutions assuming an infinite length and a short
cylinder whose entire length is insonified within the major lobe of an
acoustic source. It has been shown in Chapter 6 that exact theory can
be fit by least-squares methods to measured data derived from short-
pulse (and therefore broadband) reflection measurements. The experi-
ments result in the reflected pressure $|p_r|$ or form function $|f_\infty(ka)|$

over a frequency range. For a cylindrical geometry these two are related by the expression $|p_r/p_i| = (a/2r)^{1/2} |f_\infty(ka)|$, where r is the distance from the cylinder axis to the receiver. In addition, the exact wave-harmonic solutions for a cylindrical geometry has been extended as described in Chapter 13, to values of ka as high as 1000. It has been adequately demonstrated in Chapters 7 and 10 that "ka scaling" is a valid concept in reflection problems dealing with simple shapes. That means that a reflected pressure determination at one frequency f_1 for a cylinder radius a_1 is the same as the reflected pressure at another frequency $f_2 \neq f_1$ from a cylinder of radius $a_2 \neq a_1$, as long as ka is held constant.

PROCEDURE

In the following discussion, specific materials will be chosen for describing the method, but it is applicable in general to the same extent that the classical shell problem described earlier applies to arbitrary media. Consider a two-layer shell shown in Fig. 1 with air or water in the interior (fluid 2), water (fluid 1) on the exterior of the shell, a solid layer 2—a relatively well-defined solid such as a metal, and a solid layer 1—an acoustically lossy material. In an experiment, the incident and reflected pressure amplitudes $|p_r|$ and $|p_i|$, respectively, can be measured with demonstrated accuracy [12] in a controlled environment as shown in Chapter 9. The parameters considered as known quantities are the densities of all media and longitudinal and shear-wave speeds c_{L2} and c_{S2}, respectively, in the inner (known) shell material. The loss parameters of the inner layer would probably not be significant [8], if it is a metal, below MHz frequencies. However, if that layer is composed of a material that does have significant longitudinal and shear-wave losses, those can be determined separately and included in the theoretical calculation. The longitudinal wave speed in the external fluid (1) is known. Values for the longitudinal and shear-wave speeds c_{L1} and c_{S1} and their associated attenuation factors α_{L1} and α_{S1} can be thus derived from pressure amplitude $|p_r|$, $|p_i|$ measurements as a function of frequency, and least-squares fitting procedure provides a measure of the relative sensitivity to the fit of each of the parameters such as wave speeds and attenuations. Therefore this process can also be used as a reflection-reduction design tool to indicate the potential worth, in terms of acoustic reflection, of varying, or attempting to vary, wave speeds or losses to achieve a specific result.

It is possible, too, to describe a reflected wave resulting from incidence of, say, an infinite plane wave or any configuration of a different geometry than a cylinder such as an infinite plane interface.

This can be readily derived, with the appropriate phase, by using the material constants obtained from the cylindrical shell theory and associated acoustical field measurements.

It should be mentioned that all of the solutions referred to here for cylinders are in a plane normal to the axis of the cylinder. In spite of that limitation, because of the curvature of a cylindrical shell, if the radius is not larger than the incident-wave beamwidth, the constants derived are a result of acoustic incidence at all angles of incidence around the cylinder surface. In addition, this method of determining the echo-reduction-coating material constants is not limited to such a determination in only the far field since the theory is exact and can accommodate the near field as well. This could be a vital consideration in practical cases such as measurements in an enclosed vessel or a tank limited in size as all tanks are. Also, insonification of the cylindrical shell by a close approximation to a spherical or plane wave, although a considerable conceptional simplification, is not an absolute requirement, since the incident beam function can be included in the theory, making it no less exact.

SOME LIMITATIONS

Two shortcomings of the method will be mentioned even though one of them is common to existing methods and a means of overcoming the second is foreseen. First, the reflection-reduction material must be suitable for consideration as a homogeneous isotropic material even though it may not be, even on a macroscopic scale. If the scale of inhomogeneities in the absorbing layer is too large the difficulty will probably appear at the stage of the attempt of least-squares fitting of the exact theory to measured data. The second potential shortcoming of the method is that the reflection-reduction material cannot have wave-loss properties different from those assumed in the exact theory of the two-layer shell. For the cases considered here, $\alpha\lambda$ is constant where λ is the wavelength in the material, and α is the absorption coefficient expressed as attenuation per wavelength. So how does one detect whether the variation of reflected pressure with frequency is caused by the shell solution itself or a deviation of the coating material behavior from that assumed in the theory? If a greatly different material behavior is suspected, it is possible to resolve this potential ambiguity with additional experiments. A possible choice is the measurement of the bistatic field pressure radially around the cylinder (see Chapter 6). These observations can be calculated and will vary in a way consistent with the exact reflection theory of the two layer shell. Consistent deviations of the theory from those experiments must be attributed to an unexpected behavior of the reflection-reduction coating. The theory could then be modified to incorporate the frequency-dependent behavior of the losses in the material and the entire process repeated with those losses.

VARIATIONS ON THE METHOD

It should be noted that the reflection-reduction material can be evaluated as the innermost of the two layers of the shell even though in practice it is intended to be used as an external coating. In that case again the sound is incident on the outside of the two-layer cylinder and the outer layer must be sufficiently thin that the external reflected field be sensitive to different internal layer (solid layer 2 in Fig. 1) wave speeds and attenuations. Also, the shell can be filled with a different fluid so that the material is essentially evaluated while in contact with a different fluid without filling an entire acoustic range, such as a tank, with that fluid. Similarly, with the echo-reduction material as the inner shell layer the interior of the cylinder can be pressurized and the resultant constants for the reflection-reduction materials derived from the measurements will be gained without the requirement of a pressurized free-field facility.

SUMMARY

An acoustic reflection-reduction material can be evaluated by a combination of acoustic field measurements and theoretical computation. The material of unknown echo-reduction effectiveness is made to be one layer of a two-layer shell the interior of which may be filled with either a gas or a liquid like or unlike that surrounding the cylinder. The two-layer shell is placed in a free acoustic field and the amplitude of the resulting scattered acoustic pressure wave is measured. The amplitudes of the incident and reflected acoustic pressure wave are considered known quantities in a computer solution of the exact theory of a two-layer shell. The solution can be formulated so that it will yield values of longitudinal- and shear-wave speeds and attenuations in one of the shell layers if all other parameters of the problem are given. These include all material densities and wave speeds and attenuations in the second shell layer as well as the acoustic wave speeds in the external and internal fluids. Experimental measurements of the ratio of reflected to incident acoustic pressure amplitudes normalized by $(a/2r)^{1/2}$ to express reflection form functions versus either ka or scattering angle are least-squares fit to the exact theory by computer.

Such fitting programs indicate correlations between the various parameters as well as the sensitivity of the fit to the variation of each parameter. Thus the procedure would yield quantitative information about whether a given parameter, such as, say, shear-wave attenuation is a reasonable candidate parameter to attempt to vary to achieve a different more favorable scattering solution, whether this is achieved by anechoic material design or material selection. The procedures or method lends itself to the evaluation of materials under various conditions at the anechoic material-fluid interface such as different fluids or increased or decreased pressure, by capping the ends of the cylinder and changing the internal conditions. Insonification of those ends by the incident wave can be avoided by use of a limited acoustic beam and no different experimental facilities are needed for those conditions. Computer solutions actually yield the acoustic bulk property of the reflection-reduction material in terms of which the exact theory of the two-layer cylindrical shell is written. A specific possible such set is the complex wave number or alternatively the complex wave speed for both longitudinal and shear waves, either of which allow for acoustical wave attenuation in the material. Derived quantities such as reflection coefficients or impedances can then be evaluated.

REFERENCES

1. L. Brekhovskikh, "Waves in Layered Media," translated by D. Lieberman (Academic, New York, 1960), pp. 28-36.

2. Ref. 1, p. 31.

3. Ref. 1, p. 33.

4. L. Flax and W.G. Neubauer, J. Acoust. Soc. Am. **61**, 307-312 (1977).

5. G.C. Gaunaurd, J. Acoust. Soc. Am. **61**, 360-368 (1977).

6. H.D. Dardy, J.A. Bucaro, and L. Flax, J. Acoust. Soc. Am. **57**, S59(A) (1975).

7. I.B. Andreeva and Samovol'kin, Sov. Phys. Acoust. **22**, 361-364 (1976).

8. L. Schuetz and W.G. Neubauer, J. Acoust. Soc. Am. **62**, 513-517 (1977).

9. L. Flax, J. Acoust. Soc. Am. **61**, S37(A) (1977).

10. W.G. Neubauer, R.H. Vogt, and L.R. Dragonette, J. Acoust. Soc. Am. **55**, 1123-1129 (1974).

11. R.H. Vogt, L. Flax, L.R. Dragonette, and W.G. Neubauer, J. Acoust. Soc. Am. **57**, 558-561 (1975).

12. L.R. Dragonette, R.H. Vogt, L. Flax, and W.G. Neubauer, J. Acoust. Soc. Am. **55**, 1130-1137 (1974).

Chapter 18

REFLECTION OF A BOUNDED BEAM*

INTRODUCTION

An analytical formulation of the reflection of a scalar wave beam with bounded cross section incident from a fluid on a plane interface between the fluid and an elastic medium was first given by Schoch [1]. That theory resulted in the formulation of a lateral displacement of the reflected beam. This emergent beam was regarded as constituting the entire reflected field similar to that hypothesized and demonstrated by Goos and Hänchen [2] for light waves. An extensive treatise of this effect has been carried out by Lotsch [3]. In each case beam displacement is expressed in terms of the first derivative of the angle-dependent phase shift of the reflection. A second-order approximation in terms of the second derivative of the reflected phase was given by Brekhovskikh [4] to express the field distribution in the cross section of the reflection for a defined incident beam. In the ultrasonic case, experiments at the Rayleigh angle have indicated [5] that the so-called lateral displacement along the interface occurs at the Rayleigh wave speed. Beam-displacement measurements at the Rayleigh angle have been reported in the literature only for a limited number of materials, namely, aluminum and its alloys [6,7], stainless steel [5], and beryllium [7]. Also, the beam-displacement concept was used in reflection measurements on stainless steel [8-10]. Experimenters have observed [7-11] by means of hydrophone reflectivity measurements a pronounced minimum in the reflection from plane material surfaces that is unaccounted for by the so-called classical theory [12]. That discrepancy had been accounted for in the case of aluminum [7] by imposing a lateral displacement on the hydrophone predicted by Schoch's formulation. Later results, to be given here, have shown it to be clearly evident that such a procedure does not take the entire reflected or reradiated field into account. In general, for other materials and even for the case of aluminum, a significant acoustic field was shown to be present outside of the region that would be intercepted by a receiving crystal of the same size as the incident beam. For a plane hydrophone, oriented at

---

*These results first appeared in: J. Appl. Phys. **44**, 48 (1973).

the Rayleigh angle, with a diameter of 1 cm or larger, a uniform unique-phase signal is intercepted neither at the position of the specular beam nor at the Schoch displacement distance for a range of frequency.

A further modification to the classical theory that included the effects of attenuating waves was carried out by Becker and Richardson [8,9] whose results were confirmed with experiments [10]. That theory was written for a plane-wave assumption and the analysis of the experiment allowed for the finite beams employed by imposing the Schoch displacement on the receiving transducer. The magnitude of the minimum in reflection at the Rayleigh angle was found in that theory to be a function of frequency. This was confirmed by reflection measurements at and near the Rayleigh angle on stainless steel. The magnitude of the reflection at the Rayleigh angle was found to decrease up to 15 MHz (for stainless steel) and then increase with increased frequency above 15 MHz. That frequency was therefore called the "frequency of least reflection." At that frequency the frequency-dependent phase shift upon reflection was found to change sign with respect to the incident wave. Support of this aspect of the theory was also given by experimental observation [8-10] under the same condition as previously stated, viz., source and receiving transducer many wavelengths large of equal size and positioned at equal angles, and receiving transducer centered on the Schoch displacement distance. Although the prediction by the theory of a frequency-dependent phase is indicated to be accurate by later observations, the interpretation of a resulting "frequency of least reflection" is misleading if it is regarded that a hydrophone of the same size as the source beam is used to observe it, since significant energy can be radiated beyond the hydrophone surface. Also, the energy incident on the hydrophone would not have a unique phase over the entire face.

The fact that previous measurements validating beam-displacement theory were made with a limited range of equally sized sources and receivers, or even larger receivers, and a limited range of materials accounts for the apparent validation of a theory that is not adequate to describe the total reflected or radiated field. Previous observations have been rationalized in terms of the results that disclose that the energy redistribution can be characterized by a specular reflection and surface-wave radiation, and the entire field is not represented by the portion incident on a receiving transducer placed in the reflected field and displaced laterally according to the formulation by Schoch. Subsequent to the proposal of the model of reflections at the Rayleigh angle based on experimental observations, Bartoni and Tamir [13] presented theory that substantiated the observed behavior and model.

EXPERIMENTAL OBSERVATIONS OF REFLECTION
AT THE RAYLEIGH ANGLE

For a 1.1-cm-wide beam of 7 MHz incident at the Rayleigh angle on an aluminum (1100)-water interface Fig. 1(a) shows a diagram of the cross section of the reflected beam containing all of the reradiation from the surface laterally displaced to the right by a distance Δ = 4.5 mm as the Schoch theory would predict. The schlieren photograph of a circular beam (diameter = 1.1 cm) incident on an aluminum-water interface is shown in Fig. 1(b). The schlieren system is pulsed (see Chapter 22). The source transducer is seen at the left and the leading edge of the pulse at the right of the photograph. Slight divergence is shown in the diagram since it is unavoidable in the experiment. The fairly well delineated right edge of the reflected or reradiated energy appears out to a distance of 6 mm to the right of the right edge of the incident beam which is 1.5 mm beyond the Schoch prediction. In Fig. 1(b) energy is radiated into the specular region with no displacement. Also present is a dark strip in the reflected or reradiated energy. This occurs at a position within the radiation where specular reflection would normally be expected to be found, i.e., between the lateral limits over which the energy is incident. Equivalent observations have been made with transducers that had square faces. In the case of aluminum the differences between the experimental observation and the description of the result demanded by the theory, and depicted in the diagram of Fig. 1(a), are significant but not startling. In that case the lateral displacement was relatively small, and the distance measured in the schlieren photograph at which energy is seen to emerge from the interface, in excess of that predicted, is not large either. However, the theory is not limited to cases for which displacements are small. Aluminum oxide is a material for which both longitudinal and shear wave speeds are significantly larger than those in aluminum and whose density is almost 50% greater than aluminum. These factors combine to cause a prediction of a lateral displacement of an incident beam for aluminum oxide of 118.1 wavelengths at the Rayleigh angle of 14.8° as depicted in Fig. 2(a). That displacement is more than five times that predicted for aluminum. A schlieren photograph of an actual beam reflected from aluminum oxide at a frequency of 7 MHz is shown in Fig. 2(b). It can be seen that no separation between the incident and emergent beam occurs. A dark strip is again evident in the reradiation as it was in the case of aluminum and the brightness decreases significantly toward the right indicating a decay in pressure amplitude in that direction. Specular reflection in the normal position is evident.

(a)

(b)

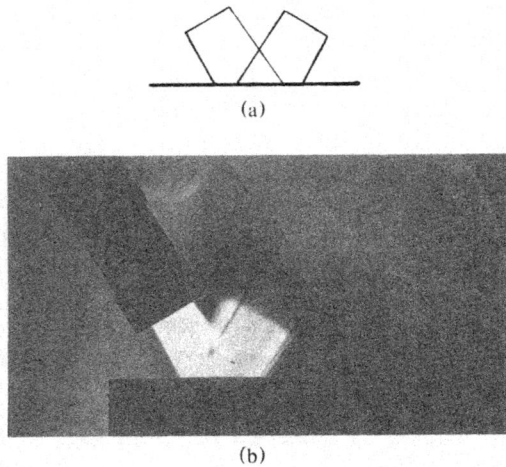

Fig. 1 — Finite beam of 7-MHz ultrasound incident from water on a plane aluminum (1100) interface at the Rayleigh angle. (a) Diagram showing the entire reflected field on the right predicted by beam-displacement theory that should result from the incident beam on the left. (b) Schlieren photograph of an actual incident beam the same size as that shown in (a) and the actual resulting reflected and radiated field.

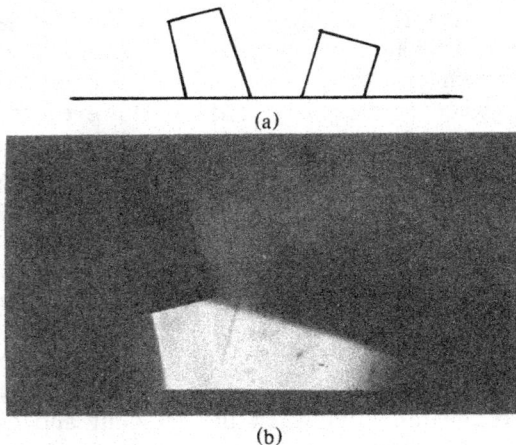

(a)

(b)

Fig. 2 — Finite beam of 7-MHz ultrasound incident from water on a plane aluminum oxide interface at the Rayleigh angle. (a) Diagram showing the entire reflected field on the right predicted by beam-displacement theory that should result from the incident beam on the left. (b) Schlieren photographs of an actual incident beam the same size as that shown in (a) and the actual resulting reflected and radiated field.

A PHYSICAL MODEL OF REFLECTION AT
THE RAYLEIGH ANGLE

A relatively simple model is put forward here that explains the observed results. It is supported by specific measurements and is consistent with the previous experimental results of others. Consider a limited beam of sound incident from a liquid on a solid-plane elastic surface. For all angles, even at and near the Rayleigh angle, energy is specularly reflected. The angle of reflection is equal to the angle of incidence and the size of the specularly reflected beam is the same as the incident beam. In addition, over the area on which energy is incident, and at the same time that specular reflection occurs, a Rayleigh wave is generated by the phase-matching mechanism for incidence *at* and *near* the Rayleigh angle. However, when energy is incident near but not exactly at the Rayleigh angle, the Rayleigh wave is still generated but to a lesser degree and if permitted to radiate, as it may in the case of an adjacent fluid, it will be radiated at precisely the Rayleigh angle.

In the case of aluminum in water, for incidence at precisely the Rayleigh angle, Fig. 3(b) shows the relative phase of the energy in the specular region to be 180° from that in the region of the Rayleigh wave radiation to the right of specular reflection. Between the two regions is a narrow dark region where the equal amplitude coincidence of energy resulting from both causes mutually cancel. Figures 3(a) and 3(c) show reflection at angles smaller and larger than the Rayleigh angle, respectively. It can be seen in Fig. 3(b) that at a frequency of 833 kHz the phase fronts may reasonably be considered to be planar in both regions. The observation of individual wavefronts and the determination of their relative phases for reflection at the Rayleigh angle was first reported in Ref. 14.

A slightly diverging incident beam was produced at 5 MHz with a transducer having a Gaussian-shaped beam [15]. For incidence at the Rayleigh angle for aluminum (1100), the resulting reflection and reradiation into water from the surface is shown in Fig. 4. No significant effect was observed that could be attributed to beam divergence. The hydrophone is shown in place with the sensitive area in the specular reflection. The hydrophone has a plane round active area with a diameter of 0.32 cm. In addition to a significant amount of specular reflection and the cancellation area along the right edge of the specular reflection apparent in the schlieren picture of Fig. 4, a bright high-amplitude region is seen with subsequent amplitude oscillations to its right. The precise cause of amplitude oscillations is not known but it

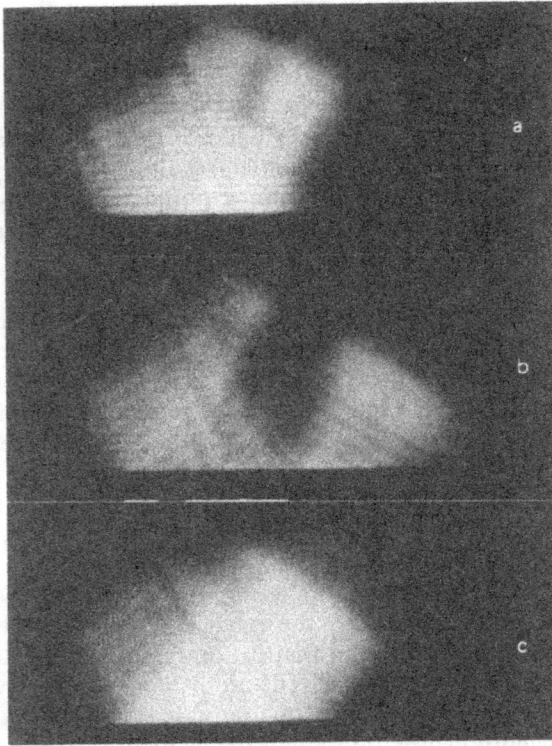

Fig. 3 — Schlieren photograph of an incident beam on the left and a reflected and reradiated beam on the right. The acoustic frequency is 833 kHz, the fluid is water, and the solid is aluminum (6061). Individual cycles are made visible by using a short light pulse. (a) Beam incidence angle $\theta = 25.9°$, (b) $\theta = 30.4°$ (the Rayleigh angle), and (c) $\theta = 35.9°$.

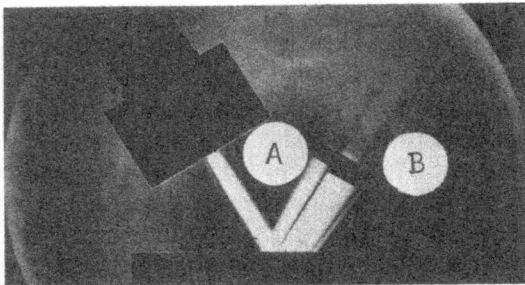

Fig. 4 — Narrow Gaussian incident beam incident on aluminum (1100) at the Rayleigh angle. A receiving transducer is shown at the left with the sensing area in the specularly reflected field. (a) labels the specular region and (b) labels the Rayleigh radiation region.

has been observed that the pattern changes periodicity inversely with source-beam width.

Representative hydrophone output voltage signals are shown in Fig. 5. The top oscilloscope trace is received in the region of the specular reflection labeled A in Fig. 4 and the lower trace is that measured in the Rayleigh-wave-radiation region labeled B in Fig. 4. The two traces have had their amplitudes adjusted so that their respective first cycles are equalized in amplitude and occur at the same time marker so that the pulses may be compared. The pulses are very much alike and can reasonably be called out of phase. The slight change in pulse shape is caused by the fact that amplitude variations exist in the region labeled B and the hydrophone is approximately the size of the bright specular reflection in Fig. 4, and since the amplitude variations to the right are narrower than that, a single and uniform wavefront is not incident on the hydrophone.

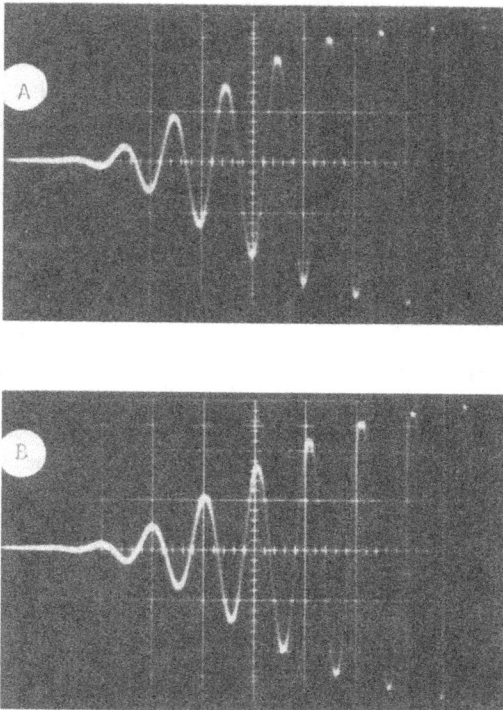

Fig. 5 — Voltage output signal of the receiving transducer, corresponding respectively to the specular region labeled (a) in Fig. 4 and in the Rayleigh radiation region labeled (b) in Fig. 4.

The much quoted formulation of Schoch [1] for incidence at the Rayleigh angle describes the lateral displacement Δ along the boundary of a limited beam as

$$\Delta = \lambda' \frac{2}{\pi} \frac{\rho}{\rho'} \left[\frac{(c_s/c')^2[(c_s/c')^2 - (c_s/c_r)^2]}{(c_s/c_r)^2[(c_s/c_r)^2 - 1]} \right]^{1/2}$$
$$\times \frac{1 + 6(c_s/c_r)^4[1 - (c_s/c_d)^2] - 2(c_s/c_r)^2[3 - 2(c_s/c_d)^2]}{(c_s/c_r)^2 - (c_s/c_d)^2}, \quad (1)$$

where ρ is the density, λ the wavelength, and c the wave speed. Primed parameters refer to the liquid, and unprimed parameters refer to the solid. The subscripts s, d, and r refer to shear, dilatational, and Rayleigh, respectively. This displacement is, therefore, directly proportional to the density ρ of the solid as well as the wavelength λ' of the dilatational wave in the liquid and inversely proportional to the density ρ' of the liquid. It is also a complicated function of the various wave speeds in both the liquid and the solid.

The shear and dilatational wave speeds, the densities, and the anticipated displacements Δ calculated from Eq. (1) are given in Table 1 for seven materials. The various materials were chosen for specific reasons for the observations reported here. Aluminum is the material on which beam displacement was first demonstrated and later extensive work has been done with 304 stainless steel. One is not at liberty to adjust material parameters such as density or wave speed at will, but it was possible to choose materials for which large displacements would be predicted and for which density and wave speed had a desirable relationship to those of aluminum. Stainless steel, molybdenum, and Kennametal K-91 all have wave speeds that are within 24% of the corresponding wave speeds in aluminum and their densities are within 6% of being three, four and five times the density of aluminum, respectively. The last three materials in Table 1 were chosen because they have significantly higher wave speeds for a range of densities. All displacements are predicted to be more than twice that of aluminum. Radiation from the Rayleigh wave generated by the incident wave was observed beyond the distance Δ predicted by the Schoch formula and was observed on materials other than those listed in Table 1.

A broad well-collimated beam of 7 MHz is shown just incident on Kennametal K-8 in Fig. 6(a) and the resultant reradiation caused by that beam is shown in Fig. 6(b). The actual maximum distance to which radiation could be observed was not measured since the end of the limited sample caused a sharp termination of the radiated wave at the far right edge. The displacement formula predicts a termination of

Table 1 — Relevant Values for Seven Materials

Material	ρ (g/cm³)	C_l (m/sec)	θ_l (deg)	C_S (m/sec)	θ_S (deg)	θ_R (deg)	Δ/λ
Al (1100)	2.7	6296	13.6	3106	28.5	30.5	21.3
SS (304)	7.9	5840	14.7	3130	28.3	30.6	58.4
Mb	10.1	6400	13.4	3369	26.1	28.3	90.0
K91[a]	13.5	6542	13.1	3847	22.7	24.8	139.8
K8[a]	14.9	6900	12.4	4200	20.7	22.6	174.5
K165[a]	5.7	8850	9.6	5410	15.9	17.5	115.0
Al₂O₃	4.0	10700	8.0	6360	13.5	14.8	118.2

[a]Material is a sintered product whose main constituent is tungsten and is made by Kennametal Inc.

(a)

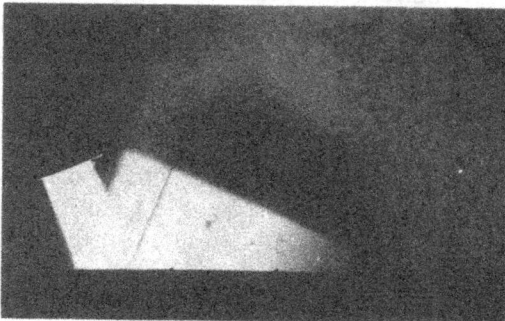

(b)

Fig. 6 — Schlieren photograph of (a) an incident ultrasonic beam of 7 MHz on a plane Kennametal K-8 interface in water and the resultant reflection and radiation at the Rayleigh angle of 22.6°. The radiation energy terminates with a sharp edge on the right at the edge of the material. The sample is 6.35 cm wide.

the radiated energy at a position about four-fifths of the sample width measured from the left edge of the sample. For the last five materials listed in Table 1, reradiated energy was observed beyond a distance of 5 cm to the right of the left edge of the incident beam for an experimental arrangement similar to that shown by Fig. 6. The characteristic cancellation strip can be seen in Fig. 6(b) as it was in all materials at this frequency. The distance along the surface from which radiation was observed was a function of the incident amplitude. The rate of amplitude diminution along the direction parallel to the boundary was observed from schlieren photographs not to be the same for all materials and was quantitatively measured with a hydrophone. Results agreed well with loaded Rayleigh wave attenuation reported by Viktorov [16] for a totally different method.

COMPARISON OF REFLECTION AT THE RAYLEIGH ANGLE AND THE MODEL

Some previous reflection measurements at the Rayleigh angle on a water-stainless-steel interface have presumed that the location of the reflected finite beam would be at a lateral displacement along the interface predicted by Schoch and therefore the transducer was placed at this position. Also, the reflected beam was supposed to have the same size as the incident beam, i.e., the source and receiving transducer were the same size and shape. Under those conditions it was observed that there was a frequency at which the reflection amplitude at the Rayleigh angle was minimum when measured at the displacement distance Δ. The frequency has been called the frequency of least reflection. The observations presented here indicate that reflected (reradiated) energy is distributed over a region larger than the incident beam and is not of uniform amplitude or phase within that region over which it is reradiated. The observation of a frequency of least reflection can be rationalized by showing that it is a result of the experimental method.

Presume a 1-cm-diam incident beam, without divergence, impinging on a plane water-solid interface. Such a beam is shown diagrammatically in Fig. 7 with the hydrophone placed in the field to the right of the incident beam at an angle equal to the angle of the source transducer (Rayleigh angle for stainless steel) and at a distance Δ predicted by the Schoch formula. A schlieren photograph in Fig. 8 shows the actual radiation at 5 MHz from stainless steel that is idealized in the diagram of Fig. 7. The regions labeled A, B, and C constitute the radiated field. The regions A and B combined normally constitute most of the specular reflection and C is the region in which the Rayleigh wave radiation predominates. Specular radiation and Rayleigh

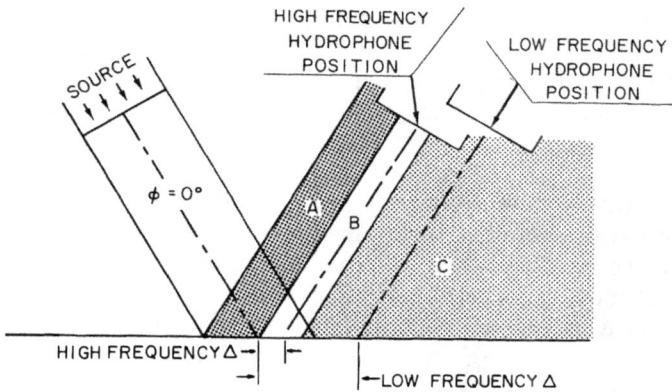

Fig. 7 — A diagram corresponding to the schlieren photograph in Fig. 8 that depicts the various regions of the radiated field: (a) is the specular reflection, (b) is the null region, and (c) is the Rayleigh radiation region.

Fig. 8 — Schlieren photograph of a 5-MHz beam in water incident from the left on a plane stainless steel (304) interface at the Rayleigh angle (30.6°).

radiation coexist in the region B. Computation of the specular reflection discloses that specular reflection from stainless steel (304) is 180° out of phase with the incident wave at exactly the Rayleigh angle up to a frequency f_a (for stainless steel f_a lies between 15 and 20 MHz). Above f_a the phase in the specular reflection is modified as a result of bulk wave attenuation to be in phase with the incident beam. Thus at and below f_a in region B a nullification would occur when the specular and Rayleigh radiation are equal in amplitude. As may be seen in Fig. 8, the delineated regions are, of course, in reality not as sharp and spatially well defined as in the diagram of Fig. 7 and are drawn as they are for purposes of description. The lateral displacements Δ, predicted by Eq. (1), and hydrophone positions at high (i.e., nearly as large as f_a) and low frequency that must be imposed on the receiving transducer to measure a frequency of least reflection are indicated in Fig. 7.

At relatively low frequency, say 5 MHz, the hydrophone is integrating the pressure that is at phase $\theta = 0°$ in region C and part of the field in the region B where the resultant amplitude is at least diminished and at most cancelled by a signal of opposite phase. At a higher frequency as Δ is decreased consistent with Eq. (1), the hydrophone has incident on it less of the pure Rayleigh radiation because of its attenuation. This results in a decreased hydrophone output at 10 MHz. The hydrophone output is a minimum at f_a, the highest frequency for which the signals in regions A and C are of opposite phase. At f_a, and for the Δ prescribed, the energy intercepted in the region A equals or is very nearly that intercepted in the region C resulting in a minimum hydrophone output. Above f_a the hydrophone output rises significantly since the pressure in all regions has been observed to have the same phase.

It must be remembered that the radiation in the region labeled C decreases to the right as the Rayleigh wave is attenuated along the interface. Theories of both Becker and Richardson [8,9] and Mott [17] that describe a frequency of minimum reflection for stainless steel also predict that below a frequency of 5 MHz the reflection increases monotonically to unity at zero frequency when received at the Rayleigh angle and at the Schoch distance. However, this is inconsistent with the observation of the generation and continuous radiation of a surface wave of the type seen in Fig. 6(b). The displacement formula would cause the hydrophone to be placed at an ever increasing distance as frequency is decreased for a fixed beam size. It would intercept a small portion of the Rayleigh radiation even at the lower frequencies when the Rayleigh wave is only slightly attenuated along the surface by radiation or loading of the adjacent liquid. The behavior of the amplitude with frequency predicted by the attenuative theory [8,9,17] is shown in Fig. 9. The monotonic dotted portion of the curve below f_b is the result given by attenuative theory. At the frequency f_b below f_a the hydrophone would have incident on it only radiation resulting from the Rayleigh wave because the distance Δ predicts that the reradiated beam would be intercepted only there. The frequency f_b is a function of hydrophone size. As frequency is decreased below f_b, the predicted hydrophone distance Δ increases directly according to Eq. (1) as the wavelength of the ultrasound. The amplitude of the radiated wave

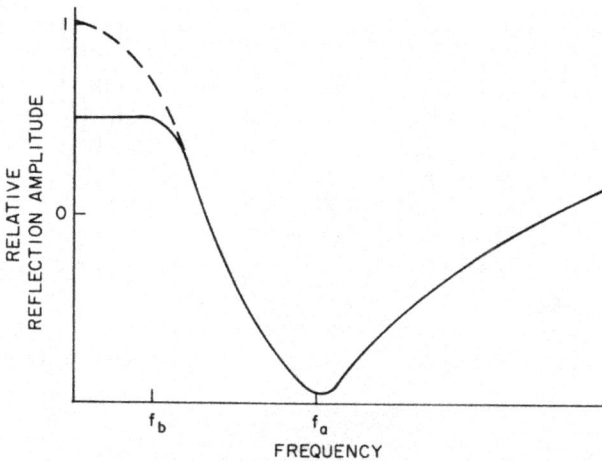

Fig. 9 — Qualitative plot of relative reflection amplitudes vs frequency. The frequency f_a indicates the "frequency of least reflection."

decreases exponentially along the surface inversely as the frequency [16,18]. Thus, the combined result of these two effects at any frequency below f_b is a constant hydrophone output resulting in the flat portion of the curve in Fig. 9.

No attempt was made to identify exactly the specific frequency f_a at which the phase in the specular reflection changed from being 180° out of phase to being in phase with the incident beam. The manifestation of this change is the disappearance of the dark strip normal to the radiated wavefront in the radiation. The source transducer outputs for this range of frequencies changed greatly with frequency which changed the schlieren-observed result, making it difficult to precisely identify the frequency at which a strip disappeared. It can be unequivocally stated, however, that schlieren observations showed f_a to be between 15 and 20 MHz for stainless steel.

EXPERIMENTAL OBSERVATIONS AT THE CRITICAL ANGLES

Evidence of a radiated wavefront similar to that observed at the Rayleigh angle has been observed at the shear critical angle in the region outside the specular reflection for some materials on which Rayleigh wave radiation was observed. The effect is difficult to isolate and distinguish from the Rayleigh wave radiation that occurs for incidence angles near the Rayleigh angle. This indicates that at the shear critical angles a surface wave is generated on the interface and radiates in a manner similar to the Rayleigh wave but to a much lesser degree. Specular reflection at an angle slightly smaller than the shear critical angle θ_{sc} for Mallory 1000 (90% tungsten, 6% nickel, 4% copper) is shown in Fig. 10(a). At exactly θ_{sc}, shown in Fig. 10(b), radiation can be observed to the right of the specular reflected beam. For an incidence angle greater than θ_{sc}, Fig. 10(c) is obtained. By comparison of Fig. 10(b) with Figs. 10(a) and 10(c) the presence of the spreading of the wave to the right can be detected. Actually, the observation was first made in the process of changing the incident angle through θ_{sc}. The continuous viewing of the change in the appearance of the reflected pulse makes the effect more visible than the static views of Fig. 10.

(a)

(b)

(c)

Fig. 10 — Schlieren photographs after reflection of a 7-MHz pulse incident from the left on a plane Mallory-1000—water interface at (a) the incident angle $\theta < 32.1°$, (b) $\theta = 32.1°$ the shear critical angle, and (c) $\theta > 32.1°$.

Evidence of surface wave generation and radiation similar to that at the shear critical angle has also been found at the longitudinal critical angle 0_{1c} and is shown in Fig. 11 for Pyrex glass at 5 MHz. This effect is present to even a lesser degree than that at the shear critical angle. Figures 11(a)-11(c) show reflected pulses for incidence angles less than θ_{1c}, equal to θ_{1c}, and greater than θ_{1c}, respectively. Again significantly more energy appears at the right edge of the reflected pulse in Fig. 11(b) which is taken at θ_{1c}.

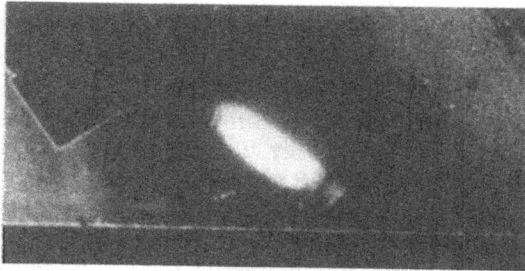

(a)

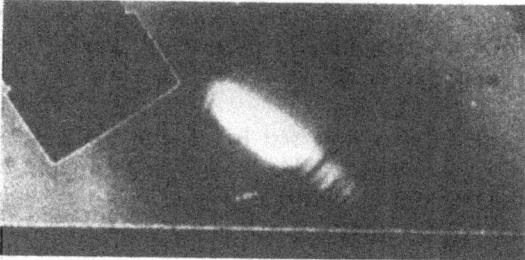

(b)

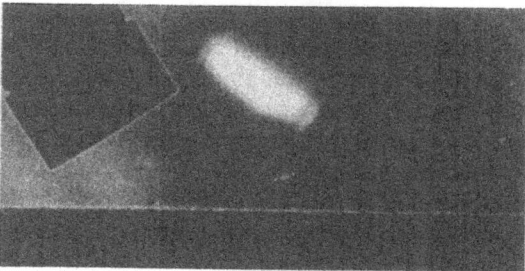

(c)

Fig. 11 — Schlieren photographs after reflection of a 5-MHz pulse incident from the left on a plane Pyrex-glass—water interface at (a) the incident angle $\theta <$ 15.2°, (b) $\theta = 15.2°$ the longitudinal critical angle, and (c) $\theta > 15.2°$.

The reader should keep in mind that all of the data, i.e., schlieren photographs in support of observations or conclusions that are reported here are not presented here. Conclusions were often derived from continuous observations made in the process of changing the incident angle or pulse length. The photographs, that are shown were chosen to demonstrate those observations and to maximize the chances of adequate reproduction.

CONCLUSIONS

The theory of Schoch is inadequate to account for observed radiation from materials for finite beam incidence. Previous apparent agreement between theory and experiment resulted from a reception of only part of the radiated field by a receiving transducer of the same size as the source beam. Radiated energy outside of the region predicted by displacement theory is difficult to measure for materials on which most measurements were previously made, such as aluminum and stainless steel. On these materials relatively small displacements are predicted by the appropriate theory. For materials for which significantly larger displacements are predicted, separations of the incident and reflected beam along the plane interface would be expected. Such separations do not occur for any of the materials examined for any size of source beam.

For the observations reported here, it is concluded that in addition to specular reflection, a Rayleigh wave is generated by the incident beam, propagated along the interface, and radiated continuously into the fluid. Extant theory does not adequately describe this surface wave generation and radiation. Interference occurs between the specular and Rayleigh radiations when they are of opposite phase and are of equal amplitude causing a null field in a portion of the specular region. The results of previous experimental observations of others can be accounted for by this description of the reflected and reradiated field of an incident finite beam. Significant Rayleigh wave generation and radiation also occurs for a range of incidence angles near the Rayleigh angle for sufficiently low frequencies at which the phase of the specular reflection is opposite that of the Rayleigh radiation which is always true [19] when attenuation of bulk waves can be ignored. The theory describes the observed reradiated field.

The theory of Becker and Richardson has shown the phase for a plane reflected wave to be frequency dependent as a result of attenuation of the bulk longitudinal and shear waves. That theory predicts a frequency for incidence at the Rayleigh angle above which the phase in the specular reflection is the same as that of the incidence wave and Rayleigh radiation. That change in phase was detected in schlieren observations by the disappearance of the dark null strip in the specular reflection for stainless steel at a frequency between 15 and 20 MHz indicating the need for considering an attenuative theory in this frequency range.

The attenuative theory also predicts a consequence of the phase change manifested as a frequency of least reflection. That effect is a result of the experimental configuration that was used to observe it which does not include the entire reradiation caused by finite beam incidence.

The theoretical consideration of the attenuation of bulk waves in the solid is indicated to be consistent with schlieren observations at frequencies above f_a. That same theory is deficient for the explanation of reflection amplitude at lower frequencies. Theory predicts the monotonic increase of amplitude to unity toward zero frequency. The model of surface wave generation and attenuation introduced in this paper demands an approach of the amplitude to a constant value as zero frequency is approached.

REFERENCES

1. A. Schoch, Ergeb. Exakten Naturwiss. **23**, 127 (1950).

2. F. Goos and H. Hänchen, Ann. Phys. (Leipz.) **1**, 333 (1947).

3. H. Lotsch, Optik (Stuttg.) **32**, 299 (1971); Optik (Stuttg.) **32**, 553 (1971).

4. L.M. Brekhovskikh, *Waves in Layered Media* (Academic, New York, 1960).

5. O.I. Diachok and W.G. Mayer, IEEE Trans. Sonics Ultrason. **16**, 219 (1969).

6. A. Schoch, Acustica **2**, 18 (1952).

7. R.F. Smith, M.S. thesis, (University of Tennessee, 1968) (unpublished).

8. F.L. Becker and R.L. Richardson, *Research Techniques in Nondestructive Testing*, edited by R. S. Sharpe (Academic, London, 1970), Chap. 4.

9. F.L. Becker and R.L. Richardson, J. Acoust. Soc. Am. **51**, 1609 (1971).

10. F.L. Becker, J. Appl. Phys. **42**, 199 (1971).

11. O.I. Diachok and W.G. Mayer, J. Acoust. Soc. Am. **47**, 155 (1970).

12. W.G. Mayer, J. Appl. Phys. **34**, 909 (1963).

13. H.L. Bertoni and T. Tamir, Appl. Phys. **2**, 157 (1973).

14. W.G. Neubauer, J. Acoust. Soc. Am. **50**, 106 (1971).

15. F.D. Martin and M.A. Breazeale, J. Acoust. Soc. Am. **49**, 1668 (1971).

16. I.A. Viktorov, *Rayleigh and Lamb Waves* (Plenum, New York, 1967).

17. G. Mott, J. Acoust. Soc. Am. **50**, 819 (1971).

18. W.G. Neubauer, J. Acoust. Soc. Am. **52**, 154 (1972).

19. A.L. Van Buren, J. Acoust. Soc. Am. **45**, 341 (1969).

Chapter 19

RAYLEIGH AND SHEAR SPEED DETERMINATION
USING SCHLIEREN VISUALIZATION*

INTRODUCTION

The low-MHz region is the frequency range in which significant internal losses in the material can be neglected and for most metals would have an upper limit in the neighborhood of 10 MHz. It has been shown in Chapter 18 that when a finite acoustic beam was incident on the solid, in the vicinity of the Rayleigh angle, a "head-wave" was generated and progressed along the interface. Figure 1(a) shows a schlieren visualization of the radiation which results from a 1-cm-wide acoustic beam incident on a molybdenum surface in water. The source is in the upper left-hand corner of the picture radiating downward, and the frequency is 5.3 MHz. The beam displacement theory of Schoch [1] predicts a displacement Δ shown in the diagram of Fig. 1(b) and is clearly inadequate to explain the observation. The observation in Fig. 1(a) and observations on other metals are explained by the following model presented in Chapter 18. At and near the Rayleigh angle, in addition to the normal specular reflection, an incident acoustic beam generates a Rayleigh wave in the solid which radiates into the water as it progresses along the surface. At exactly the Rayleigh angle the specular reflection and the Rayleigh wave radiation are 180° out of phase. This model, based on experimental observations, is in agreement with a theoretical model of Rayleigh-angle phenomenon by Bertoni and Tamir [2]. The dark strip in Fig. 1(a) is caused by the mutual cancellation of specular and Rayleigh radiation where they have equal amplitude.

It is this null strip which makes the Rayleigh angle easily and accurately observable and thus makes possible accurate measurements of Rayleigh phase velocity. Specular reflection occurs from the region directly illuminated by the incident beam, and outside of this region Rayleigh wave radiation alone is responsible for the observed wave front.

---

*These results were first reported in: W. G. Neubauer and L. R. Dragonette, J. Appl. Phys. **45**, 618-622 (1974).

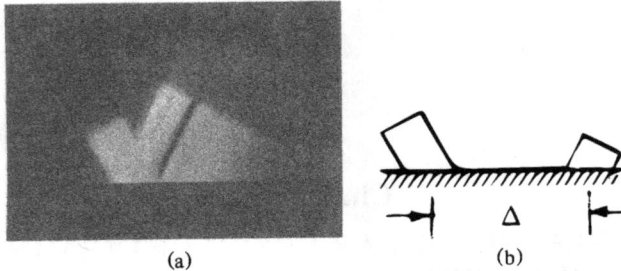

(a)   (b)

Fig. 1 — (a) Schlieren visualization of the reflection of a finite
beam from a molybdenum-water interface at the Rayleigh angle.
The incident beam is on the left-hand side. (b) Diagram of a
displaced beam according to the prediction of Schoch theory for
the incident beam in (a).

MEASUREMENTS

The materials were solid circular cylindrical samples with radii of
6.35 cm and with 7.62-cm lengths. The circular ends are flat to within
±0.001 cm, and these flat ends are the metal surfaces used in the
experiments. The sound source is a 1.9-cm-diam lead-zirconium-
titanate transducer. No receiving transducer is needed for the velocity
measurements. A transducer similar to the source with a mask having
a 1-mm wide slit was used as a receiver in attenuation measurements.

The Rayleigh phase velocity c_r is related to the Rayleigh angle θ_r
by [4]

$$c_r = c_w/\sin \theta_r, \tag{1}$$

where c_w is the speed of sound in water. The speed of sound in fresh
water is known to at least 0.01% from Chapter 23. Thus, the limitation
on how accurately c_r can be obtained is limited by the measurement of
θ_r. The generation of a head wave by an incident acoustic beam does
not by itself give an accurate measure of θ_r. As shown theoretically in
Viktorov [3] and experimentally by Neubauer [1], Rayleigh waves are
generated at many angles and are strongly generated over a wide range
of angles in the vicinity of the Rayleigh angle. The dark null strip seen
in Fig. 1(a) is, however, a sensitive function of the Rayleigh angle,
since it is at this exact angle that the Rayleigh radiation and specular
reflection are 180° out of phase. Figures 2 and 3, respectively, show
the radiation observed when stainless steel and molybdenum interfaces
are insonified at and near the Rayleigh angle. In both cases the dark
strip in the middle frame at Rayleigh-angle incidence is easily distin-
guishable from the radiation at angles 0.1° away. With real-time obser-

Fig. 2 — Reflection of a finite beam from a water-stainless steel interface for incidence at (a) 0.1° less than the Rayleigh angle, (b) at the Rayleigh angle, and (c) 0.1° greater than the Rayleigh angle.

Fig. 3 — Reflection of a finite beam from a water-molybdenum interface for incidence at (a) 0.1° less than the Rayleigh angle, (b) at the Rayleigh angle, and (c) 0.1° greater than the Rayleigh angle.

vation of the radiation using a TV monitor, the Rayleigh-angle measurements could be repeated to 0.10° and the limitation was the
mechanical measurement system and not the method. Table 1 gives
measured Rayleigh angles and calculated phase velocities for several
materials that were measured. The velocity errors given in the table are
based on the ±0.1° limitation in angle measurement imposed by the
angular measurement system used here. This is not necessarily the
limitation of the experimental method.

Table 1 — Rayleigh Phase Velocities at 20°C

Material	Experimentally determined θ_r (deg)	Rayleigh phase velocity (m/sec)
stainless steel	31.62	2827 ± 8
aluminum	31.36	2848 ± 8
molybdenum	28.61	3095 ± 8
K-91	25.40	3456 ± 13
K-8	23.00	3794 ± 16
K-165	17.26	4996 ± 28
aluminum oxide	15.10	5691 ± 38

SHEAR VELOCITY ESTIMATES

In addition to the accurate and direct measurement of Rayleigh
phase velocity obtained from this schlieren technique, estimates of the
shear velocity may be made which are accurate to within a few percent.
These estimates can be made by using only the acoustic surface
reflection measurements discussed above. The determination of a bulk
property using a surface property can be very advantageous in materials
where direct shear velocity measurements are quite difficult. The shear
velocity estimates can be made by using the approximation found in
Mason [5]. This approximation is shown graphically in Fig. 4, which
shows the ratio of Rayleigh velocity c_r to shear velocity c_s plotted vs
Poisson's ratio σ. The velocity ratio is slowly varying with Poisson's
ratio so that σ need not be known with great accuracy. For example, a
value $\sigma = 0.25 \pm 0.10$ on the curve of Fig. 4 corresponds to a range in
c_r/c_s of $0.90 \leqslant c_r/c_s \leqslant 0.935$. This is an error of about ±2% in the
estimate of c_r/c_s. Over the entire curve, the ratio c_r/c_s varies only from
0.86 to 0.96, so that a value $c_r/c_s = 0.91$ can be chosen if nothing is
known about σ and still give a result within usable limits for many
applications.

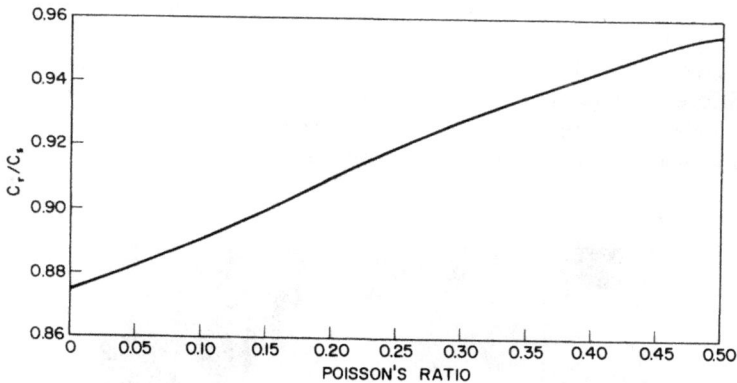

Fig. 4 — Ratio of Rayleigh velocity to shear velocity plotted
vs Poisson's ratio.

Table 2 gives a comparison between shear velocities estimated from the schlieren measurements of c_r and shear velocities measured by direct shear wave propagation through the same samples.

Table 2 — Comparison of Shear Velocity Estimates
with Direct Measurements

Material	c_s estimates (m/sec)	c_s direct propagation measurements (m/sec)[a]	% difference
aluminum	3030	3143	3.6
stainless steel	3040	3148	3.4
molybdenum	3328	3347	0.5
K-91	3757	3853	2.5
K-8	4135	4172	0.8
K-165	5490	5410	1.5
aluminum oxide	6254	6300	0.7

[a]The direct propagation measurements were made by the Ocean Materials Criteria Branch at NRL on the same samples used in the schlieren results. These direct measurements are accurate to within ±1%.

A similar technique, one of measuring elastic constants by the identification of the Rayleigh angle through observation of the well-defined dark strip in the reradiated field, can be applied to plates. This method of estimating shear wave speed on plates is usable for plates three Rayleigh wavelengths thick, it is probably usable for plates even thinner. A finite beam incident on a nickel plate approximately three Rayleigh wavelengths thick is shown in Fig. 5. The sound pulse has a center frequency of 5.8 MHz. Values for Rayleigh speed measured on five different plate materials, as well as the estimated shear speed in those materials, are given in Table 3. Comparison with handbook values is also given where these are available [6].

Fig. 5 — Reflection of a finite beam at the Rayleigh angle
from a nickel plate 3.0 Rayleigh wavelength thick.

Table 3 — Shear Wave Speeds Inferred from
Schlieren Measurements on Plates

Material	Rayleigh speed (m/sec)	Shear speed (m/sec)	Handbook values of shear speed (m/sec)	% difference
Nickel	2780	2973	3000	0.9
Armco iron	2935	3166	3240	2.3
Mu metal	2749	2988	...	...
Iconel	2757	2996	...	...
Permalloy	2757	2996	...	...

ATTENUATION RESULTS

In the low-MHz region, in which these experiments were performed, the attenuation of the Rayleigh wave caused by radiation into the water is given by [7]

$$\alpha_r = \rho_w c_w / \rho c_r \lambda, \qquad (2)$$

where α_r is the absorption coefficient of the Rayleigh wave, ρ_w is the density of water, ρ is the density of the solid, and λ is the Rayleigh wavelength. As demonstrated in Fig. 2 for stainless steel, even when the Rayleigh wave is not transmitted over a large distance along the interface, the identification of the null strip remains a sensitive identifier of the Rayleigh angle. Attenuation measurements were made on the materials listed in Table 1. As mentioned previously, the receiving transducer used in attenuation measurements was similar to the source transducer and was masked to have an active face 1 mm wide. The Rayleigh angle was first determined by using the null strip method described. A receiving transducer was then placed in a fixed position and the source transducer was moved parallel to the liquid-solid interface. A typical plot of amplitude normalized to the maximum value vs distance parallel to the interface for reflection from molybdenum is given in Fig. 6. The initial peak is the specular peak. The dip

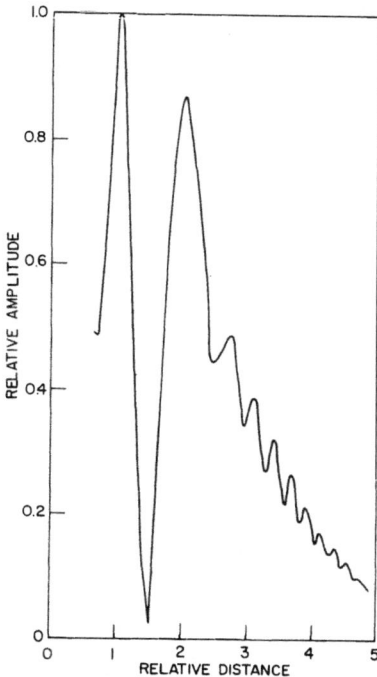

Fig. 6 — Relative amplitude of the reflection of a finite beam from a molybdenum-water interface at the Rayleigh angle, plotted vs the relative distance parallel to the interface along the surface.

in the curve corresponds to the position of the cancellation of specular and Rayleigh radiation, and the second peak is the Rayleigh peak which is followed by an exponential decay. Figures 7(a) and 7(b), respectively, show a linear and logarithmic plot of the decay of the Rayleigh amplitude as a function of distance parallel to the interface in cm for stainless steel. Measured values of α_r are listed in Table 4 and are compared to values calculated from Eq. (2). This equation predicts that α_r is directly proportional to frequency, and this was verified experimentally on aluminum oxide at frequencies from 4.46 to 6.19 MHz. The amplitude measurements given in Figs. 6 and 7 show oscillations about the exponential decay of the radiated wave. Hydrophone measurements of the amplitude of these fluctuations about the exponential decay vary with receiving hydrophone aperture, and these differences with aperture are discussed in Ref. 8. However, a consistent value of α_r is obtained from a least-squares fit of a plot of the logarithm of the amplitude vs the distance.

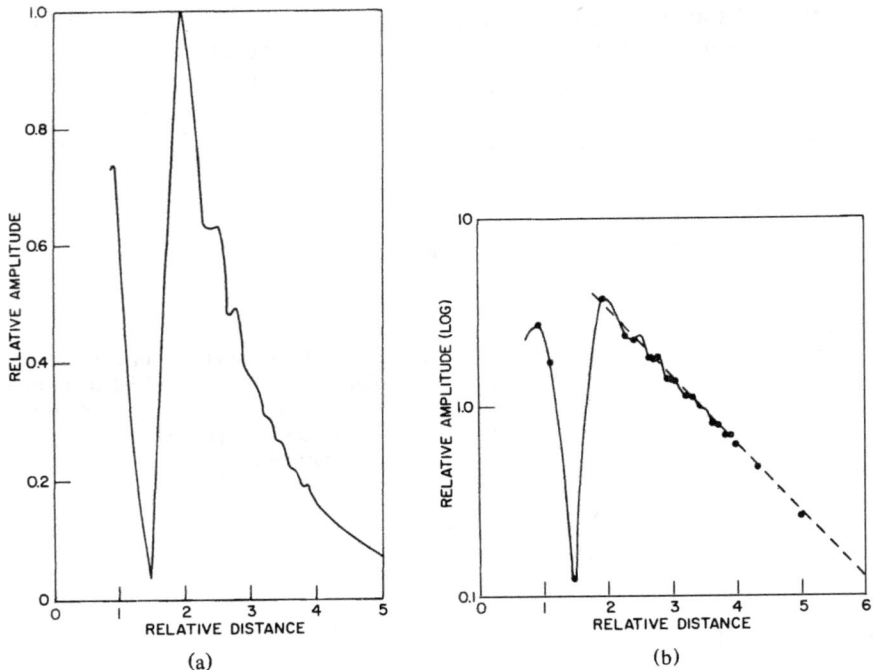

Fig. 7 — (a) Relative amplitude of the reflection of a finite beam from a stainless steel-water interface at the Rayleigh angle, plotted vs the relative distance parallel to the liquid-solid interface. (b) Relative amplitude in (a) plotted logarithmically vs relative distance parallel to the liquid-solid interface.

Table 4 — Measurement of Rayleigh Attenuation α_r in Np/cm

Material	Frequency (MHz)	Experimental α_r	Calculated α_r
stainless steel	3.8	0.82	0.89
aluminum	5.00	3.43	3.41
molybdenum	3.3	0.58	0.51
K-8	4.72	0.35	0.33
K-91	3.8	0.42	0.35
K-165	6.50	0.67	0.64
aluminum oxide	4.46	0.46	0.52
aluminum oxide	5.00	0.55	0.58
aluminum oxide	6.19	0.71	0.71

CONCLUSION

Measurements of the Rayleigh phase velocity in the low-MHz region can be directly and easily made by using a schlieren technique. When a finite acoustic beam is incident from water on a plane solid interface, head waves are generated at and near the Rayleigh angle. These head waves are due to the generation of a Rayleigh wave. In addition to normal specular reflection, the Rayleigh wave radiates into the water as it progresses along the interface. At exactly the Rayleigh angle, the specular reflection and the Rayleigh wave radiation are 180° out of phase. Over a small region where they are of equal amplitude, the specular and Rayleigh radiations mutually cancel and thus form an easily observable dark null strip in the radiation. This measured angle is related to Rayleigh phase velocity by a simple equation. The attenuation of the Rayleigh wave in the frequency range used was shown to be primarily due to radiation loss into the water. Even for materials with a high attenuation, the null strip was easily observed and phase velocity was easily obtained. The direct measurements of Rayleigh phase velocity were also used to estimate the shear velocities of the materials by utilizing a theoretical approximation. The least accurate estimate of shear velocity by this method disagreed with measured shear velocities by 3.6%. The technique was attempted on plates of the order of three Rayleigh wavelengths thick, and agreement, when comparisons could be made, was excellent. Thus, given an arbitrarily shaped body with a flat surface, a simple method exists for determining its Rayleigh phase velocity accurately and for estimating its shear velocity to within a useful accuracy.

REFERENCES

1. A. Schoch, Ergeb. Exakten Naturwiss **23**, 127 (1950).

2. H.L. Bertoni and T. Tamir, Appl. Phys, **2**, 157 (1973).

3. I.A. Viktorov, *Rayleigh and Lamb Waves* (Plenum, New York, 1967).

4. W.G. Neubauer, J. Appl. Phys. **44**, 48 (1973).

5. W.P. Mason, *Physical Acoustics and the Properties of Solids* (Van Nostrand, Princeton, N.J., 1958), p. 21.

6. *Handbook of Chemistry and Physics,* 52nd ed. (Chemical Rubber Co., Cleveland, Ohio, 1971-1972), p. E-41.

7. K. Dransfeld and E. Salzmann, in *Physical Acoustics, Vol. VII,* edited by W.P. Mason and R.N. Thurston (Academic, New York, 1970), pp. 260-261.

8. W.G. Neubauer, in *Physical Acoustics, Vol. X,* edited by W.P. Mason and R.N. Thurston (Academic, New York, 1973), p. 61.

Chapter 20

SCHLIEREN VISUALIZATION OF "HEADWAVES" ON PLATES*

INTRODUCTION

A schlieren technique of measurement and a path trace method of analysis have been used [1] to identify the wavefronts radiated from solid and hollow aluminum cylinders illuminated by short acoustic pulses in water. The schlieren technique was found to be helpful, and in some cases necessary, towards understanding pressure pulse hydrophone measurements of periodic pulses in the shadow zone of cylinders and cylindrical shells [1-5]. Two distinct guided waves with circumferential properties were found on cylindrical shells [1]. The position and group velocity of the radiated wavefronts resulting from these two effects were shown to be consistent with appropriate combinations of multiple shear and longitudinal reflections between the shell surfaces. Illumination by a plane wave at a single angle of incidence is not possible in the case of the shell because of the cylindrical geometry and the finite beam width. Methods for determining the angles at which incident energy contributed to the observed effects are discussed in Ref. 1. Illumination at a single incidence angle can be more nearly achieved in the case of a flat plate.

Guided modes in aluminum plates are isolated here by illuminating the plates with short acoustic pulses at specific angles over a range from 0° to 35°. This range includes the longitudinal and shear critical angles, 13.5° and 28.5°, respectively, and the Rayleigh angle, 31.0°. In general, only those guided modes which are not strongly attenuated by radiation into the water will be isolated and considered. Both the trace velocity and the group velocity of the modes are easily obtained from schlieren visualizations. A formula giving the group velocity of the observed radiated wavefronts or "headwaves" along the plate is derived independently of the type of wave motion in the plate. The effects generated are compared with previous experimental results on plates in water [7, 8]. The utility of a path trace technique for predicting the group velocity of the observed "headwaves" is determined for the plate geometry. Comparisons between the plate and previous [1] cylindrical shell results are made, and the association of Lamb modes, with the guided waves seen in shells, is discussed.

---

*These results were first reported in: Louis R. Dragonette, J. Acoust. Soc. Am. **51** 920 (1972).

EXPERIMENT

The schlieren system utilized here is fully described in Chapter 22 and in Ref. 9. A diagram of the system is given in Fig. 1. The sources are lead-zirconium-titanate transducers and are, in all cases, above the plate and radiating downward. The radiated field in the water is visible above and below the plate. The plate, in some cases, rested at each end on a 1/4-in.-thick rubber cushion glued to a brass block. In other' cases, the plate rested at each end on a brass rod supported by a device capable of accurately rotating the plate. The aluminum plates used in the schlieren measurements were made of 6061 aluminum and had dimensions of approximately 12.7 × 7.6 cm. The plate thickness ranged from 0.002 to 0.241 cm. At the frequencies used, these thicknesses (D), expressed in wavelengths, varied from $0.023\lambda_p$ to $1.89\lambda_p$, where λ_p is the compressional wavelength in aluminum, or from $D = 0.046\lambda_s$ to $D = 3.84\lambda_s$, where λ_s is the shear wavelength. The values for shear wavespeed ($V_s = 3.136 \times 10^5$ cm/sec) and compressional wavespeed ($V_p = 6.370 \times 10^5$ cm/sec) were measured directly from circular cylindrical samples of 6061 aluminum. These samples were 2.54 and 10.16 cm thick, and 7.62 cm in diameter. The speed of sound in water at 20°C, 1.4853×10^5 cm/sec, was obtained from Ref. 10, and is given in Chapter 23.

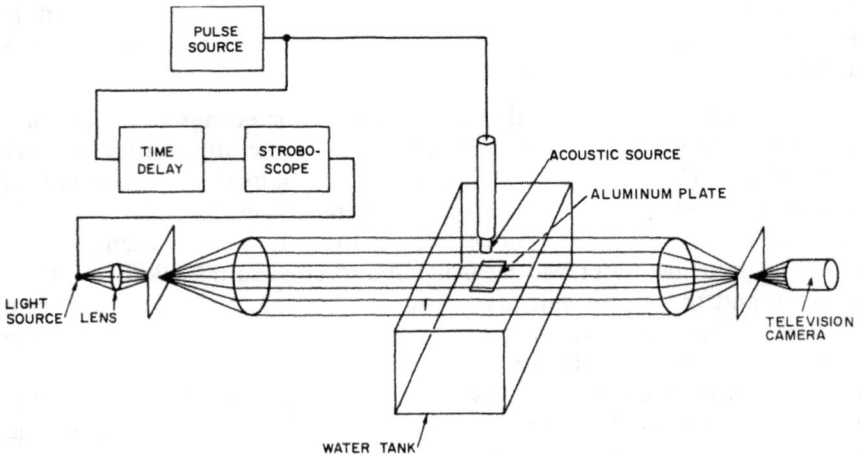

Fig. 1 — A diagram of the experimental system.

Figure 2 shows a schlieren observation of the radiated field in the water above and below a 0.130-cm-thick plate illuminated by a 3-μsec acoustic pulse. At the frequency 7.2 MHz, this thickness corresponds to 1.47λ_p. In the successive frames of Fig. 2, specific angles of incidence, between 0° and 26.5°, are shown. Frames (a)-(c) (incidence angles 0°, 2.0°, and 4.0°) show specular reflection above the plate and transmission below it. In addition, a definite brightening appears behind the specular and transmitted pulses which seems to maximize between 2.0° and 4.0°. No detailed interpretations of this brightening are attempted since the effect does not separate sufficiently from specular reflection to make any useful study of its properties possible; however, it could correspond to a highly attenuating plate mode. The only effects which will be considered in detail are those which persist far enough along the plate, so that separation and isolation from specular reflection are achieved. The first occurrence of such an effect is seen in frame (e) (8.0°). At 10.0°, in frame (f), only specular reflection and transmission are observed. In frames (g) and (i) (12.0° and 13.2°), the appearance of two distinct wavefronts which have progressed along the plate are observed and in frame (h) (12.5°), both are seen simultaneously. From 16.0° through 22.0° [frames (j)-(l)], only specular reflection and transmission are present. Finally, the occurrence of another progressing wavefront is observed in frames (n) and (o) (26.0° and 26.5°). Only specular reflection and transmission were observed as rotation of the plate was continued through 35.0°. The additional wavefront, seen in frames (g)-(o), is a reflection from the transducer of the reflected pulse. The effects seen in Fig. 2 are observed separately in Figs. 3 through 7.

When the 0.130-cm-thick plate is illuminated by an acoustic pulse at an incidence angle of 12.5°, the result shown in Fig. 3 is obtained. The driving pulse is approximately 3-μsec long at a frequency of 7.2 MHz. The successive frames of Fig. 3 show the sound beam striking the upper surface of the plate and at later times the formation and progression along the plate of two distinct wavefronts, each with its own characteristic inclination angle and speed. Separation in time between the frames is obtained by varying the time delay between the triggering of the sound source and the strobe-light source. In this way, the sound pulse and its effect can be frozen at any instant, from the time the sound pulse leaves the transducer until it and its effects disappear below the dynamic range of the system. Both of the wavefronts, seen in Fig. 3, are equally well observed above or below the plate. The two wavefronts can be isolated by changes in the incidence angle and are observed separately in Figs. 4 and 5 (12.0° and 13.2°). The strong

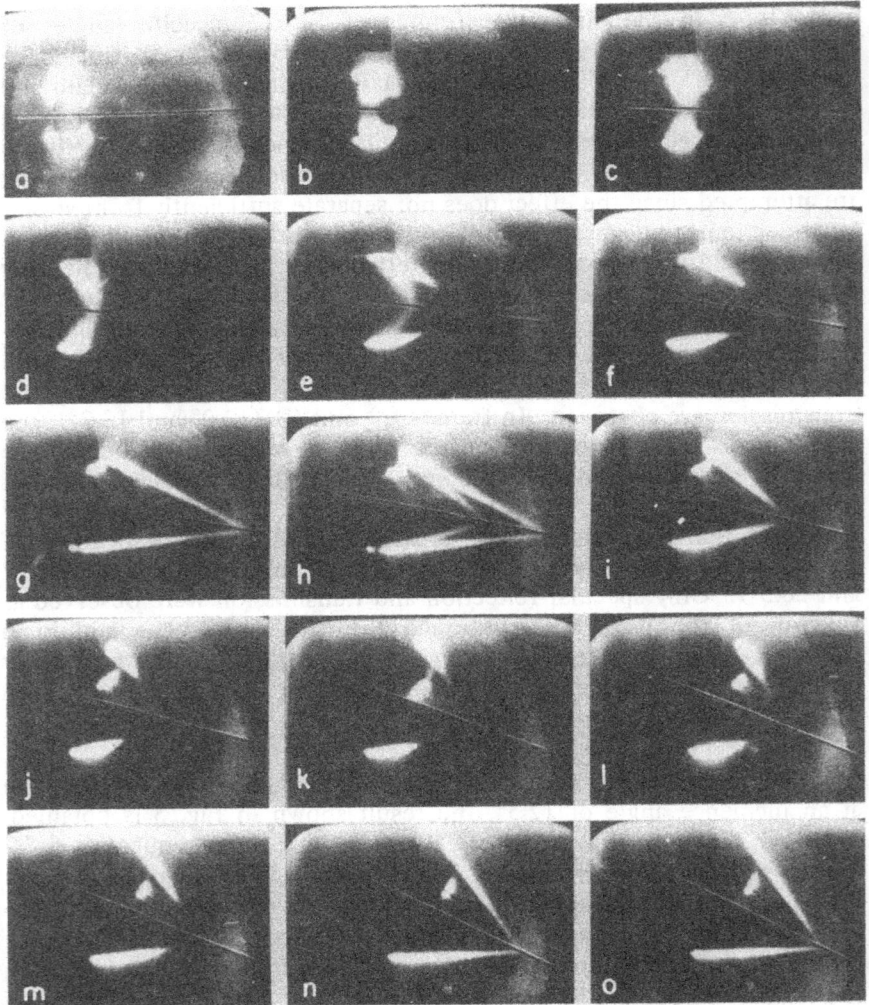

Fig. 2 — Schlieren photographs taken as a 0.130-cm-thick plate is rotated from normal to the direction of propagation of an incoming acoustic pulse through 26.5°. Pulse length is 3 μsec; frequency is 7.2 MHz. Frame (a) is 0.0°, (b) 2.0°, (c) 4.0°, (d) 6.0°, (e) 8.0°, (f) 10.0°, (g) 12.0°, (h) 12.5°, (i) 13.2°, (j) 16.0°, (k) 20.0°, (l) 22.0°, (m) 24.0°, (n) 26.0°, and (o) 26.5°.

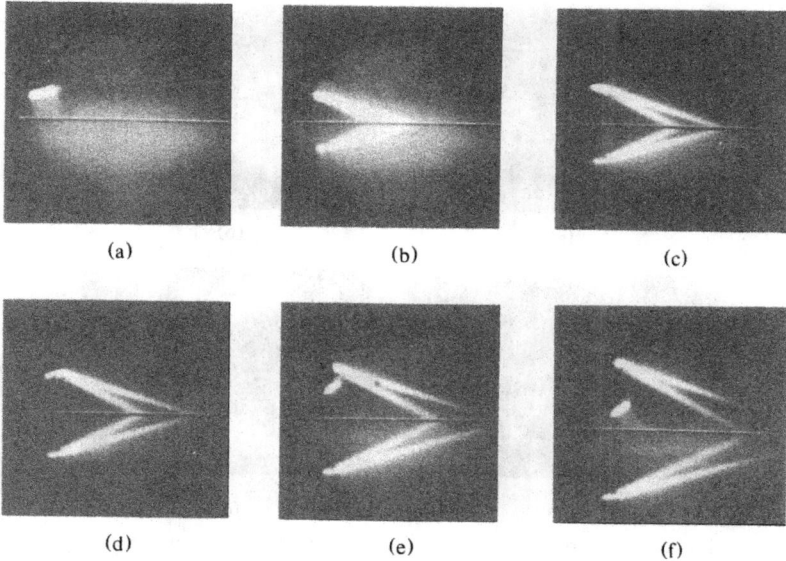

(a)   (b)   (c)

(d)   (e)   (f)

Fig. 3 — A time sequence of schlieren photographs showing the effects of illuminating a 0.130-cm-thick plate with an incoming acoustic pulse at $\theta_i = 12.5°$. Pulse length is 3 μsec; frequency is 7.2 MHz.

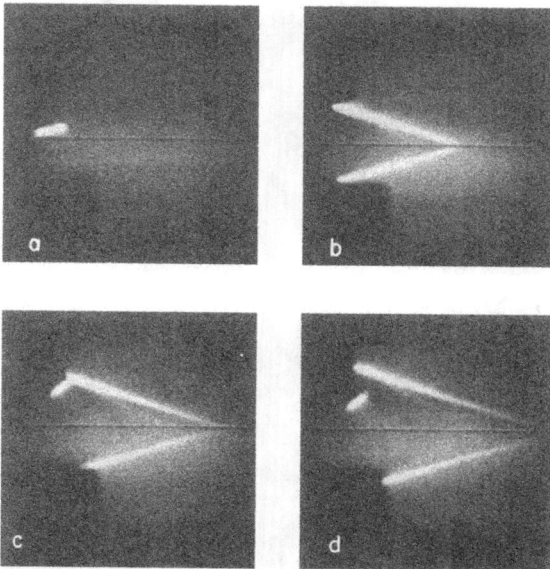

Fig. 4 — A time sequence of schlieren photographs showing the effect of illuminating a 0.130-cm-thick plate with an incoming acoustic pulse at $\theta_i = 12.0°$. Pulse length is 3 μsec; frequency is 7.2 MHz.

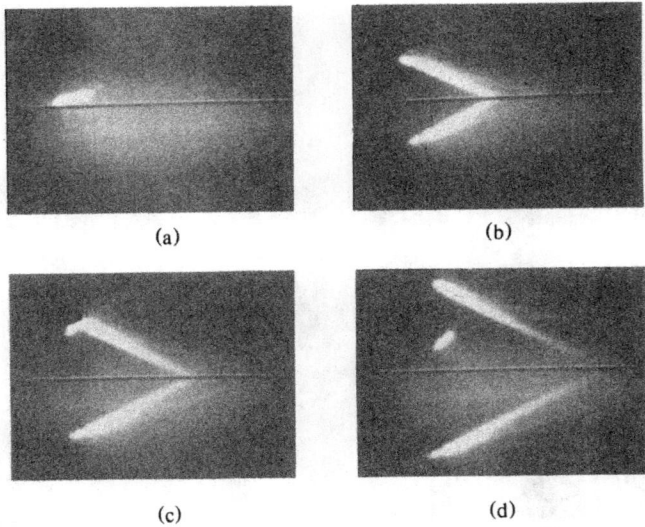

(a)   (b)

(c)   (d)

Fig. 5 — A time sequence of schlieren photographs showing the effect of illuminating a 0.130-cm-thick plate with an incoming acoustic pulse at $\theta_i = 13.2°$. Pulse length is 3 μsec; frequency is 7.2 MHz.

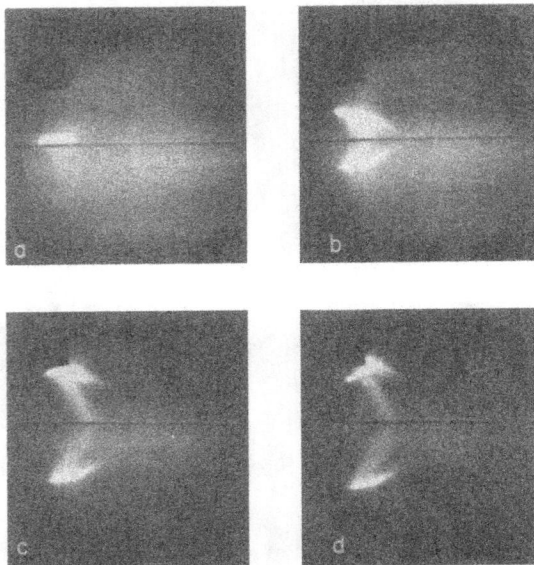

Fig. 6 — A time sequence of schlieren photographs showing the effect of illuminating a 0.130-cm-thick plate with an incoming acoustic pulse at $\theta_i = 8.0°$. Pulse length is 3 μsec; frequency is 7.2 MHz.

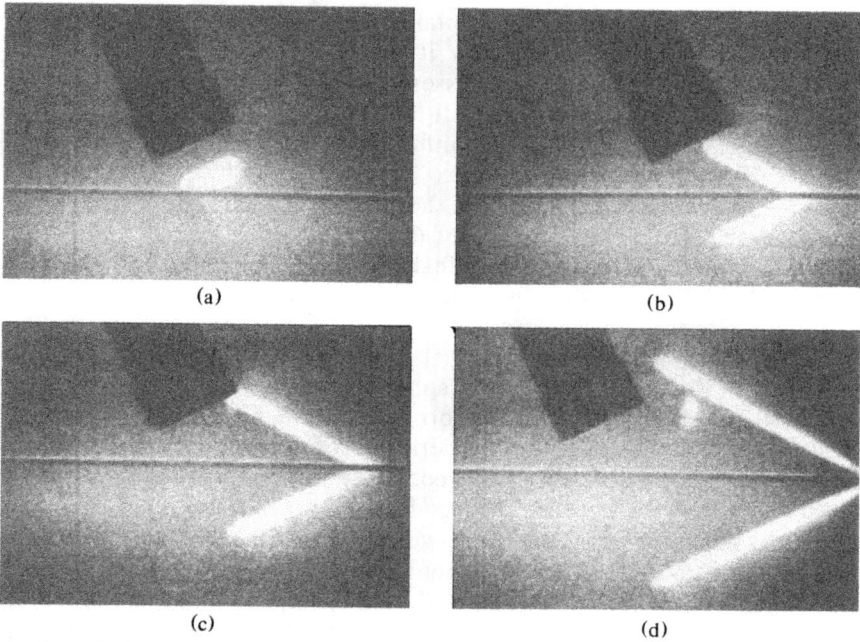

Fig. 7 — A time sequence of schlieren photographs showing the effect of illuminating a 0.130-cm-thick plate with an incoming acoustic pulse at $\theta_i = 26.5°$. Pulse length is 3 μsec; frequency is 7.2 MHz.

persistence of each of the radiated wavefronts over a small range of incidence angles is consistent with the amount of beam spreading expected with the source transducer. At incidence angles of $8.0°$ and $26.5°$, the final two effects observed in Fig. 2 were maximized and time-sequence schlieren photographs of the progressions of the radiated wavefronts along the plate are seen in Figs. 6 and 7, respectively. The angles of incidence of the incoming acoustic wave in Figs. 2-7 are measured with a possible systematic error of $\pm 0.5°$. Measurements were made from Figs. 3-7 of the angle of inclination, θ_r, between the radiated wavefront and the plate surface (i.e., the angle at the intersection of the wavefront and the plate). The measurements show that θ_r is not equal to θ_i, the original angle of incidence of the sound pulse. In all cases, however, the specular reflection makes an angle θ_i with the plate. The radiated wavefront angles are measured when the "headwave" has progressed far enough along the plate that the specular reflection is easily avoided. Differences between θ_r and θ_i are especially obvious in Fig. 6, where sound is incident at $\theta_i = 8.0°$, and θ_r is $60.0°$, and, in Figs. 4 and 5, where a change of $1.2°$ in incidence angle leads to generation of two wavefronts, whose inclination angles differ by 10.0°. Measurements of θ_r, taken from different frames in the time-sequence photographs, show it to be a constant for each wavefront, as that wavefront progresses along the plate. Theoretically, Snell's law predicts that energy will leave the plate at an angle equal to the angle of incidence. This is confirmed by the plate results of Worlton [7, 8]. The apparent paradox between the expected result, that energy should leave the plate at the same angle at which sound was incident, and the determinations of θ_r, made from the schlieren photographs, is in reality only a manifestation that the group velocity and phase velocity of the effect involved are different.

On the single assumption that energy making up the wavefronts observed in the schlieren photographs does leave the plate at an angle equal to the incidence angle, a formula for the group velocity of the wavefront along the plate can be derived from geometric considerations alone. The derivation does not depend upon any consideration of the type of motion within the plate. In Fig. 8, the lines AC and BD represent positions of a radiated wavefront above a plate OD at times t_1 and t_2. Positions A and B represent positions of a ray which has left the plate at an angle θ_i and traveled with the speed of sound in water, V_w, to position A in time t_1, and from position A to B in time $t_2 - t_1$. The radiated wavefront moves a distance CD along the plate in the time, $t_2 - t_1$. The group speed V_g is related to the speed of sound in water, V_w, by $V_g/V_w = CD/AB$, which can be expressed as

$$V_g = V_w \cos (\theta_i - \theta_r)/\sin \theta_r. \qquad (1)$$

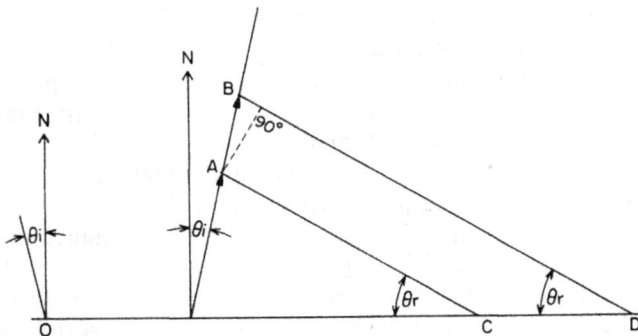

Fig. 8 — Diagram, representing the position of a radiated wavefront at successive times t_1 and t_2, from which group velocity formula is derived.

Table 1 gives values of V_g which were calculated from the measured values of θ_r. The latter were obtained from the schlieren photographs presented. The radiated wavefront angle, determined from photographs shown in Fig. 6 (entry No. 1), was measured with a precision of $\pm 5.0°$. The thinner, sharper wavefronts in the schlieren photographs, corresponding to entries 2 through 9 in Table 1, were measured with a precision of $\pm 2.0°$. Included in Table 1 are results obtained from schlieren photographs of waves generated on plates thinner compared to a wavelength. These schlieren photographs are seen in Figs. 9 through 12, and correspond to entries 5 through 10 in Table 1. Figures 9(a) through 9(c) show wavefronts which occur when a 0.130-cm-thick plate is tilted at 12.0°, 12.5°, 13.4°, and 26.0° with respect to the direction of propagation of an incoming acoustic pulse. The pulse is 5 μsec long and the frequency is 5 MHz. At this frequency, the plate thickness corresponds to $1.02\lambda_p$. The three separate wavefronts isolated in Figs. 9(a), 9(c), and 9(d) were the only observed progressions along the plate, in a range of rotation from 0° to 35.0°. Figures 10-12 show the single strongly persistent wavefronts observed on plates of thickness 0.051, 0.025, and 0.002 cm, respectively. In all three cases, the pulse length is 3 μsec, the frequency is 7.2 MHz, and the angle of incidence is 13.0°. The 0.002-cm plate is not 6061 aluminum. It is made from commerically available aluminum foil. Entries 8-10, in Table 1, show the results obtained from these three figures. Values for the incidence angle and wavefront angle for the thin foil are only best estimates, since the surface of the foil could not be flattened sufficiently for precise measurements to be made.

Some direct measurements of V_g, which utilized a receiving hydrophone, were also made in a separate experiment. A source and receiving hydrophone were set with angles of incidence and reception the same. The receiver was moved at a constant height above the plate and the velocity is given by $V_g = d/t$, where d is the distance that the receiver moved and t is the time delay in the received pulse. The results of these measurements are also included in Table 1. Pulse shape changes and pulse lengthening effects, noted by others [8, 11], made meaningful direct measurements of V_g, for the situation described in entry No. 1 of Table 1 impossible. No hydrophone velocity measurements were attempted in the two thinnest cases because of the difficulties in obtaining and mounting long plates that thin. A summary of the experimental results, tabulated in Table 1, shows that almost all of the persistent "headwaves" were generated at angles within 2.0° less than the longitudinal and shear critical angles (13.5° and 28.5°,

Table 1 — Group Velocities of Observed Wavefronts

Entry No.	Plate Thickness	Incident and Radiated Angles Corresponds to Schlieren Photograph Summarized θ_i	θ_r	$V_g \times 10^5$ cm/sec from Eq. (1)	$V_g \times 10^5$ cm/sec Measured with Receiver
1	$1.47\lambda_p$	8.0°	60° ± 5°	$1.0 \begin{array}{l} +0.2 \\ -0.1 \end{array}$	...
2	$1.47\lambda_p$	12.5°	17° ± 2°	5.1 ± 0.6	4.8 ± 0.3
3	$1.47\lambda_p$	12.5°	27° ± 2°	$3.2 \begin{array}{l} +0.2 \\ -0.3 \end{array}$	2.9 ± 0.2
4	$1.47\lambda_p$	26.5°	30° ± 2°	3.0 ± 0.2	3.0 ± 0.2
5	$1.02\lambda_p$	12.0°	19° ± 2°	$4.5 \begin{array}{l} +0.5 \\ -0.3 \end{array}$	4.5 ± 0.3
6	$1.02\lambda_p$	13.4°	29° ± 2°	3.0 ± 0.2	3.0 ± 0.2
7	$1.02\lambda_p$	26.0°	29° ± 2°	3.0 ± 0.2	3.0 ± 0.2
8	$0.58\lambda_p$	13.0°	18° ± 2°	$4.8 \begin{array}{l} +0.6 \\ -0.5 \end{array}$	4.6 ± 0.3
9	$0.28\lambda_p$	13.0°	18° ± 2°	$4.8 \begin{array}{l} +0.6 \\ -0.5 \end{array}$	...
10	$0.023\lambda_p$	13°[a]	16°[a]	5[a]	...

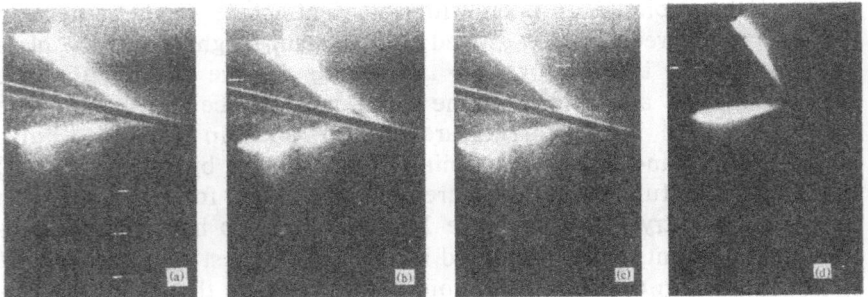

Fig. 9 — Schlieren photographs showing effects generated on a 0.130-cm-thick plate at (a) $\theta_i = 12.0°$, (b) $\theta_i = 12.5°$, (c) $\theta_i = 13.4°$, and (d) $\theta_i = 26.0°$. Pulse length is 5 μsec; frequency is 5 MHz.

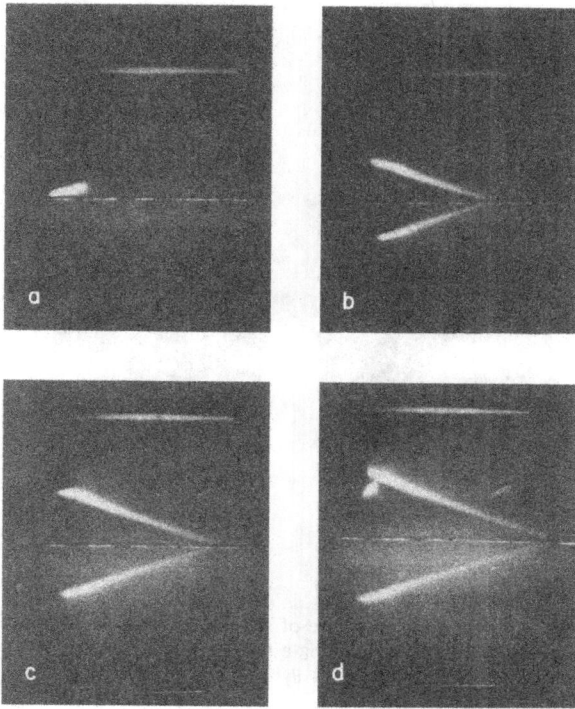

Fig. 10 — A time sequence of schlieren photographs showing the effect of illuminating a 0.051-cm-thick plate with an
incoming acoustic pulse at $\theta_i = 13.0°$. Pulse length is 3
μsec; frequency is 7.2 MHz.

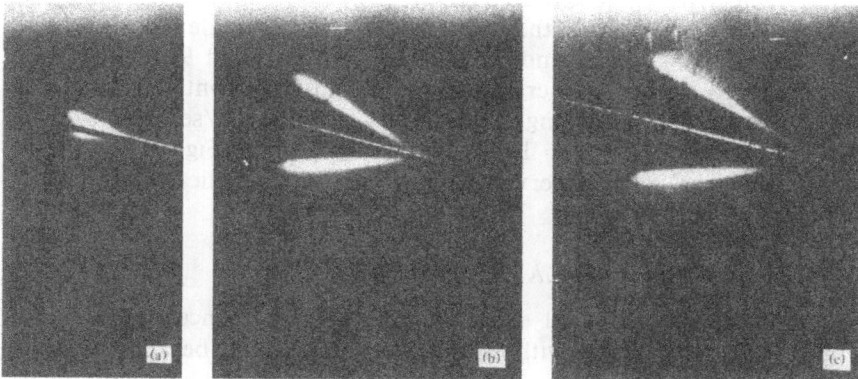

Fig. 11 — A time sequence of schlieren photographs showing the effect of illuminating a
0.025-cm-thick plate with an incoming acoustic pulse at $\theta_i = 13.0°$. Pulse length is 3
μsec; frequency is 7.2 MHz.

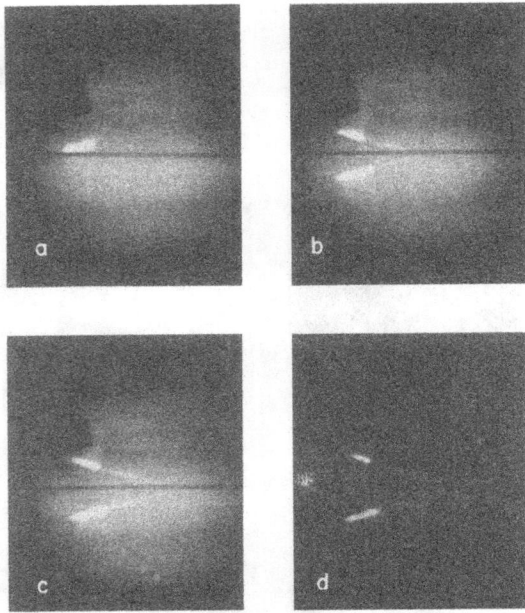

Fig. 12 — A time sequence of schlieren photographs showing the effect of illuminating a 0.002-cm-thick plate with an
incoming acoustic pulse at $\theta_l = 13.0°$. Pulse length is 3
μsec; frequency is 7.2 MHz.

respectively). Wavefronts with $V_g \approx 4.5 \times 10^5$ cm/sec were generated
at incidence angles within 1.5° less than the longitudinal critical angle
on plates whose thicknesses ranged from $0.023\lambda_p < D < 1.47\lambda_p$. The
effects are isolated in Figs. 4, 9(a), 10, 11, and 12. Wavefronts with
group velocity, $V_g \approx 3 \times 10^5$ cm/sec, were generated for incidence
angles within 2.5° less than the shear critical angle, on plates of
thicknesses $D = 1.47\lambda_p$ and $D = 1.02\lambda_p$ [Figs. 7 and 9(d)], but were
not observed on the thinner plates. A second wavefront generated near
the longitudinal critical angle with $V_g \approx 3 \times 10^5$ cm/sec was seen on
plates of thicknesses $D = 1.02\lambda_p$ and $D = 1.47\lambda_p$ [Figs. 5 and 9(c)].
No "headwaves" were observed between the shear critical and the Rayleigh angles.

COMPARISON WITH LAMB THEORY

Theoretical [12] and experimental [7, 8] evidence indicates that
Lamb theory for plates with stress-free surfaces can be applied to the
case of water-loaded plates. Lamb [6] derived two transcendental equations, which can be written in terms of the frequency-plate-thickness
product fD and the phase velocity V of the Lamb modes. The first of
these equations,

$$\frac{\tan h\{(\pi fD/V)[(V_s^2 - V^2)/V_s^2]^{1/2}\}}{\tan h\{(\pi fD/V)[(V_p^2 - V^2)/V_p^2]^{1/2}\}}$$

$$= \frac{4\{[(V_p^2 - V^2)/V_p^2]^{1/2} \cdot [(V_s^2 - V^2)/V_s^2]^{1/2}\}}{[(2V_s^2 - V^2)/V_s^2]^2}, \quad (2)$$

is satisfied if the plate motion is symmetrical with respect to a plane parallel to, and midway between, the plate surfaces. The second equation,

$$\frac{\tan h\{(\pi fD/V)[(V_s^2 - V^2)/V_s^2]^{1/2}\}}{\tan h\{(\pi fD/V)[(V_p^2 - V^2)/V_p^2]^{1/2}\}}$$

$$= \frac{[(2V_s^2 - V^2)/V_s^2]^2}{4\{[(V_p^2 - V^2)/V_p^2]^{1/2} \cdot [(V_s^2 - V^2)/V_s^2]^{1/2}\}}, \quad (3)$$

is satisfied when motion is asymmetrical with respect to this median plane. The number of propagating symmetrical modes possible at a given fD is determined by the number of real values of V which satisfy Eq. (2). Likewise, the number of propagating asymmetrical modes possible is determined by real values of V which satisfy Eq. (3) at a given fD value. Numerical solutions of Eqs. (2 and 3), in the form of curves relating phase velocity of Lamb modes to the frequency-plate-thickness product are presented for various materials in Refs. 7 and 8. A plot of the first four symmetric and asymmetric modes for 6061 aluminum was computed and is seen in Fig. 13.

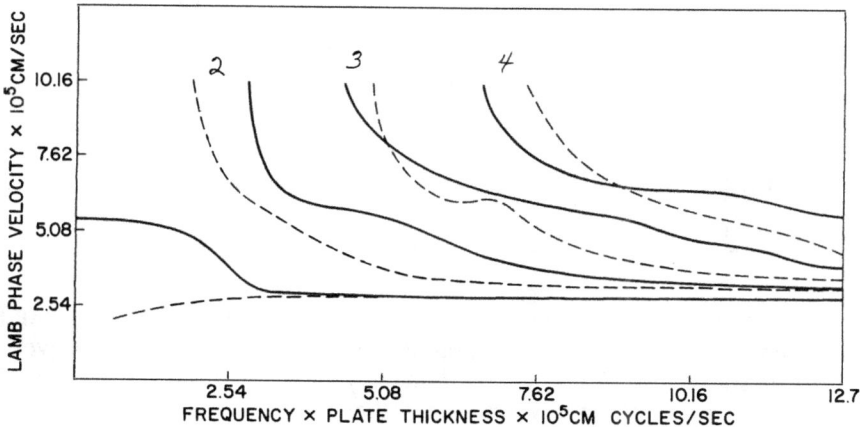

Fig. 13 — Lamb phase velocity vs fD for the first four symmetric (———)
and asymmetric modes (— —).

The angles of incidence at which the effects, seen in the schlieren photographs, are generated are compared in Table 2 to theoretically calculated incidence angles at which Lamb modes can be generated. The relationship between the angle of incidence θ_i and the phase velocity V, of a Lamb mode, is given by [7]:

$$V = V_w/\sin \theta_i. \tag{4}$$

Real values of V, satisfying Eqs. (2 and 3), are computed at the fD values corresponding to the experimental situations presented, and these are used in Eq. (4) to obtain the theoretical values of θ_i at which Lamb modes can be generated.

Table 2 — Comparison Between Incidence Angles at Which Effects are Strongly Observed and Angles Predicted by Lamb Theory

Entry No.	$fD \times 10^5$ cm Cycles/sec	θ_i Measured	θ_i Lamb Theory	Lamb Mode
1	9.4	8.0°	9.2°	5th symmetric
2	9.4	12.0°-12.5°	12.7°	4th symmetric
3	9.4	12.5°-13.2°	13.5°	4th asymmetric
4	9.4	26.5°	27.2°	2nd asymmetric
5	6.5	12.0°-12.5°	12.7°	3rd symmetric
6	6.5	12.5°-13.4°	14.4°	3rd asymmetric
7	6.5	26.0°	25.7°	2nd asymmetric
8	3.7	13.0°	13.5°	2nd symmetric
9	1.8	13.0°	17.0°	1st symmetric
10	0.14	13.0°	16.5°	1st symmetric

Many more Lamb modes are theoretically possible than were actually seen in the schlieren photographs for a given fD value; however, with this short pulse schlieren method, the presence of an effect is considered unequivocal only when the radiated wavefront progresses far enough down the plate so that its separation from specular reflection is undeniable. In most cases, identification of a mode which is highly attenuated (that is, radiates a great deal of energy into the water over a short distance) would be ambiguous using this short pulse method.

Though no radiation is involved in Lamb's theory, Worlton [7] made estimates of those modes that would be least attenuated due to radiation, based on the ratio of normal to horizontal particle motion at the plate surface for a specific mode. By this means, based on Lamb's theory, it is possible to estimate those modes that would propagate furthest along the plate at a given fD value. Calculations based on this method indicate that the wavefronts observed in the schlieren photographs are those expected to be least attenuated.

Table 2 shows good agreement between the experimentally observed results, and predicted angles of incidence for Lamb modes, when fD is at least 3.7×10^5 cm cycles/sec. The agreement is poorer for the two thinnest plates. The poorer agreement between experiment and prediction on the thinner plates suggests that water loading becomes a more significant factor for small values of fD where only the first symmetric and asymmetric Lamb modes are possible. Osborne and Hart [12] made exact calculations of the effect of water loading on phase velocity vs fD for the first symmetric Lamb mode on stainless steel plates. These calculations did not indicate as large a modification of phase velocity as seen experimentally at the small fD values. Their observations that, in general, water loading had little effect on Lamb modes were confirmed for sufficiently large fD.

Of special note are the effects summarized in entries 2, 5, 8, 9, and 10 of the Tables. These effects are similar in that they are all generated within 1.5° of the longitudinal critical angle, and have the same group velocity to within experimental error. Despite their similarity, they do not involve a single Lamb mode, but correspond to the fourth, third, second, and first symmetric Lamb modes depending on the plate thickness. Thus for aluminum plates, even as thin as $D = 0.02\lambda_p$, a strongly persistent wavefront with $V_g \approx 4.5 \times 10^5$ cm/sec, will be generated by an acoustic pulse incident within 1.5° less than the longitudinal critical angle. This phenomenon jumps Lamb modes depending on the plate thickness.

PATH TRACE COMPARISON

Path trace analyses for "headwaves" applied to the effects generated at angles of incidence slightly less than the shear critical angle [Figs. 7 and 9(d)] show that "headwave" position and angle can be accounted for by internal reflections of shear waves between the plate surfaces. Here group velocity is used in the same manner considered by Brillouin [13] in his discussions of guided waves in pipes. With the aid of Fig. 14(a), V_g can be calculated directly from the formula

$$V_g = V_s \sin \beta, \tag{5}$$

where β is the angle between the refracted, and subsequently reflected, shear ray and the normal to the plate surface. The formula is derived by noting from Fig. 14(a) that a refracted ray travels a path AC with velocity V_s, and that the component of V_s along the plate surface is $V_s \sin\beta$. The value obtained for V_g at an angle of incidence of 26.5° ($\beta = 70.4°$) is $V_g = 3.0 \times 10^5$ cm/sec, which is in excellent agreement with the schlieren observations of Fig. 7, and the direct hydrophone measurements, both of which gave $V_g = (3.0 \pm 0.2) \times 10^5$ cm/sec. The plate thickness in Fig. 7 was $3.0\lambda_s$, but owing to refraction a shear ray travels a distance $9.0\lambda_s$ between surfaces.

In Fig. 14(b), the radiated field above plates of thicknesses d and $2d$, illuminated at an incidence angle of 26.5°, is constructed. The upper surfaces of the two plates are colinear in the figure, and contributions to the radiated field above the plates from a ray incident at 26.5° at position A are determined. These contributions are calculated by path tracing of the internal shear reflections in the plates and the radiation into the water at each reflection of a shear ray from the plate surface. All path lengths, directions, and times of travel in the plate and in the water were calculated using Snell's law. Each time an internal shear reflection strikes the surface, sound energy enters the water at 26.5°, the original incidence angle. In the time taken for the internal shear reflections to reach position B, rays leaving the thinner plate have reached positions labeled ($\times$). Rays leaving the plate of thickness $2d$ have reached positions labeled ($\bigcirc$) in the same time. The fact that internal shear reflections in both plates reach position B at the same time is expected since the group velocity formula, Eq. (5), is independent of plate thickness. The radiated wavefronts from the two plates are observed to be colinear, and to make an angle of 30.0° with the plate surface. This is in agreement with the schlieren observation of Fig. 7 that, for $\theta_i = 26.5°$, $\theta_r = 30.0°$. Thus, a ray trace gives quantitative agreement with both group velocity and position of the wavefront observed in Fig. 7. From Eq. (5), it can be seen that the group velocity calculated from a path trace analysis does not depend on plate thickness. It depends on the refracted angle β which is determined from the original angle of incidence. While the group velocity is not dependent on plate thickness, Fig. 14(b) shows that the continuity of the radiated wavefront is. As seen in Fig. 14(b), the effect of doubling the plate thickness was to reduce to half the number of contributions to the radiated wavefront of the incident ray, over the same travel distance along the plate. If this description of the effect, seen at 26.5°, is physically correct, a combination of a plate sufficiently thick or an

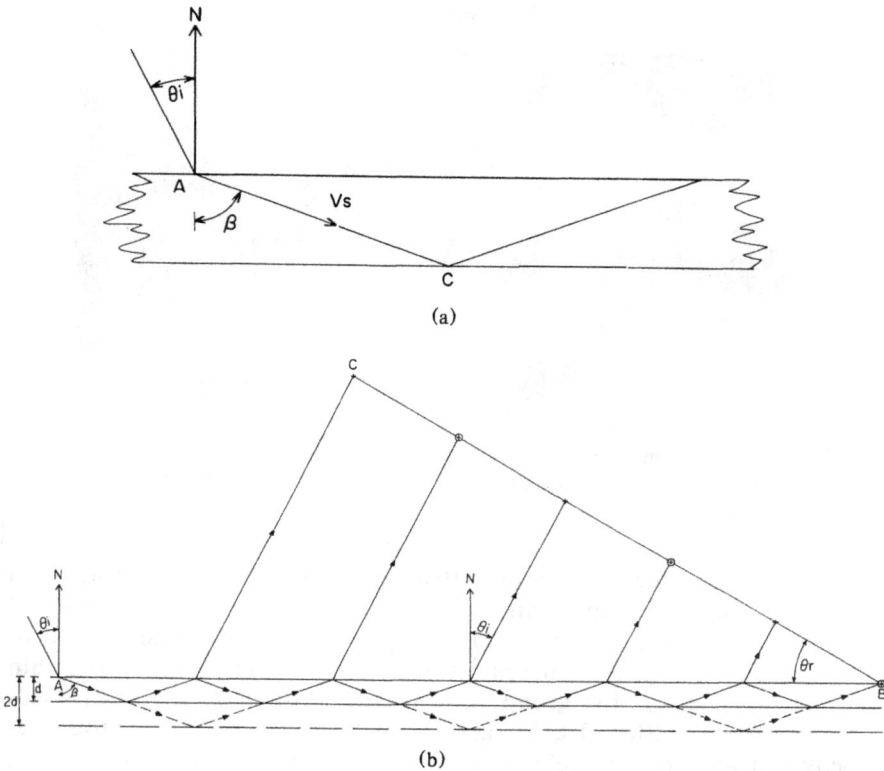

Fig. 14 — (a) Internal path taken by a guided wave consisting of all shear rays, from which group velocity is calculated. (b) Radiated wavefront above superimposed plates of thicknesses d and $2d$, in which a guided wave consisting of all shear paths is present.

incident plane wave sufficiently narrow, should allow gaps in the radiated wavefront to be observed. In Fig. 15, this effect is demonstrated by a resulting wavefront, which occurs when a 5-MHz pulse is incident at 26.5° on a plate 0.241-cm thick. Calculations of the expected spacings agree with the observations in Fig. 15 to within the accuracy with which the spacing of the gaps can be measured from the photograph. The wavefront angle and group velocity remain the same as those observed on the thinner plate at the same incidence angle. This plate is $1.9\lambda_p$ thick.

At angles of incidence where both shear and compressional waves are possible, the ray trace paths are not readily singled out. Radiated wavefronts made up of many different combinations of internal reflections may occur. For sound incident, water to aluminum, at

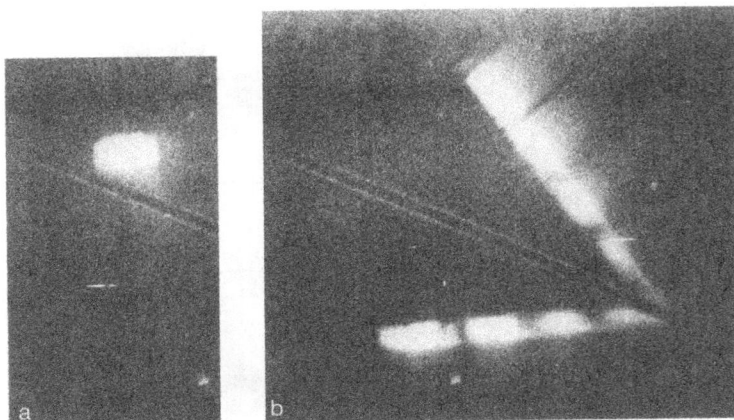

Fig. 15 — A time sequence of schlieren photographs showing the effect of illuminating a 0.241-cm-thick plate with an incoming acoustic pulse at $\theta_i =$ 26.5°. Pulse length is 5 μsec; frequency is 5 MHz.

angles within a few degrees less than the longitudinal critical angle, a guided wave consisting of an equal number of shear and longitudinal propagation paths in the plate appears probable from a consideration of the reflection coefficient curves [14-16]. Such a path predicts, to within experimental error, the group velocity and position of the effects isolated in Figs. 4, 9(a), 10, 11, and 12, which have the same group velocity but are not uniquely determined by a single Lamb mode. The other wavefront seen in the region of incidence near the longitudinal critical angle [Figs. 5 and 9(c), entries 3 and 6, Table 2] would have to be explained by a combination of internal reflections consisting of more shear than longitudinal paths. A path consisting of three times as many shear paths as longitudinal paths is needed to give the observed group velocity. There is, however, no obvious way of predicting *a priori* such a path description. The result is not, however, inconsistent with a path trace explanation. A propagation path consisting of all shear reflections, for the effect generated at 8.0° (Fig. 6), predicts a group velocity $V_g = 0.9 \times 10^5$ cm/sec and a wavefront angle $\theta_r = 64.7°$. This compares favorably with the results seen in Table 1 (entry No. 1).

COMPARISONS WITH CYLINDRICAL SHELLS

Comparisons between the plate result and results on cylindrical shells [1] will be limited to shells with $b/a = 0.9$ for reasons discussed below. A refracted ray leaving one surface in a plate will strike the other surface at the refracted angle. For a cylindrical shell this is not true, and in fact a refracted ray can miss the inner surface entirely, taking chord paths from outer surface to outer surface. Radiation from energy taking chord paths predominates for shells thicker than those with an inner to outer radius ratio, b/a, of 0.9 [1].

A path trace analysis was strongly suggested by the reflection coefficient curves for a water-aluminum boundary in two regions of incidence on aluminum plates. The group velocity of the wavefronts generated near the shear critical angle were described by a path involving only shear reflections (called S-S), and the faster wavefronts generated near the longitudinal critical angle had group velocities described by a path consisting of an equal number of shear and longitudinal segments (called L-S). The L-S and S-S path descriptions also described the two radiated wavefronts observed on cylindrical shells [1], which were also generated near the longitudinal and shear critical angles. Of the two wavefronts observed on cylindrical shells, only the wavefront generated near the longitudinal critical angle persisted to shell thicknesses much smaller than a wavelength, and the L-S path continued to describe the position and group velocity of the observation. This is in agreement with the results on very thin plates presented in Figs. 11 and 12, where only one wavefront persists.

A bent plate with both flat and curved portions was used to determine the persistence of the wavefronts generated on plates when they encountered a curved surface. The effects seen in Figs. 4, 5, and 7 were generated on the flat portion of a bent plate of the same thickness. The persistence of the wavefronts around the curved portion of the plate is seen in Fig. 16. Figure 16(a) is included for reference to show the size and approximate position of the incoming pulse. In Figs. 16(b) and 16(c), respectively, the effects generated at incidence angles near the longitudinal and shear critical angles, and described by L-S and S-S paths, are observed after they have reached the curved portion of the bent plate. Both are seen to have persisted strongly around the curvature. In Fig. 16(d), the second wavefront generated near the longitudinal critical angle (corresponding to entries 3 and 6, Table 2) is also seen to persist around the curved surface, but is more highly attenuated as seen in the fluid above the plate.

Neubauer and Dragonette [1] previously pointed out that the two predominant guided wavefronts observed on cylindrical shells could not be related to only the first symmetric or first asymmetric Lamb mode. Comparisons between the experimental results on cylindrical shells and the results presented here on plates indicate that the faster of the two circumferential waves seen on shells corresponds to the plate effect

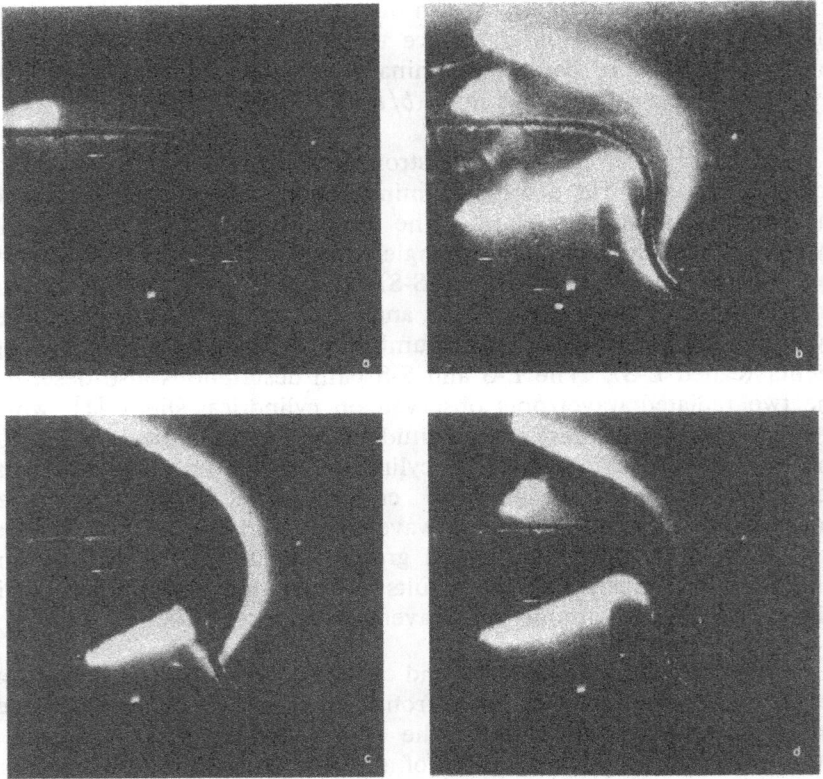

Fig. 16 — Schlieren photographs showing the effect of plate curvature on the persistence of three of the waves generated on a 0.130-cm-thick plate. The pulse length is 3 μsec; the frequency is 7.2 MHz. (a) An incident pulse; (b) The wavefront isolated at 12.0°; (c) The wavefront maximized at 26.5°; (d) The wavefront isolated at 13.2°.

which jumps symmetric modes as thickness is increased, and the slower circumferential wave corresponds most closely to the second asymmetric mode.

CONCLUSION

Radiated wavefronts were observed when aluminum plates were illuminated at specific incidence angles by short acoustic pulses in water. The effects least attenuated by radiation into the water occurred at incidence angles within a few degrees less than the longitudinal and shear critical angles. The propagation of these "headwaves" along a plate has not been previously reported and would be difficult to describe with hydrophone measurements alone. With a single schlieren photograph, it is possible to determine both the group velocity along the plate of the effects generated, and the trace velocity, even when these are quite different. The measured angle of inclination between the radiated wavefronts and the plate surface was related to the group velocity of the wavefront along the plate by a formula derived independently of the type of motion within the plate.

Comparisons between the incidence angles at which the effects were generated and incidence angles calculated from Lamb theory for unloaded plates showed good agreement for plates as thin as $0.58\lambda_p$. Agreement between Lamb theory and experiment became poorer when plate thickness was reduced below $0.3\lambda_p$, suggesting that water loading was becoming more significant. There are more Lamb modes theoretically possible than there were effects observed at any given plate thickness. Only those effects which persisted far enough along the plate so that specular reflection was avoided were considered by this method. Estimates of the attenuation following a procedure used by Worlton indicated that the modes observed were those that would be least attenuated by radiation into the water. Those effects on plates whose position and velocity were describable by path tracing corresponded to the two circumferential waves found on cylindrical shells of $b/a \geqslant 0.9$ [1]. The faster circumferential wave on cylindrical shells is similar to the plate effect, generated near the longitudinal critical angle, which jumps modes as thickness is increased. The slower cirumferential wave on shells is most closely associated with the second asymmetric Lamb mode, generated near the shear critical angle.

REFERENCES

1. W.G. Neubauer and L.R. Dragonette, J. Acoust. Soc. Am. **48**, 1135-1149 (1970).

2. C.W. Horton, W.R. King, and K.J. Dierks, J. Acoust. Soc. Am. **34**, 1929-1932 (1962).

3. K.J. Dierks, T.G. Goldsberry, and C.W. Horton, J. Acoust. Soc. Am. **35**, 59-64 (1963).

4. W.G. Neubauer, J. Acoust. Soc. Am. **44**, 1150-1152 (1968).

5. W.G. Neubauer, J. Acoust. Soc. Am. **45**, 1134-1144 (1969).

6. H. Lamb, Proc. Roy. Soc. (London) **93**, 114-128 (1917).

7.   D.C. Worlton, J. Appl. Phys. **32**, 967-971 (1961).

8.   D.C. Worlton, Hanford Atomic Products Operation Rep. No. HW-60662 (1959).

9.   W.G. Neubauer and L.R. Dragonette, J. Acoust. Soc. Am. **49**, 410-411 (1971).

10.   W.G. Neubauer and L.R. Dragonette, J. Acoust. Soc. Am. **36**, 1685-1690 (1964).

11.   T.N. Grisby and E.J. Tajchman, IRE Trans. **UE-8**, 26-33 (1961).

12.   M.F.M. Osborne and S.D. Hart, J. Acoust. Soc. Am. **17**, 1-18 (1945).

13.   L. Brillouin, *Wave Propagation and Group Velocity* (Academic New York, 1960), pp. 144-148.

14.   W.G. Mayer, J. Appl. Phys. **34**, 909-911 (1963).

15.   W.G. Mayer, J. Appl. Phy. **34**, 3286-3290 (1963).

16.   R.E. Bunney, "Circumferential Waves on Cylinders," PhD dissertation, Dept. Phys., Colorado State Univ. (1968).

Chapter 21

SCHLIEREN VISUALIZATION TO DETECT FLAWS ON PLATES*

INTRODUCTION

The use of Lamb [1] waves for nondestructive testing applications has been the subject of numerous papers [2-5]. Generally these applications involve the hydrophone reception of a Lamb mode or a Lamb wave reflected from a flaw.

Surface waves may be launched on a plate immersed in a fluid by illuminating the plate at specific angles of incidence with short acoustic pulses. A wave thus launched propagates along the plate and reradiates energy back into the fluid as it propagates. The reradiated energy leaves the plate at an exit angle equal to the original angle of incidence. This reradiated wave is called a "headwave," and the detection of a "headwave," in a region beyond that where normal specular reflection occurs, serves as an indicator for the presence of a surface wave.

Schlieren visualization of "headwaves" generated when thin aluminum plates in water were illuminated by short acoustic pulses were presented in Chapter 19. "Headwaves" were observed when the plates were insonified at specific angles of incidence, and their properties compared with predictions of Lamb theory. In addition, the position and group velocities of the "headwaves" were described in terms of a path trace technique. Those waves not strongly attenuated by radiation into the water are of particular interest in nondestructive test applications. Such waves are found to be generated by acoustic pulses incident within 2° less than the longitudinal and shear critical angles on plates made of metals and glasses. The angle at which the "headwaves" are generated, and their phase and group velocities, can be predicted by the known elastic properties of the materials, and the distortion of these "headwaves" when flaws are present in the material is easily observable in real time by the schlieren method.

---

*Some of this material first appeared in "Materials Evaluation," Oct. 1974.

EXPERIMENT

The schlieren system used is fully described in Chapter 22. The sound source used is a lead zirconium titanate transducer pulsed at a center frequency of 6.6 MHz. The active face of the transducer has a diameter of 0.75 in. The plate material in all of the photographs that will be shown is 6061 aluminum, which has a compressional wave speed of 6370 m/sec and a shear wave speed of 3136 m/sec. The longitudinal and shear critical angles are 13.5° and 28.2°, respectively. The plates are supported at each end by brass rods. Both the source and the plates can be accurately rotated. The radiated sound field in the water above and below the plates will be visible.

Figure 1 shows the "headwave" which results when an aluminum (6061) plate 0.130-cm thick is illuminated by a 5-μs-long acoustic pulse at an incident angle of 12.0°. At the center frequency of 6.6 MHz, the plate thickness, d, is 1.5 λ_p, where λ_p is the compressional wavelength in aluminum. The incident pulse is seen in Fig. 2(a), and the successive frames of Fig. 1 show the generation and propagation of a "headwave" along the upper and lower surface of the plate at later times. Separation in time between the frames is obtained by varying the time delay between the triggering of the sound source and the light source. In this manner the sound pulse and its effects can be frozen and observed at any instant, from the time the sound pulse leaves the transducer until it and its effects attenuate below the dynamic range of the system. Observations are made in real time on a TV monitor, and the photographs seen here were made from the TV screen. The wave seen in Fig. 1 was generated within 2° less than the longitudinal critical angle on aluminum plates as thin as $d = 0.023 \lambda_p$. The phase velocity, V, of the plate mode seen in Fig. 1 is given by:

$$V = V_w/\sin \theta_i \tag{1}$$

where θ_i is the angle of incidence of the sound wave, and V_w is the velocity of sound in water. The group velocity, V_g, of the wave is given by [6]

$$V_g = V_w \frac{\cos (\theta_i - \theta_r)}{\sin \theta_r} \tag{2}$$

where θ_r is the angle of intersection of the plate surface and "headwave" as seen in the photograph. The group velocity of this "headwave" was measured to be $V_g \approx 4.5 \times 10^5$ cm/sec. This velocity can be predicted by assuming that the observed radiation results from a guided wave in the plate which takes an equal number of shear and longitudinal paths between the plate surfaces. This type of guided wave

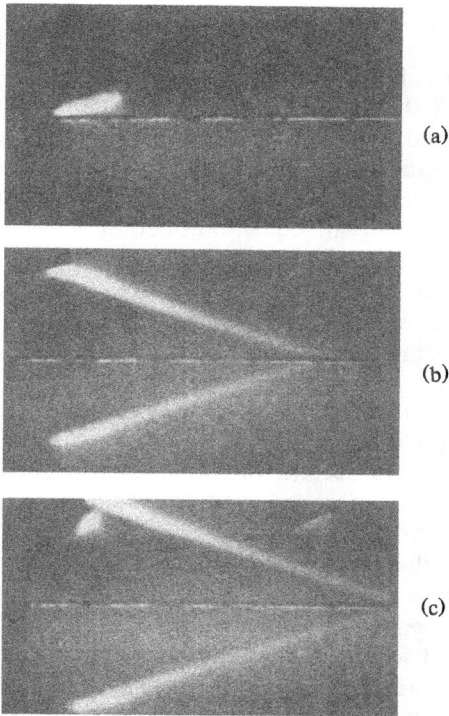

Fig. 1 — A time sequence showing the generation and propagation of a "headwave" on a 0.130-cm-thick plate when sound is incident near the longitudinal critical angle.

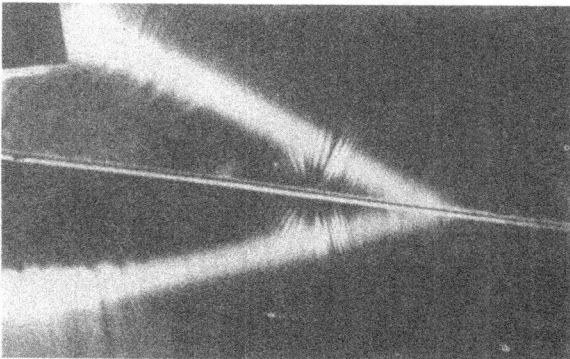

Fig. 2 — The distortion of the "headwave" seen in Fig. 2 by a 0.025-cm (deep and wide) scratch on the upper surface of a 0.130-cm-thick plate.

is consistent with reflection coefficient calculations on aluminum, and with previously shown results on cylindrical shells [6]. This "headwave" was generated on all aluminum samples studied. These samples ranged in thickness from $0.023\lambda_p \leqslant d \leqslant 2\lambda_p$, and the "headwaves" had the same properties on all the samples, i.e., they were generated within 2° less than the longitudinal critical angle. Their group velocity was $V_g \approx 4.5 \times 10^5$ cm/sec, and their explanation was consistent with a guided path consisting of an equal number of shear and longitudinal legs between the plate surfaces. The strong generation of a guided wave near the longitudinal critical angle is not limited to aluminum. Such waves were generated on samples of nickel, armco iron, inconel, stainless steel, and other metals and glasses.

Figure 2 shows the distortion which has occurred in a "headwave" generated under conditions similar to those in Fig. 1, except that the plate had a 0.025 cm (deep and wide) flaw on its upper surface. The radiated wavefront has proceeded past the position of the flaw, and the distortion the wavefront is easily observed. The distortion obtained from a scratch on the bottom surface of a 0.050-cm-thick aluminum plate is seen in Fig. 3. The flaw is 0.012-cm deep and wide. The plate thickness here is $d = 0.6\lambda_p$, and the "headwave" has the same properties and is generated within 1.5° of the longitudinal critical angle under the same conditions (except for plate thickness) as that seen in Figs. 1 and 2. The distortion of the wavefront begins when the flaw is initially encountered in frame (4c) and clearly identifies the presence of the flaw.

The plate seen in Fig. 4 is 0.130-cm thick and has a 0.040-cm-diam hole drilled 0.254-cm deep into its side between its surfaces. This size hole was chosen since this was the smallest drill easily available. The encounter of the "headwave," seen in Figs. 1-3, with a flaw between the plate surfaces is seen in Fig. 4 to distort the "headwave" in a similar manner.

Within 2° less than the shear critical angle, a strongly persistent "headwave" with a group velocity $V_g \approx 3.0 \times 10^5$ cm/sec was generated on aluminum plates as thin as $d = 1.0\lambda_p$, and was not observed on thinner plates. This wave was also generated on the other metals previously mentioned and glasses. The group velocity observed can be predicted by assuming a guided wave consisting of multiple shear reflections between the surfaces. The generation of this "shear headwave" on a 0.130-cm plate is seen in Fig. 5. The interactions of the wavefront, seen in Fig. 5, with flaws are similar to the effects seen in Figs. 2-4.

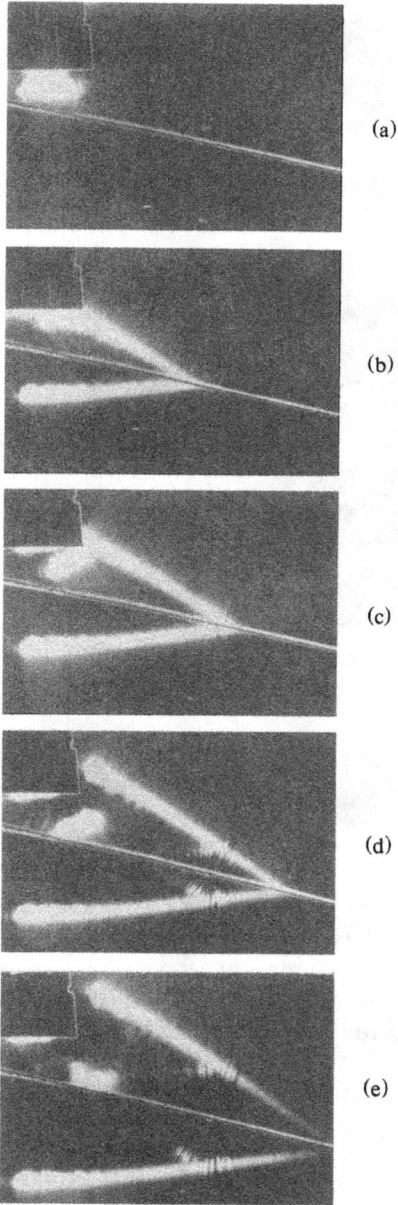

Fig. 3 — The distortion of a "headwave" by a 0.012-cm (deep and wide) scratch on the bottom surface of a 0.051-cm-thick plate.

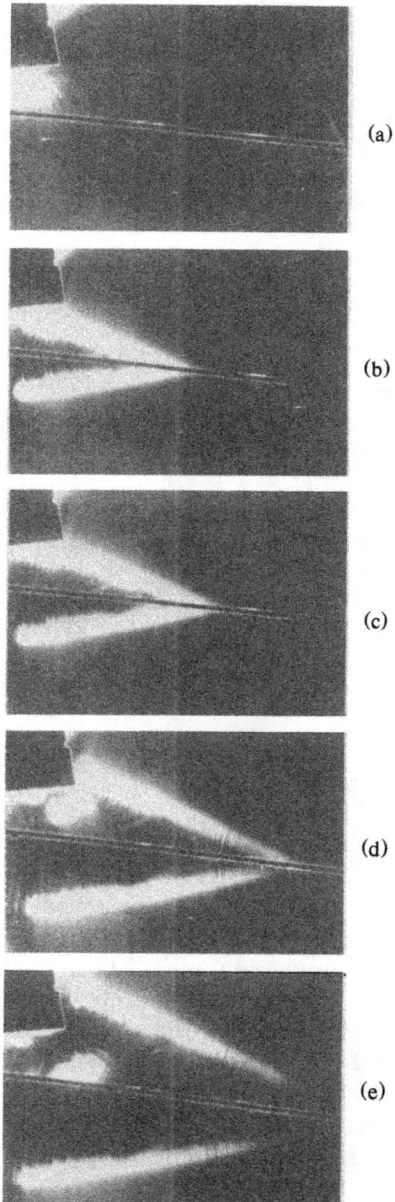

Fig. 4 — The interaction of a "headwave" with a flaw between the surfaces of a 0.130-cm-thick plate.

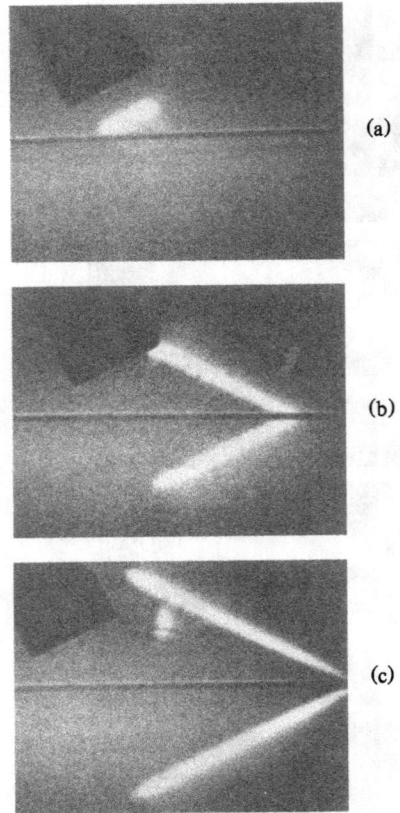

(a)

Fig. 5 — The generation of a
"headwave" by a sound pulse incident
near the shear critical angle on a 0.130-
cm-thick plate.

(b)

(c)

When a flaw is contained within the incident sound beam, detection by schlieren visualization is immediately obvious. Figure 6 shows
the illumination of a 0.005-cm-deep scratch on the top surface of a
0.130-cm-thick plate. The angle of incidence is 27.5°. Frame (a)
shows the incident pulse. In frame (b) the near shear angle "headwave"
has been generated, and its progression along the plate as well as its
reflection by the flaw in the opposite direction are observed. A semicircular pattern is seen in frames (c) and (d). This is caused by the flaw
which is at the center of the semicircle. With a flaw on the bottom surface of a plate, the semicircular pattern appears below the bottom surface, as seen in Fig. 7. Here the flaw is a 0.025-cm (deep and wide)
scratch on the bottom surface of a 0.130-cm-thick plate. The center
frequency is 7.2 MHz; the angle of incidence is 27.5°. Frame (a) shows
the generation and reflection of the near shear angle "headwave," and
the subsequent frames show clearly the semicircular pattern formed

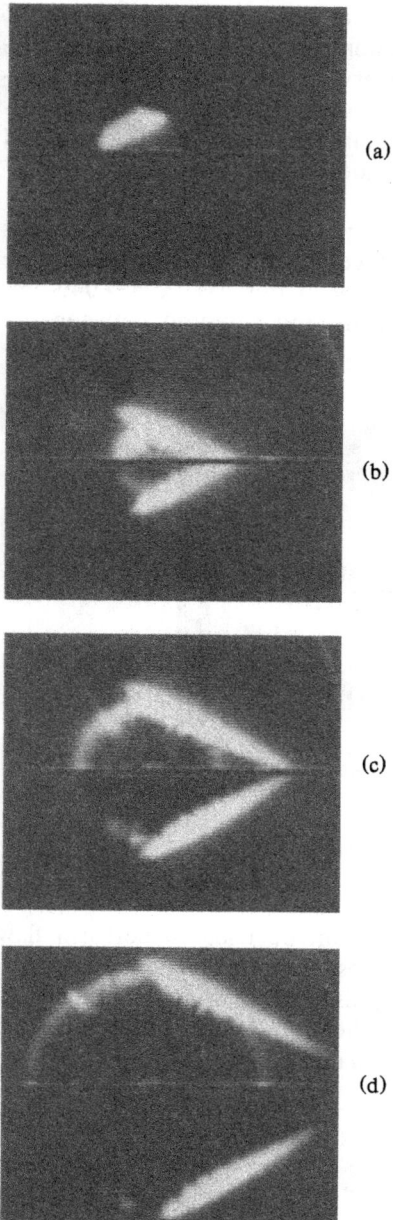

(a)

(b)

(c)

(d)

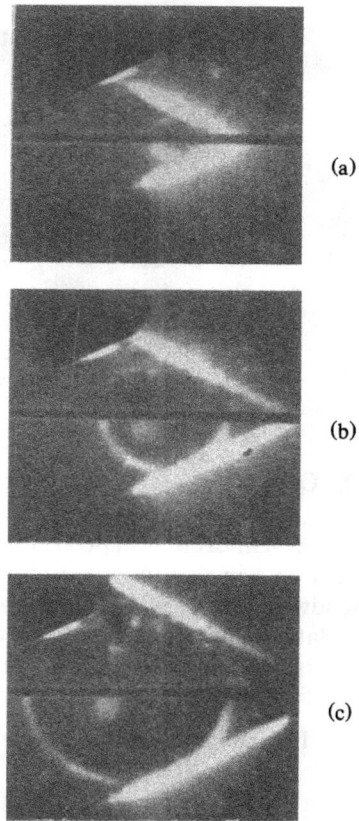

(a)

(b)

(c)

Fig. 6 — The observation of effects, caused by a surface flaw, contained within the incident beam. The plate is 0.130-cm thick; the scratch is 0.025-cm deep and wide.

Fig. 7 — An observation under conditions similar to Fig. 7, with the flaw on the nonilluminated side of the plate.

below the bottom plate surface. This effect is not limited to angles of incidence at which "headwaves" are generated. The semicircular patterns seen in Figs. 6 and 7 are easily observed and separated from specular reflection at all angles of incidence from 5° through the shear critical angle, the only necessary criterion being that the flaw is contained within the physical dimensions of the incident beam.

The distorted "headwave" seen in Fig. 8 is characteristic of results obtained when a flaw between the surfaces of the plate is within the dimensions of the incident beam. The flaw in this case is a 0.040-cm-diam hole drilled into the side of the plate and centered between the plate surfaces.

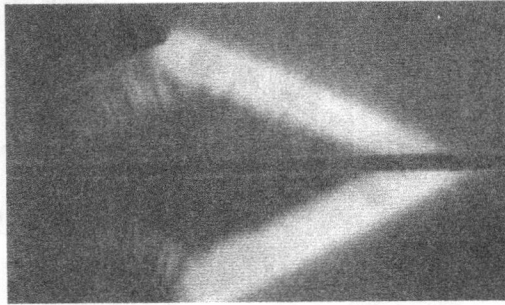

Fig. 8 — The pattern obtained when a flaw between the plate surfaces is contained within the incident beam.

CONCLUSION

A schlieren visualization technique can be used to perform flaw detection inspections on flat samples of various thicknesses. "Headwaves" are generated by acoustic pulses incident at specific angles on plates in water. Distortion of these "headwaves" in the presence of flaws is directly observed by schlieren visualization.

The advantages of a schlieren technique are:

1. Detection of energy scattered by a flaw is not limited in direction by a receiver. Scattered energy in all directions perpendicular to the light source can be accomplished with the schlieren.

2. The existence of the flaw and its position along the surface are visually observed in real time.

3. A flaw not large enough to generate a significant reflected echo may still distort the radiated "headwave" and be detectable.

REFERENCES

1. Horace Lamb, "On Waves in an Elastic Plate," *Proceedings of the Royal Society (London)*, **93**, 114 (1917).

2. D.C. Worlton, J. Appl. Phys. **32**, 967 (1961).

3. T.N. Grisby and E.J. Tajchmam, IRE Transactions on Ultrasonics Engineering, **UE-8**, 26 (1961).

4. C.L. Frederick and D.C. Worlton, *Nondestructive Testing (Materials Evaluation)*, **20**, 51 (1962).

5. M. Luukkala and P. Meriläinen, Ultrasonics, **11**, 218 (1973).

6. W.G. Neubauer and L.R. Dragonette, J. Acoust. Soc. Am. **48**, 1135 (1970).

Chapter 22

EXPERIMENTAL FACILITIES

INTRODUCTION

Throughout this volume various experimental measurements have been presented; mostly acoustic field measurements of pulsed sine waves in water and air, and schlieren visualization of pulsed acoustic fields. In some cases, detailed experimental procedures were described, but there has been only little physical description of the facilities suitable for performing the experiments. Of course, a variety of equipment had been used over the period of almost 20 years that this work spans. It would serve little purpose to describe old outdated equipment and facilities, so mostly recent facilities will be described. In the case of acoustic pool facilities the new ones are far more elaborate and capable than earlier facilities because of the direct connection to a computer. However, this advanced system is also suitable for measurements that were carried out in earlier simpler systems.

GENERAL CONSIDERATIONS

The measurement of acoustic reflection in a closed body of water or an air enclosure, such as a tank or room, is always encumbered with one ever present factor: reflection from the boundaries (tank walls and water-air interface or room walls). Additionally, in a water volume, difficulties, sometimes unexplainable, with electrical grounding seem omnipresent. The latter becomes a constant enemy conspiring to obscure measurement values. It often seemed that the greatest care had been taken to eliminate ground loops one day so that signals were adequately noise free and the next day a signal is found walking through the trace on a bed of grass offering a new grounding challenge. It has happened that a hydrophone immersed in a water tank was picking up aircraft pilot conversations from the airport across the river. The lesson, surely a common rule in other areas of physical measurement is — while the system is working well, take all the data possible.

Some experimenters have sought to make enclosures that absorb the energy incident on them. Such anechoic treatments of boundaries are very common and quite successful in air. In water, in addition to the significant expense, anechoic linings are relatively narrowband and often have undesirable directional properties. Reflection measurements such as those for spheres in Chapters 7 and 8 demand a very low electrical background for the signal. In that case uniform (plane wave) insonification is required and a far field measurement is sought. To achieve a reasonably plane incident wave, usually a large distance is required between source and reflector, so that a $1/r$ diminution with distance of acoustic intensity is suffered. On reflection, the same or a similar distance is required resulting in an additional $1/r$ diminution of the intensity with range. The reflection results from the interception of only a small portion of the source field. For a sphere and most other double curved objects the absolute reflection is not large. A representative example will demonstrate the magnitude of signal involved. For a sphere (see Chapter 7),

$$p_r = \frac{a}{2r} \, p_0 |f_\infty|, \tag{1}$$

where p_0 is the incident pressure at the position of the center of the sphere. Take an average of unity for $|f_\infty|$ which is the high-frequency rigid-sphere value. Consider a sphere with radius a, and impose a 10-diam range $r = 10(2a)$) for the receiver to satisfy far field conditions. Now

$$\frac{p_r}{p_0} = \frac{a}{2(20a)} = 0.025, \tag{2}$$

or the reflection is 32 dB below the incident pressure level. It can be seen that a 30 dB echo reduction tank-wall coating would cause serious competition for reflection from a sphere. Of course, some reflectors return larger signals and some smaller signals and even an elastic sphere can return significantly higher signals at some frequencies. However, to achieve the measurement of small magnitudes of $|f_\infty|$ such as those in Fig. 4 of Chapter 7, every dB of receiver preamplifier gain and reduction in background noise that can be garnered are required.

It may seem that as far as the enclosure (tank, pool, etc.) is concerned, bigger is better. As a practical matter this turns out not to be the case. With larger size several things become a great deal more difficult. First, there is the matter of maintenance of the water volume. That medium is ideally made to be of uniform temperature, still, quiet, and constant in these and other properties. If optical sighting is required, such as for distance measurements, optical clarity is also necessary. Of course, the frequency range and pulse lengths used in the experiments are a vital consideration, with larger volumes required for lower frequencies. At all frequencies the shortest achievable pulse lengths in water or the actual physical size of reflector are usually the major limiting factors for tank size.

Estimates of tank size required for specific conditions result from a complicated interaction of many parameters. Limitations of tank size resulting from reflection interference from walls can be generalized to some extent and used as a basis for judgment about tank size. When source and receiver are equidistant from a reflecting surface, the modification of the received, directly transmitted pulse by energy which has suffered a reflection and is at least in part received simultaneously with the direct pulse, may be predicted by geometrical considerations. Consider a sound source located at S_1 (Fig. 1) and a receiver at S_2, both at a distance d from the nearest reflecting surface, i.e., the water-air interface or a tank wall. The interference with the direct pulse takes place when acoustic energy simultaneously arrives via the path $S_1 a S_2$ and the path $S_1 b S_2$. This is the "Lloyd mirror effect." Let

t_1 = transit time of pulse via path $S_1 b S_2$,

t_2 = transit time of pulse via path $S_1 a S_2$, and

t_p = duration of pulse.

To avoid interference,

$$t_2 \geqslant t_1 + t_p.$$

Since

$$t_1 = r/c$$

and

$$t_2 = 2q/c, \quad \text{where } q = \sqrt{d^2 + r^2/4},$$
$$2\sqrt{d^2 + r^2/4} = r + ct_p$$

or

$$r \leqslant \frac{2d^2}{ct_p} - \frac{ct_p}{2}.$$

Thus r is the maximum separation of S_1 and S_2 for a given depth and pulse duration where no interference will occur.

For the case where S_1 and S_2 are not equidistant from the nearest surface, an interference relation can be derived in which the point of

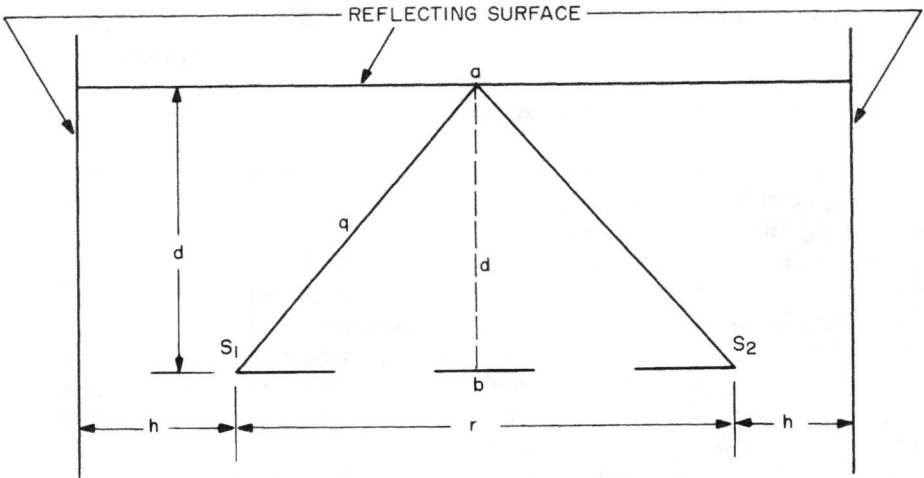

Fig. 1 — Geometry for reflection interference from side-wall with
source and receiver equidistance from the reflecting surface.

reflection must be experimentally determined. This becomes unnecessarily complicated and makes it advisable in most cases to maintain both S_1 and S_2 as close to equidistant as possible from a single surface so that this distance is smaller than that to any other reflecting surface parallel to it.

Reflection from the plane surface behind the source and normal to the line between source and receiver can interfere with the directly transmitted pulse at the receiver. To avoid this interference, the following relationship between the distance h from the source to the surface should be maintained

$$h > \frac{ct_p}{2}.$$

Reflection from the surface behind the receiver is also capable of interfering with the directly transmitted pulse; so the above relationship should also be maintained where h is the distance from receiver to the surface behind it. These considerations will be true for a perfectly square pulse emitted by the source, reflected, and propagated in the water and also for the case of measurement on a pulse at some specific fixed time along its length. To assume interference for no part of the pulse, the effective pulse lengthening due to the finite bandwidth of the source, medium, reflector, and receiver must be taken into account. The number of cycles needed to allow the pulse that is introduced into the water to reach a desired percentage of its steady-state value can be determined from the quality factor Q of the source. The degree of interference between the direct and reflected pulse will of course be a function of the relative magnitudes of energy emitted in the direction of the source and reflector. For a highly directional source and receiver, experimentation in an interference region may be permissible to the extent that the reflected signal is smaller to a desired degree than the directly received signal.

The transient pulse analysis of reflection data described in Chapter 8 affords a significant advantage over long pulse experiments by using a pulse as short as possible to maximize bandwidth and maximize also the free volume in which interference-free measurements can be made.

TANKS AND POOL DESCRIPTIONS

One tank of a size that has proven useful had a capacity of approximately 250 gals which was built to the normal standards of a fish aquarium of that size. Its dimensions were 76 cm × 183 cm × 76 cm deep. A frame was placed on top of the tank which allowed the positioning of acoustic sources, receivers, and reflectors. A side view of the tank showing the frame and the positioning of transducers and cylinders in the water is shown in Fig. 2. Smaller glass-walled tanks have been used for reflected-pulse hydrophone measurements and for schlieren visualization experiments. The relatively small glass-sided tank was used for measuring the circumferential-pulse sequences of Chapter 3. No pulses reflecting from the side walls interfered with the desired pulses in most cases. If and when such wall pulses did exist, they were identified and avoided for very limited experimental configurations. The glass tank was too small to allow a sufficient distance between the cylinder and hydrophone at all angles around the cylinder so scattering-cross-section experiments were done in a larger tank. A photograph of that larger tank is shown in Fig. 3. The tank has cypress walls and bottom normally 7.6-cm thick and contained a water volume 152 cm × 305 cm × 152 cm deep. Location of acoustical elements in a volume 1-m square by 10-cm deep in the center of the tank's volume was accomplished optically with microalignment telescopes mounted at right angles to each other on two adjacent sides of the tank (see Fig. 3) and a vertical cathetometer that viewed through a porthole in one end of the tank. Elements were placed in the tank by a system of tracks, rails, and carriages mounted on top of the tank. The wooden walls of the tank into which water soaked to a certain depth afforded some reduction in acoustic reflection from them depending on the degree to which water soaked into the wood. This occurred presumably because of the gradual impedance change over the wood thickness.

Fig. 2 — A side view of the glass-walled tank showing the framework
which supported the transducers and diffracting cylinder.

Fig. 3 — A view of the wooden tank showing placement and
location equipment on top of the tank.

A much larger wooden tank was also used. A photograph of the work area showing the tank protruding above the deck around the tank is seen in Fig. 4. This redwood tank is set in a hole in the ground and has dimensions of 9-m diam and 6-m deep. Typical swimming-pool water-maintenance techniques were used to keep the water pure.

Fig. 4 — The work area around the top of the 30-foot-diam redwood tank.

The pool of a pool-type research nuclear reactor became available after the reactor was deactivated at NRL in 1970. The pool is approximately 8 m × 10 m × 6 m deep at the shallowest point. A photograph of the pool area at NRL is shown in Fig. 5. Sources, receivers, and reflectors are suspended from bridges that span the pool (see Fig. 6). The bridges travel the length of the pool on rails that are flat and level to better than 1 mm but, are seldom moved during the course of an experiment. The pool contains approximately 150,000 gal of filtered and deionized water. The water purification system was originally designed and used with the reactor to remove from the water moderator all particles capable of becoming radioactive. Therefore, the water is very clear and has an extremely low electrical conductivity. This is often a distinct advantage in that electrical leakage paths are minimized, especially if transducers lose their watertight integrity. The purity of the water also helps retardation of corrosion of immersed metals.

The incident acoustic pulse is either measured directly or a replica of the incident pulse is obtained by the sphere calibration procedure

Fig. 5 — The large pool used for acoustic reflection
and scattering measurments.

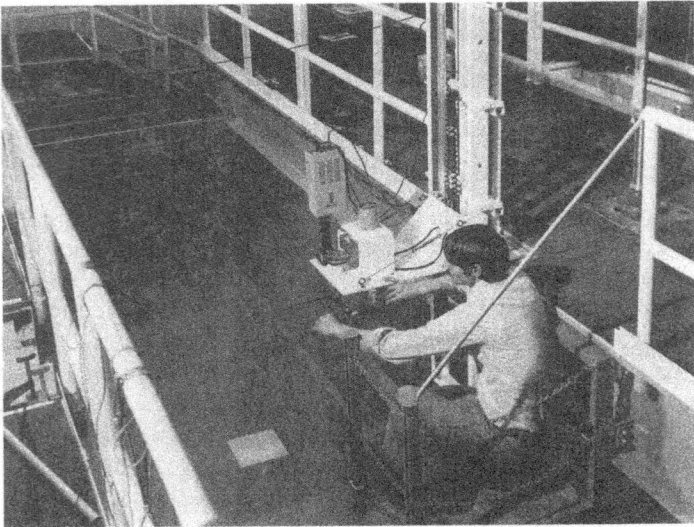

Fig 6 — The target suspension platform attached to one bridge
that spans the large acoustic pool.

described in Chapter 16. Alignment and relative aspect of source, receiver, and target center are usually, but not necessarily, fixed in a horizontal plane. The target is suspended from a stepping motor assembly so that it can be rotated about its center. Target aspect angle is varied by the stepping motor which are controlled by the computer. The resolution of the motor that rotates the target is 1/30th of 1°. Measurements of reflection amplitude vs target aspect are routinely acquired in the pool at 1/2° intervals over 360°. After the target has been rotated by the stepping motor, a waiting period is imposed before data collection to permit damping of any oscillation of the target in its suspension. Care has been taken to design a suspension system that minimizes such oscillations. The support members for targets are usually 0.076-mm (0.003-in.)-diam-tungsten wires.

Data collection is performed by the computer through a Biomation model 8100 transient recorder. The entire data collection system is shown diagrammatically in Fig. 7. The computer-operation area is shown in Fig. 8. The transient recorder digitizes incoming signals into 2's complement 8 bit data (7 bits plus sign). The computer can adjust the input gain of the digitizer so that each return to be measured is digitized with the maximum dynamic range possible for that particular signal. The range of adjustment is from 0.05 volts to 50 volts for full scale digitization giving a 40 dB range in which individual echoes can vary and still be sampled with the same relative accuracy. The recorder digitizes 2048 samples of the incoming signal at a rate set either by the operator or the computer. Sampling rates may be varied from one sample every 10 sec to one sample every 10 nsec. The sampling rate used depends on the frequency content and time duration of the signals. The digitizing process begins on a trigger signal which can be delayed by some integral number of sample intervals. This delay can be set by the computer to optimally position the signal that will be recorded in the time window. Once the signal has been stored in the recorder's memory, the computer can read the data and store it on permanent medium such as a disk or magnetic tape for later processing.

AIR SYSTEM

An air acoustics system is shown schematically in Fig. 9. The experimental technique in air is similar to that in water, but the sources, receivers, and frequencies are different. The air acoustics system allows the simulation of rigid boundary conditions for the target. For the measurements in Chapter 8, the acoustic system employed LTV electrostatic sources 5.08 cm and 15.24 cm in diam. The targets are hung in a large room 9 m × 30 m × 15 m high. Variable length gated-sine-wave pulses were produced by using a Hewlett Packard model

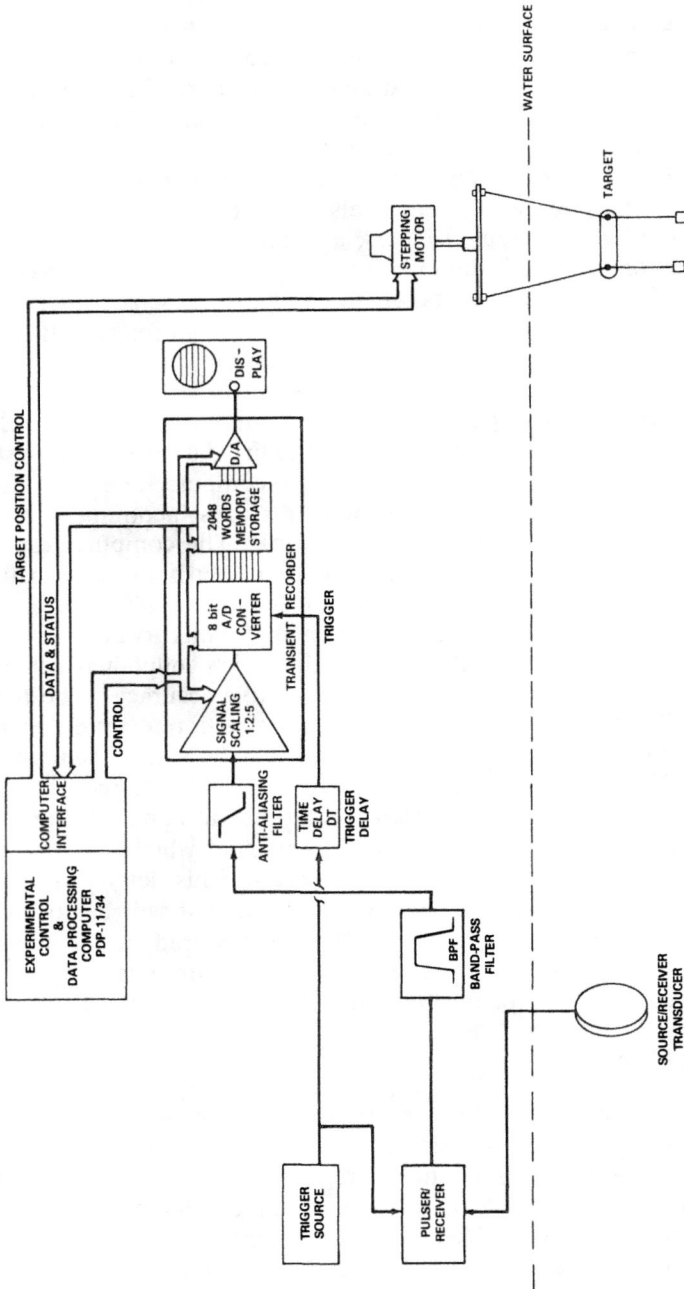

Fig. 7 — A diagram of the instrumentation used with the acoustic reflection facilities.

Fig. 8 — The computer used to control experimental conditions and analyze the data for the large pool (Fig. 6).

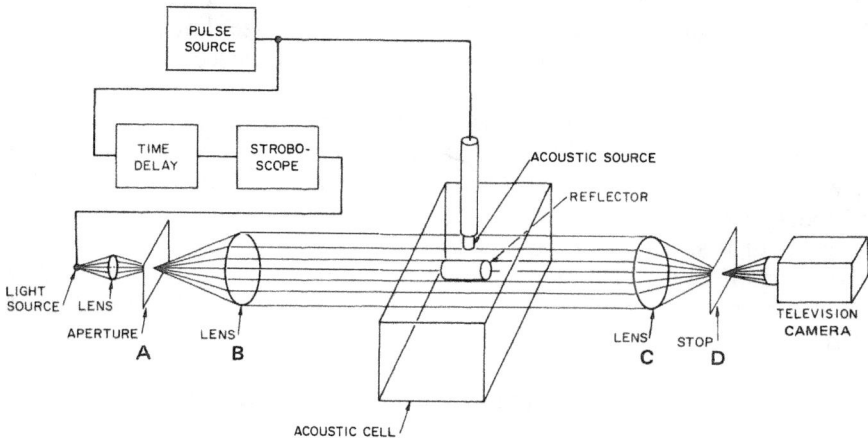

Fig. 9 — Schematic diagram of an air-acoustic system.

214A pulse generator and a Sanders switch to gate the continuous sine wave output of a Hewlett Packard model 5110 frequency synthesiser. The gated-sine-wave output was amplified by a Krohn-Hite model DCA 50 or ENI model 240L amplifier whose output drives the electrostatic speaker. The signal received by the 0.635-cm-diam microphone was filtered by a Krohn-Hite model 312 bandbass filter and amplified with a Bruel and Kjaer type 2107 frequency analyzer. The microphone was mounted on a stand and placed by hand at the desired aspect angle. A positioning of $\pm 2°$ is possible in the air measurements. The received signal is amplified and analyzed by the same equipment as described for the pool facility. The water pool and air facility were adjacent to each other.

SCHLIEREN SYSTEM

On several occasions [1,2] motion pictures of schlieren visualizations of ultrasonic wave interactions with cylinders have been shown. These movies demonstrated the great usefulness of the schlieren technique in identifying the individual components of the diffracted field of the cylinders. In fact, such identification was not possible using the standard pulse measurement technique of placing a hydrophone in the shadow zone of the cylinder, and interpreting the received pulse train displayed on an oscilloscope. The schlieren system that will be described can be used to examine ultrasonic wave interactions with bodies of various materials and shapes, with a size limitation of the viewed field imposed by the diameters of the lenses used.

The great utility of the schlieren system and numerous queries about it prompts the description of the system and its components and the method of producing motion pictures with it. Although schlieren visualization of sound waves is not a new procedure, practical information about it seems not to be readily available. The basic description of the fundamental phenomena involved is described by others, [3] and only a brief description will be given here. A diagram of the dark field schlieren system is shown in Fig. 10.

The following description of the schlieren system used to visualize sound waves would not satisfy the optical science purist, but is sufficient for the assembly and operation of a schlieren system for that purpose. A parallel light beam is established between the lenses B and C in Fig. 10. This parallel light beam has all rays within it parallel to each other. The system is a telecentric system that is generated by a small stop at A, which approximates a point source. The parallel rays are focused by lens C in the plane D in Fig. 10 to a point. A stop is

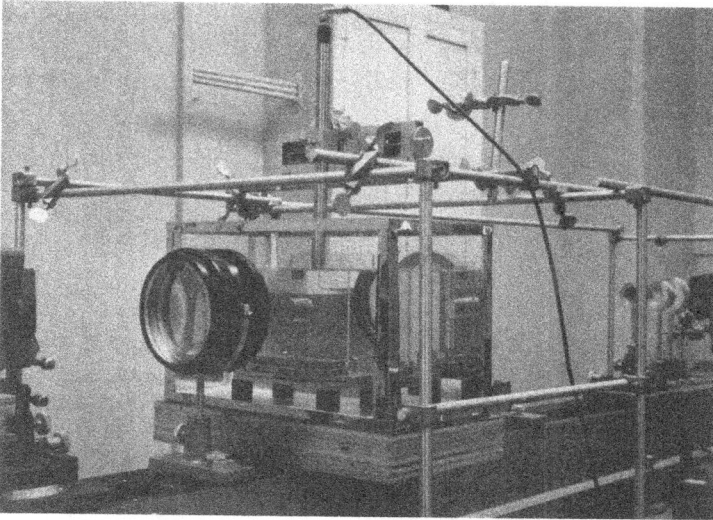

Fig. 10 — A photograph of the broad-band schlieren system.

placed at the focal point of the lens C that obliterates all of the parallel rays between lens B and lens C. The purpose of the lens before the aperture A is merely to focus the light source on the aperture. A source of pulses imposes a burst of sine waves on the acoustic source which is placed so that the acoustic wave travels normal to the direction of the parallel light rays in the light beam. A time delay is introduced between the pulse source and the light source. At the delayed time the stroboscope is driven to illuminate the acoustic pulse in the water tank. In the absence of sound, no image is seen in the image plane. However, when the acoustic source is turned on, that acoustic source causes deviation of some of the light rays in the parallel light bundle. After these rays pass through the lens C they are caused to miss the stop in the plane D in Fig. 10 and pass through a lens between the plane D and the image plane. This lens focuses the image plane at the center of the acoustic pulse. There results an image of the sound that has been emitted from the acoustic source. The stroboscope is timed so that the pulse is illuminated at a time after it has propagated a desired distance from the acoustic source or after it has interacted with the diffracting field. This is done repetitively, faster than the flicker rate of the eye, giving an apparent fixed image. If the time delay between the sound emission and light flash is continuously varied, for instance increased, the acoustic pulse will appear to be emitted from the acoustic source and will appear to propagate toward the diffracting cylinder at the rate of time delay increase and subsequently will appear to interact with it and cause the reradiated field. It is not necessary to use the exact instruments that are described here, as many of them have equivalents and would produce the same result. The instruments are named here as an example of ones that could be used to assemble a working schlieren system.

A useful source of illumination for schlieren visualization is a laser. A helium-neon laser can be used as a steady-state source or it may be intensity modulated to some extent. Usually the on-to-off amplitude ratio needed for good strobed schlieren pictures is high and not achievable by direct internal modulation. A laser light source obviates the need for achromatic lenses in the schlieren system. However, it puts greater demands on such elements as the glass walls of the water tank. Because of the coherence of the laser much glass, especially thin glass plates as well as plastics, cause the optical field to be striated and nonuniform. Of course, any visible-light laser is usable for this purpose. An argon-iron laser has been used for experiments at NRL. The laser beam was diverted by an acousto-optic modulator through an aperture. This aperture must be at a distance from the modulator large enough so that no residual edge of the undeflected beam passes through the aperture during the time that the light is off in the schlieren system. It was found that an extinction ratio of about 10,000 to 1 was desirable, and at times necessary. The light modulator must have a fast rise time and a pulse duration that is short compared to the travel time of a particular point on the acoustic wave if the wave is to appear stationary. In water, sound travels at approximately 1500 nsec/mm, so a pulse length of less than 100 nsec would be required. In addition, if so called phase resolution of the acoustic wave or pulse is desired, a light pulse short compared to the period of the highest acoustic frequency is needed. The period of a 5 MHz wave is 200 nsec, so a pulse of approximately 20 nsec would be desirable.

A photograph of a broad-band light-source system is shown in Fig. 11. The light source is a General Radio Stroboslave (model 1539-A). Even with this relatively slowly switched light source, phase resolution of 1 MHz was possible using it. The condensing lens system focuses the source image, which is in this case a zenon arc sufficiently bright that an aperture of 0.66-cm diam can be used at A (Fig. 10) to limit the parallel light rays to a reasonably high-quality telecentric system between B and C. The zenon light source is a broadband light source and produces a schlieren picture with the full spectrum of colors. In Fig. 9 the lens at B has a 12.7-cm diam with a 62.9-cm focal length. All lenses are high-quality achromatic lenses. The aperture A is at the focus of the lens at B. The distance from B to C is 38 cm. The tank between B and C is a stainless-steel-frame glass aquarium (40 cm × 26.6 cm × 25.4 cm deep). The original two largest sides of the tank were replaced with 0.03-cm-thick plate glass in order to minimize the distortion caused by bowing of the sides. Lens C is a pair of 12.7-cm lenses that have 27.3-cm combined focal length. The stop at D, which is at the focus of lens C, is a circle (0.076-cm diam) of black photographic masking tape stuck to the center of a slide-glass cover. India ink stops (on slide-glass covers) can be used as well. Large black circles photographically reduced in size to the obtained diameter afford the highest quality stops but can be costly. The cover glass is mounted on an aluminum plate with a 2-cm hole in it. The stop is centered by means of a microscope mechanical stage mounted vertically. A television camera (Fairchild model TC-177 with high-resolution RCA 8507A vidicon) is mounted behind the stop. An $f/0.95$ lens is used on the television camera but others are usable. The TV camera or any other recording camera must be focused at the position in the tank where the center of the acoustic pulse will appear. This is conveniently done with a finely graduated scale placed in the plane where the acoustic pulse will be but with external ambient light. Since the conventional television scan rate is used, an ordinary television monitor can be used to view the picture. Of course, at this point one may choose to record directly from the television camera on a video tape recorder. To take advantage of the high-resolution vidicon, a higher-resolution monitor can be used, as it was for the picture shown in Fig. 11. To make motion pictures, an Auricon motion picture camera with a TVT shutter was placed 1 m in front of the television screen. The camera is synchronized to the television scan rate by means of the 60 Hz power frequency. The movie was taken at $f/0.95$ with Eastman Plus-X film with an ASA rating of 80. The television image was adjusted for brightness and contrast to give the best directly viewed picture. Pictures are taken in a dark room. The dynamic range of this system is 40 dB measured between maximum voltage across a 6.6-MHz transducer driven by an Arenberg pulsed oscillator (model PG 650-C) and the voltage across the transducer to produce the minimum detectable wave on the television screen. The transducer is a Branson type ZT lead-zirconate-titanate disk in a 2-cm housing.

Motion of the acoustic pulse is introduced by changing the time delay between the acoustic pulse output and the strobe synchronization input (light pulse). Although an internal delay is available on the Arenberg oscillator driver, an external pulse delay (Hewlett-Packard 214A) was used in most cases to ensure the time-delay range that was from 0 to 100 μsec. The entire apparatus was set up on a Gaertner 6-ft optical bench. A similar system has also been set up on a smaller 4-ft Cenco bench with 7.6-cm-diam lenses at B and C. Similar results were obtained but adjustments were significantly more difficult. Careful initial alignment is important for optimum results. It is necessary to align

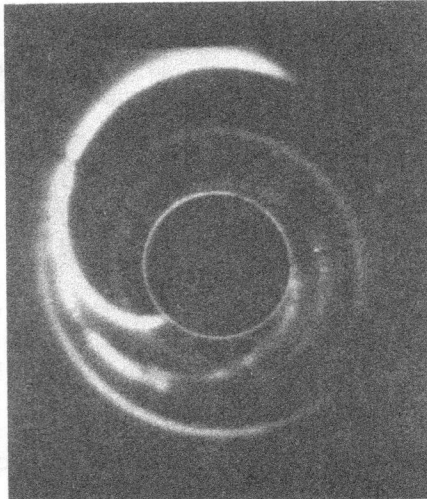

Fig. 11 — Schlieren photograph of a 3-μ-sec pulse reflected from a solid aluminum cylinder and subsequent radiation.

the cylinder elements in the same direction as that of the parallel light bundle and then it is necessary to align the transducer acoustic beam axis normal to both the cylinder elements and the parallel light bundle. After the cylinder is aligned in the parallel light beam, the acoustic pulse is observed as it is specularly reflected from the cylinder surface so that is has maximum intensity on the television screen both on incidence and after reflection. To get a schlieren photograph of a 5-cm-long (Fig. 12) cylinder, without showing its supports, the cylinder was glued with its elements perpendicular to a piece of 0.64-cm plate glass with Eastman 910 adhesive. The plate glass was larger than the 12.7-m lenses and was mounted vertically on a 0.64-cm thick horizontal brass base. The entire structure was set on a tiltable table that was placed in the bottom of the tank with adjustments accessible through the water for both tilt and rotation.

REFERENCES

1. W. G. Neubauer and L. R. Dragonette, J. Acoust. Soc. Am. **47**, 62(A) (1970).

2. W. G. Neubauer and L. R. Dragonette, J. Acoust. Soc. Am. **48**, 101(A) (1970).

3. L. Bergmann and H. Hatfield, *Ultrasonics* (Wiley, New York, 1938).

Chapter 23

SPEED OF SOUND IN FRESH WATER*

INTRODUCTION

In the course of free field measurements of the acoustic scattering by finite bodies in water, it was felt that the measurement of a fundamental quantity would unearth systematic errors in the placement-and-location equipment, electronics, and the measurement of ambient conditions. Toward this end, the measurement of the free field sound speed was undertaken since methods of time, distance, and temperature measurement were already available in the laboratory. In the experiment, a pulse was transmitted from a fixed source to a single receiver at two different distances along the same radius from the source. Sound speed was determined by measuring the time difference related to the distance difference. The resulting values for sound speed proved to be the most accurate that then existed. Soon after these results first appeared, corroborating measurements by different methods were published [1].

APPARATUS

The water medium was contained in the cypress tank described earlier in Chapter 22. A diagram of the electronic apparatus is shown in Fig. 1. The 100-kHz laboratory standard frequency was multiplied first by 5 and then by 20, causing the resultant 500-kHz and 10-MHz signals to be synchronized. The 10 MHz was scaled down by a factor of 10^5 and triggered a pulse generator, which in turn triggered the cathode-ray oscilloscope sweep each 10 msec. The pulse generator produced a $0.4\text{-}\mu\text{sec}$ pulse that, after amplification and series tuning, was applied to the acoustic source. The acoustic signal was received and was amplified and displayed on one trace of a 4-trace oscilloscope. The other three traces displayed on successive sweeps, the 100-, 500-kHz, and 10-MHz signals, all synchronized with respect to each other. Since each trace utilized the same electron gun and sweep circuitry in the

---

*Part of these results first appeared in: W. G. Neubauer and L. R. Dragonette, J. Acoust. Soc. Am. 36, 1685-1690 (1964).

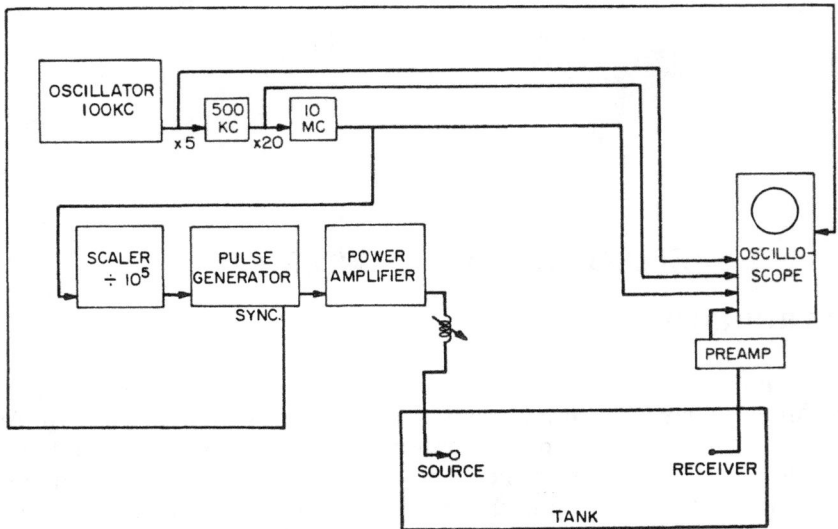

Fig. 1 — Diagram of electronic apparatus

oscilloscope as well as the same external synchronization, no timing error could be introduced. The pulse-generator jitter was unnoticeable. The composite signal displayed on the oscilloscope is shown in Fig. 2.

The acoustic source and receiver were fixed on rails that spanned the tank and moved on tracks that were mounted on opposite sides of the tank. An accurately graduated bar was slung in a framework just over the water surface. A vernier fastened to the receiver rod moved along the scale graduations when the distance between acoustic source and receiver was changed. The distance between two positions of the vernier on the scale represented the actual distance between two respective positions of the pressure-sensing element of the receiver, because at each position the rail to which the receiver was clamped was leveled along the line between source and receiver. This was accomplished with a coincidence level rigidly clamped to the rod as close as possible to the position on the rail where the 1.905-cm $\left(\frac{3}{4}\text{-in.}\right)$ receiver rod was clamped. The level had a sensitivity of 1 sec, causing a predictable uncertainty in the correspondence between actual distances between two positions of the sensing element in the water and the respective distance between two positions of the receiver as indicated on the scale above the water.

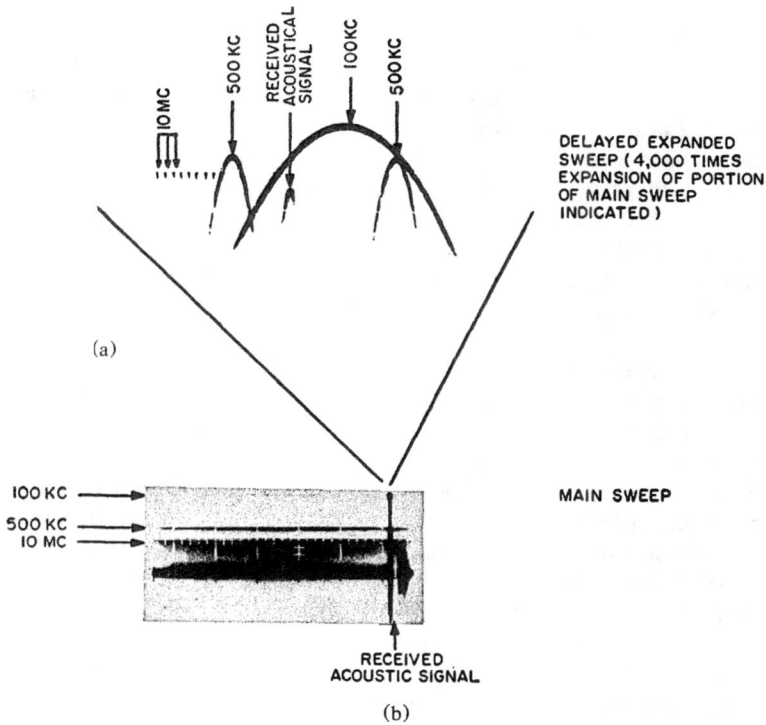

Fig. 2 — Four-trace oscillograph of 100, 500 kHz, 10 MHz and the first peak of the acoustic signal. (a) Delayed sweep. (b) Main sweep. It was necessary to darken with ink some of the 10-MHz peaks to indicate their existence in (a), since they did not survive the photographic reproduction.

EXPERIMENTAL PROCEDURE

Interference of the pulse at the receiver was avoided in all measurements by time separation of the direct transmission and any reflection. The rail holding the receiver was clamped and leveled at a position along the measuring bar, establishing a distance (d_s) between source and receiver. At each setting of the receiver, the temperature was read with a thermometer, which is described later. A time position with respect to the 10-MHz signal was read on the oscilloscope and identified relative to the 500- and 100-kHz signals (see Fig. 2). The time position of the signal was determined at the center of the cathode-ray tube in order to minimize parallax errors in reading. At each time determination, a photograph of the oscilloscope face was taken. The photograph (Fig. 2) shows an expanded portion of the entire sweep obtained by the use of a delayed sweep. The accuracy of the delay time of the delayed sweep was not used for timing but only to allow the convenient observation of the signals. The rail holding the receiver was then moved to the other end of the bar without touching the bar, and another distance (d_l) was then established in the same way as was d_s. To ensure that the bar had not been accidentally moved while the receiver was being moved, a graduation of the measuring bar was viewed through a fixed telescope. To ensure that the receiver-carrying rail moved along the tank and was maintained parallel at both distances d_s and d_l, the distances of the ends of the rail were measured with respect to a fixed rail.

The delayed sweep on the oscilloscope was then moved with respect to the same main sweep on which both signal positions were displayed at different times. The amount that the delay was increased or that the delayed sweep was moved along the main sweep was determined by counting the number of cycles of 100 kHz on the main sweep. Each interval of 100-kHz peaks was subdivided by 500-kHz peaks, each of which were subdivided by 10-MHz peaks. In this way, the acoustic signal was located with reference to the three displayed frequencies. The time interval Δt thus established in terms of periods of 100 kHz, plus periods of 500 kHz, plus periods of 10 MHz, is related to the distance difference $(d_l - d_s = \Delta d)$ or acoustic-signal path. Sound speed is thus determined, since $c = \Delta d / \Delta t$.

DISCUSSION

Since a transient signal is employed in the experiment, an exact analysis of the measurement would necessitate the application of a Fourier transform to find the frequency components of the pulse. The inherent differences between the transmission speed of the components and of the group would be in evidence. Despite the existence of such differences in the interpretation of sound speed, on the basis of pulse similarities, which are discussed, and the use of a difference method, the results are interpreted in terms of a steady-state solution.

The elementary solution to the scalar wave equation most closely corresponding to the conditions of the experiment that was done is that of a radial wave propagating outward from a point source in continuous, simple harmonic vibration, in an infinite homogeneous medium. Such a pressure wave would propagate with a phase velocity c and would diminish in amplitude as the reciprocal of the radial distance (r) from the point source. The wavelength and phase velocity are related by $c = f\lambda$ where f is the frequency of vibration of the source and λ is the wavelength. The phase velocity can also be expressed as $c = \Delta d / \Delta t$, where Δd is the radial distance that a wave is propagated and Δt is the time taken for the propagation.

In practice, the conditions that are explicit are not satisfied exactly nor are some that are implicit. Such an implicit condition is that of detection of an acoustical quantity such as pressure at a true point rather than a finite area or volume. No specific quantitative evaluation of each condition is yet possible that will allow a direct relationship between the quantity c expressed in the elementary solution to the wave equation and the experimentally determined quantity c. Therefore, no exact numerical limits of uncertainty resulting from these experimental conditions can be attributed to the measured sound speed. Subjective judgment of each condition separately and an estimate of their composite effect is the only recourse. In the experiment, the following conditions prevailed.

The acoustic field for some of the measurements was produced by a 290-deg spherical cap [2] that approximates a 1.27-cm-diam-spherical source. The radial pressure pattern at the 230-kHz resonance of the source has a central lobe of approximately 200 deg. Near the axis of this lobe, the pressure at a constant radius is constant to approximately $\pm 2\%$ over an angle of about 30 deg. The broad, single-lobe pattern implies that the cap was vibrating in its fundamental mode, which approximates the vibration of a sphere to a desired degree if a significantly small region near the axis is considered. Since the pattern is not truly spherical, a region near the source is present, even over small angles, in which the pressure behavior is measurably different from that of a true spherical radiator. This "near field" region has been measured as extending no further than 5 cm from the center of the source since, beyond this distance, the pressure has been measured as decreasing with the inverse of the radial distance to $\pm 2\%$.

In sound-speed determination, the pressure field was sampled with an acoustic probe [2] that has as its sensing element a 0.16-cm ceramic disk. The acoustic probe has a radial pressure pattern with a 300-deg lobe, measured for reception, that approaches the spherical pattern characteristic of a true probe. The angle subtended at the source by the largest sensing part of the probe at its nearest approach to the source in these measurements was 0.46 deg. For the greatest distance between source and receiver, an angle of 0.03 deg was subtended by the probe. The active disk-element radius is approximately 0.16 λ, whereas the probe housing is approximately 0.25 λ, at the source resonant frequency.

If, because of the finite curvature of the wavefront at the receiver, there was an inaccuracy of pulse location in time, it would be due to the difference in onset at d_s and d_l of the pulse on the sensitive element. This difference in onset would result in a slower rise time of the received signal at d_s than at d_l. In such a case, an observer would have made a time determination at d_s that was later on the sweep than it should have been and the Δt associated with the Δd determination would have been too small, resulting in a velocity derived from the experimental values, which would have been too large. That is, the true sound speed would have been below that determined experimentally. No such difference in initial rise time of the pulse was noticed, and therefore the influence of the effect was considered negligible. However, had a rise-time effect been present below the voltage sensitivity of the equipment and, therefore, had existed and gone unnoticed, disagreement between results, which are given, and confined-field measurements would be greater rather than smaller. A noise limitation would have resulted in an error of the location of the axis departure or any other characteristic of the received signal. However during these measurements noise was observed to be not significant; i.e., the effect on time location was less than the accuracy of the time reading.

Free-field or free-wave conditions at a field point were satisfied to a desired degree by using short pulses and causing time separation of reception at a field point of the pulse from the source and all reflections. Also, a pulse repetition rate was used that was long enough so that all reflections in the tank were below the noise level of the receiver-system electronics at the time that the source was pulsed again. The necessity for the use of pulses destroys the conformity of the experimental conditions with those assumed in the elementary solution to the wave equation in two respects. There is a lack of both a truly continuous wave and a true line-spectrum sinusoid. The bandwidth of the source causes a problem in determining sound speed since $c = f\lambda$ and in the actual transmitted pulse no unique value of either f or λ can be measured. A range of frequencies is present at the beginning of the pulse, where measurements are made. If, however, the medium in which the pulse is transmitted can be considered nondispersive, i.e., $c(f,\lambda)$ is a constant, a unique value of c is subject to definition to the degree that dispersion may be ignored. If dispersion were present, the various frequency components of the pulse would be transmitted with different speeds, and, at different distances from the source, a composite pressure resultant would be detected, and the pulse shape would necessarily change. A similar effect would be present if the attenuation of the medium over the entire frequency spectrum of the pulse could not be ignored as negligible. Since the pulse shape would indicate the presence of both of these factors, pulse shape at the short transmission was compared with that at the long transmission for each combination of transducers used in the measurements.

The initial part of received acoustic pulse is shown in Fig. 3 as it was received at distances d_s and d_l. Within the experimenters' ability to locate in time any characteristic part of this pulse, e.g., initial axis departure or first or second peak, the two received pulses are indistinguishable from each other. Therefore, the resultant determination of sound speed is no further in error than that error which is the result of the time measurement itself. Pulse-shape comparisons were carried out with different received pulse shapes as well as with different distances, with identical results. Pulse-shaped comparisons were carried out with 0.63-cm-radius transducer disks and again the results were the same even though such a disk has approximately a 25-deg central lobe in the radial pressure pattern. Figure 4 shows the pulses that were compared using these larger disks. In each situation used to determine sound speed, the entire transmitted pulses were also compared by photographically superimposing them. Over the entire pulse, no detectable differences were observed. What has been measured then can be concluded to be the speed of sound emitted by a true spherical source detected with a true probe within the experimental accuracy at a frequency represented by $f = c\lambda$, where f may be any frequency within the relatively small bandwidth involved in the resonance of the source. The resonant frequency of the 0.63-cm-radius disk is nominally 1 MHz.

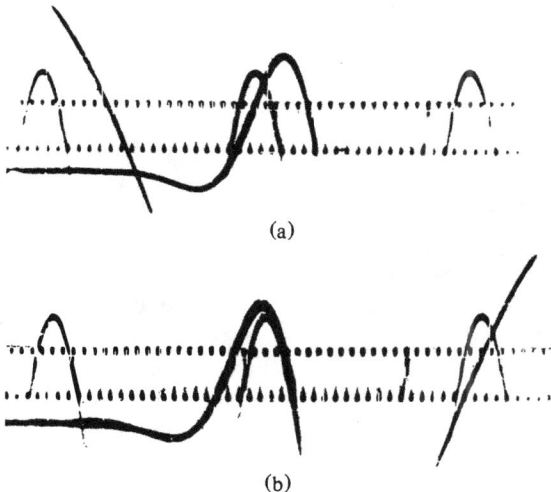

(a)

(b)

Fig. 3 — Initial portion of received acoustic pulse at (a) d_s and at (b) d_l, using the spherical-cap source and probe.

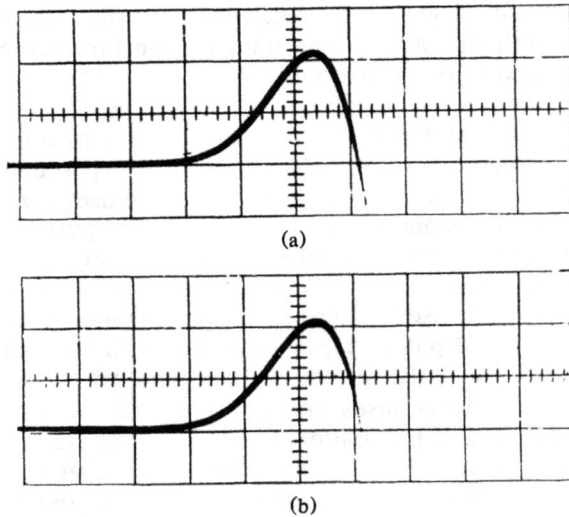

(a)

(b)

Fig. 4 — Received acoustic pulse using 0.63-cm-radius
disk source and receiver at (a) d_s and at (b) d_l.

RESULTS

Figure 5 is a plot of measurements made with the acoustic sources
and receivers described. The measured values appear in Table 1.
Measurements were carried out by three experimenters independently.
As to whether this sound speed is what is normally called phase, signal,
or group velocity, it is all three to within the accuracy of the experimental results.

These results also represent the plane-wave sound speed since the
elementary plane-wave solution to the one-dimensional wave equation
has the same form as the spherical-wave case, disregarding the amplitude and allowing the distance to the source to be sufficiently large that
the spherical divergence or wavefront curvature is essentially zero over
the receiver sensing area. Again, the pulse comparison supports this
conclusion. Actually, in the experiments that were carried out, the limitations resulting from the rigid definitions of conditions of the simple
solutions were sufficiently overcome to result not only in a spherical
wave but also a plane wave. The plane wave could not be as good an
approximation as the spherical wave, but the difference was indistinguishable at the distances used and within the accuracy of the measured
values.

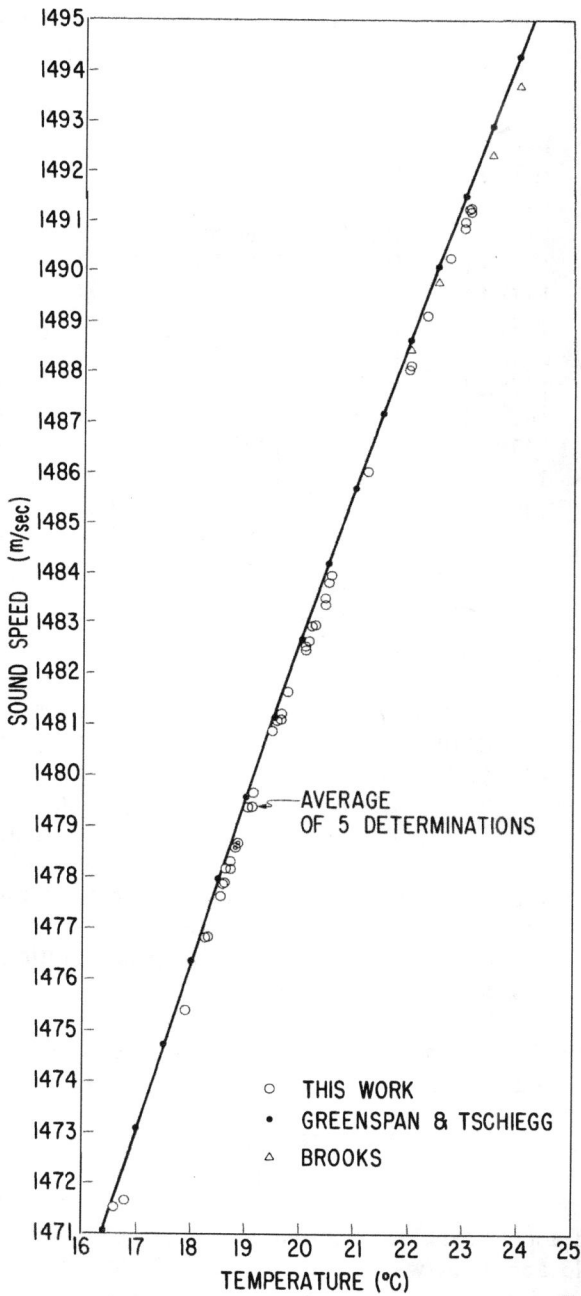

Fig. 5 — Measured values of sound speed vs temperature.

Table 1 — Measured Values of Sound Speed vs Temperature

$|\theta(°C)|$	c(m/sec)	$|\theta(°C)|$	c(m/sec)|	$|\theta(°C)|$	c(m/sec)
16.6[a]	1471.51	19.01	1479.22	20.42	1483.35
16.8[a]	1471.65	19.04	1479.36	20.42	1483.48
17.9[a]	1475.38	19.11	1479.47	20.51	1483.81
18.26	1476.81	19.14	1479.64	20.55	1483.95
18.31	1476.82	19.47	1480.85	20.56	1483.95
18.55	1477.61	19.51	1480.98	21.22	1486.02
18.59	1477.86	19.57	1481.04	21.97	1488.04
18.62	1477.88	19.63	1481.07	22.01	1488.11
18.64	1478.16	19.64	1481.19	22.30	1489.11
18.72	1478.31	19.75	1481.61	22.72	1490.25
18.72	1478.31	20.06	1482.51	22.98	1490.86
18.81	1478.58	20.07	1482.44	22.99	1490.99
18.81	1478.57	20.12	1482.63	23.07	1491.24
18.85	1478.65	20.18	1482.92	23.09	1491.19
18.85	1478.67	20.24	1482.94	23.09	1491.26

MEASUREMENT ACCURACY

In order to determine the total maximum error due to the uncertainties in the measurement of the independent variables, the total differential of $c(\Delta d, \Delta t, \theta)$ was computed [3]. Factors that were considered as contributing to the error in the determination of Δd were bar-scale calibration, bar-material coefficient of thermal expansion, scale and vernier reading error (parallax), source and receiver alignment, and receiver rail leveling and parallelism. The total error in Δd caused by these factors was $+0.0075$, -0.0067 cm. Since the maximum error in c due to a 0.001-cm error in Δd is 0.015 m/sec, the resulting total maximum error in c is $+0.113$, -0.101 m/sec.

The laboratory standard frequency of 100 kHz was known to be accurate and constant to better than one part in 10^9, contributing no significant error in Δt. Since the 500-kHz and 10-MHz signals are derived directly from the 100-kHz signal, they also are accurate and constant to one part in 10^9. If the scaling of the 10 MHz were in error, causing a lack of a constant synchronization interval, only the 10-MHz signal would appear synchronized while the signals on the three other sweeps would appear to wander across the cathode-ray-tube face. The only factor that was significant in the contribution to the error of a determination of Δt was the experimenter's time-position reading error on the scope face, relative to the standard frequencies displayed. The experimenter could determine the position of the received acoustic signal at each distance position to the nearest peak or axis crossing of the 10-MHz signal. In measuring Δt, two such determinations were necessary and therefore the maximum error in Δt was ± 0.025 μsec. Since the maximum error in c due to a 0.01-μsec error in Δt is 0.022 m/sec, the maximum total error in c related to time reading was ± 0.055 m/sec.

The temperature was read with a fast-response, 61-cm-long bomb-type fuel calorimeter that was graduated at each 0.01°C between the temperatures of 18° and 28°C. This thermometer was calibrated by comparison with a similar thermometer that had been calibrated by the National Bureau of Standards. The thermometer was always read with the entire thermometer in the transmission path before and after each pair of distance readings. Also, in each case the temperature was taken with the bulb toward the receiver as well as toward the source. Sound-speed determinations were not made when a horizontal gradient as great as 0.01°C was noticed. The partial derivative of c with respect to θ was taken as 3.28 m/sec/°C from the report of Greenspan and Tschiegg [4]. Temperature was determined in the transmission path to an accuracy of ± 0.01°C, which uncertainty resulted in a maximum error in c of ± 0.033 m/sec.

WATER PURITY

Greenspan and Tschiegg [4] report that "Several measurements made on local tap water give results about 30 ppm higher than for distilled water." Thus, since the 1800-gal water tank used in the measurements reported here was filled with tap water in the same city in which the difference was measured, an error of c of -0.045 m/sec was assumed. The results do not represent a single body of water since the temperature was changed by adding hot or cold water from the tap.

The maximum total error in c as a result of the sum of the errors associated with all of the independent variables and the impurity of the water was $+0.20$, -0.23 m/sec.

SOUND SPEED TABLES

Sound speed in water plays a vital role in the computation and measurement of reflection and scattering by bodies and surfaces. A tabulation of sound speed as a function of temperature for pure water is therefore given here. Confident values for sound speed for all temperatures within the originally measured range were derived by curve fitting techniques. Forty-two values of sound speed were measured directly. To these points a best straight-line fit was made as well as a quadratic fit.

Polynomial fits, in general, have the form

$$c = \sum_{n=0}^{m} A_n T^n.$$

To permit the confident extrapolation of our free field data to temperature values above and below the measured values, we assumed the constants for $0 \leqslant n \leqslant 5$ that were derived from confined-field sound-speed values measured over a temperature range 0°C to 100°C by Greenspan and Tschiegg [4]. This assumption is justified on the basis of the standard deviations derived for all fits of the data.

Within the temperature range (from 18.26°C to 23.09°C) for a straight-line fit of the measured free field values, $A_0 = 1.422522 \times 10^3$ and $A_1 = 2.9824$. The standard deviation σ for this data is 0.14 m/sec calculated by the formula

$$\sigma = \sum_{i=1}^{m} \frac{(y_i - x_i)}{N - (m + 1)}$$

where x_i = the experimental values of sound speed, y_i = the values calculated from the straight-line fit, N = the number of data points, and m = highest order term of the polynomial. A quadratic fit yielded:

$A_0 = 1.398732 \times 10^3$

$A_1 = 5.2905$

$A_2 = -5.5639 \times 10^{-2}$

with $\sigma = 0.09$ m/sec.

Using the values of Greenspan and Tschiegg [4]:

$A_1 = 5.03358$

$A_2 = -5.79506 \times 10^{-2}$

$A_3 = 3.31636 \times 10^{-4}$

$A_4 = -1.45262 \times 10^{-6}$

$A_5 = 3.0449 \times 10^{-9}.$

A new value of A_0 was calculated at 1.402330×10^3. This means that the shape of the Greenspan and Tschiegg curve was retained (see Ref. 4) for our free field data and a new ordinate intersection was derived in the plot of sound speed vs temperature at atmospheric pressure. A calculation of σ for our data relative to the resulting curve yielded a value of ± 0.10 m/sec.

Values of sound speed are given in Table 2 to the nearest 0.1 m/sec from 15.00°C to 30.99°C for temperature intervals of 0.01°C. This range is intended to be useful for general laboratory use at atmospheric pressure especially in open tanks for free-wave conditions.

The probable error for all values of sound speed are not greater than ± 0.10 m/sec. Experimental values for sound speed reported by others [4-7] are within a σ of values for that temperature given in the table. The values are plotted and joined by a solid line in Fig. 6. Experimentally measured values are shown on the curve.

Similar values of sound speed expressed in ft/sec over a temperature range from 60.00°F to 91.99°F, for temperature intervals of 0.01°F are given in Table 3. Here the probable error is not greater than ± 0.32 ft/sec. As was done for the data of Table 2, an accompanying plot is given in Fig. 7.

REFERENCES

1. H.J. McSkimin, J. Acoust. Soc. Am. **37**, 325-328 (1965).

2. The characteristics of the source and probe have been described in W.G. Neubauer, J. Acoust. Soc. Am. **34**, 312-318 (1962).

3. A more detailed description of these factors is given in W.G. Neubauer and L.R. Dragonette, "Microacoustic System Analysis by the Measurement of Free-Field Sound Speed," Naval Res. Lab. Report No. 6113 (1964).

4. M. Greenspan and C.E. Tschiegg, J. Res. Natl. Bur. Std. (US) **59**, 249-254 (1957).

5. R. Brooks, J. Acoust. Soc. Am. **32**, 1422-1425 (1960).

6. V.A. Del Grosso, "The Velocity of Sound in Sea Water at Zero Depth," Naval Res. Lab. Report No. 4002 (1952).

7. W.D. Wilson, J. Acoust. Soc. Am. **31**, 1067-1072 (1959).

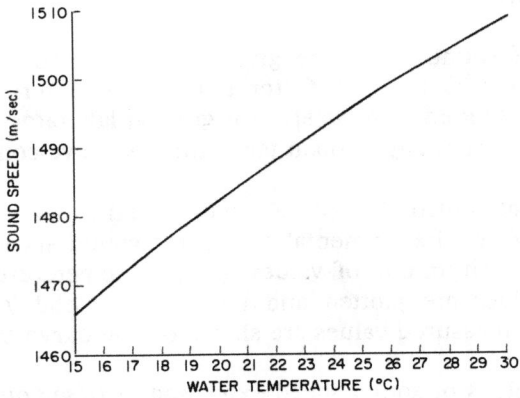

Fig. 6 — Sound speed in m/sec vs temperature
in degrees centigrade

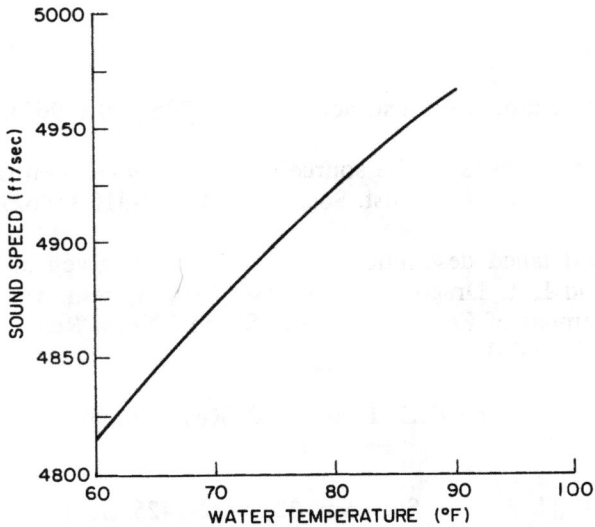

Fig. 7 — Sound speed in f_t sec vs temperature in
degrees Fahrenheit.

Table 2 — The Speed of Sound in Water in Meters/Sec as a Function of Temperature in Degrees Centigrade 23.00 to 26.99

TEMP	0	1	2	3	4	5	6	7	8	9
23.0	1491.09	1491.12	1491.15	1491.18	1491.21	1491.24	1491.26	1491.29	1491.32	1491.35
23.1	1491.38	1491.41	1491.43	1491.46	1491.49	1491.52	1491.55	1491.57	1491.60	1491.63
23.2	1491.66	1491.69	1491.71	1491.74	1491.77	1491.80	1491.83	1491.85	1491.88	1491.91
23.3	1491.94	1491.97	1492.00	1492.02	1492.05	1492.08	1492.11	1492.14	1492.16	1492.19
23.4	1492.22	1492.25	1492.28	1492.30	1492.33	1492.36	1492.39	1492.41	1492.44	1492.47
23.5	1492.50	1492.53	1492.55	1492.58	1492.61	1492.64	1492.67	1492.69	1492.72	1492.75
23.6	1492.78	1492.80	1492.83	1492.86	1492.89	1492.92	1492.94	1492.97	1493.00	1493.03
23.7	1493.05	1493.08	1493.11	1493.14	1493.17	1493.19	1493.22	1493.25	1493.28	1493.30
23.8	1493.33	1493.36	1493.39	1493.41	1493.44	1493.47	1493.50	1493.53	1493.55	1493.58
23.9	1493.61	1493.64	1493.67	1493.69	1493.72	1493.75	1493.77	1493.80	1493.83	1493.86
24.0	1493.88	1493.91	1493.94	1493.97	1493.99	1494.02	1494.05	1494.08	1494.10	1494.13
24.1	1494.16	1494.19	1494.21	1494.24	1494.27	1494.29	1494.32	1494.35	1494.38	1494.40
24.2	1494.43	1494.46	1494.49	1494.51	1494.54	1494.57	1494.60	1494.62	1494.65	1494.68
24.3	1494.70	1494.73	1494.76	1494.79	1494.81	1494.84	1494.87	1494.90	1494.92	1494.95
24.4	1494.98	1495.00	1495.03	1495.06	1495.09	1495.11	1495.14	1495.17	1495.19	1495.22
24.5	1495.25	1495.28	1495.30	1495.33	1495.36	1495.38	1495.41	1495.44	1495.47	1495.49
24.6	1495.52	1495.55	1495.57	1495.60	1495.63	1495.65	1495.68	1495.71	1495.74	1495.76
24.7	1495.79	1495.82	1495.84	1495.87	1495.90	1495.92	1495.95	1495.98	1496.00	1496.03
24.8	1496.06	1496.09	1496.11	1496.14	1496.16	1496.19	1496.22	1496.25	1496.27	1496.30
24.9	1496.33	1496.35	1496.38	1496.41	1496.43	1496.46	1496.49	1496.51	1496.54	1496.57
25.0	1496.59	1496.62	1496.65	1496.67	1496.70	1496.73	1496.75	1496.78	1496.81	1496.83
25.1	1496.86	1496.89	1496.91	1496.94	1496.97	1496.99	1497.02	1497.05	1497.07	1497.10
25.2	1497.13	1497.15	1497.18	1497.21	1497.23	1497.26	1497.29	1497.31	1497.34	1497.37
25.3	1497.39	1497.42	1497.45	1497.47	1497.50	1497.53	1497.55	1497.58	1497.60	1497.63
25.4	1497.66	1497.68	1497.71	1497.74	1497.76	1497.79	1497.82	1497.84	1497.87	1497.90
25.5	1497.93	1497.95	1497.97	1498.00	1498.03	1498.05	1498.08	1498.11	1498.13	1498.16
25.6	1498.18	1498.21	1498.24	1498.26	1498.29	1498.32	1498.34	1498.37	1498.39	1498.42
25.7	1498.45	1498.47	1498.50	1498.53	1498.55	1498.58	1498.60	1498.63	1498.66	1498.68
25.8	1498.71	1498.73	1498.76	1498.79	1498.81	1498.84	1498.87	1498.89	1498.92	1498.94
25.9	1498.97	1499.00	1499.02	1499.05	1499.07	1499.10	1499.13	1499.15	1499.18	1499.20
26.0	1499.23	1499.26	1499.28	1499.31	1499.33	1499.36	1499.39	1499.41	1499.44	1499.46
26.1	1499.49	1499.51	1499.54	1499.57	1499.59	1499.62	1499.64	1499.67	1499.70	1499.72
26.2	1499.75	1499.77	1499.80	1499.83	1499.85	1499.88	1499.90	1499.93	1499.95	1499.98
26.3	1500.01	1500.03	1500.06	1500.08	1500.11	1500.13	1500.16	1500.19	1500.21	1500.24
26.4	1500.26	1500.29	1500.31	1500.34	1500.37	1500.39	1500.42	1500.44	1500.47	1500.49
26.5	1500.52	1500.54	1500.57	1500.60	1500.62	1500.65	1500.67	1500.70	1500.72	1500.75
26.6	1500.77	1500.80	1500.83	1500.85	1500.88	1500.90	1500.93	1500.95	1500.98	1501.00
26.7	1501.03	1501.05	1501.08	1501.11	1501.13	1501.16	1501.18	1501.21	1501.23	1501.26
26.8	1501.29	1501.31	1501.33	1501.36	1501.39	1501.41	1501.44	1501.46	1501.49	1501.51
26.9	1501.54	1501.56	1501.59	1501.61	1501.64	1501.66	1501.69	1501.71	1501.74	1501.76
TEMP	0	1	2	3	4	5	6	7	8	9

Table 2 (Continued) — The Speed of Sound in Water in Meters/Sec as a Function of Temperature in Degrees Centigrade 27.00 to 30.99

TEMP	0	1	2	3	4	5	6	7	8	9
27.0	1501.79	1501.82	1501.84	1501.87	1501.89	1501.92	1501.94	1501.97	1501.99	1502.02
27.1	1502.04	1502.07	1502.09	1502.12	1502.14	1502.17	1502.19	1502.22	1502.24	1502.27
27.2	1502.29	1502.32	1502.34	1502.37	1502.39	1502.42	1502.44	1502.47	1502.49	1502.52
27.3	1502.54	1502.57	1502.59	1502.62	1502.64	1502.67	1502.69	1502.72	1502.74	1502.77
27.4	1502.79	1502.82	1502.84	1502.87	1502.89	1502.92	1502.94	1502.97	1502.99	1503.02
27.5	1503.04	1503.07	1503.09	1503.12	1503.14	1503.17	1503.19	1503.22	1503.24	1503.27
27.6	1503.29	1503.32	1503.34	1503.36	1503.39	1503.41	1503.44	1503.46	1503.49	1503.51
27.7	1503.54	1503.56	1503.59	1503.61	1503.64	1503.66	1503.69	1503.71	1503.74	1503.76
27.8	1503.79	1503.81	1503.83	1503.86	1503.88	1503.91	1503.93	1503.96	1503.98	1504.01
27.9	1504.03	1504.06	1504.08	1504.10	1504.13	1504.15	1504.18	1504.20	1504.23	1504.25
28.0	1504.28	1504.30	1504.33	1504.35	1504.37	1504.40	1504.42	1504.45	1504.47	1504.50
28.1	1504.52	1504.55	1504.57	1504.59	1504.62	1504.64	1504.67	1504.69	1504.72	1504.74
28.2	1504.77	1504.79	1504.81	1504.84	1504.86	1504.89	1504.91	1504.94	1504.96	1504.98
28.3	1505.01	1505.03	1505.06	1505.08	1505.11	1505.13	1505.15	1505.18	1505.20	1505.23
28.4	1505.25	1505.27	1505.30	1505.32	1505.35	1505.37	1505.40	1505.42	1505.44	1505.47
28.5	1505.49	1505.52	1505.54	1505.56	1505.59	1505.61	1505.64	1505.66	1505.69	1505.71
28.6	1505.73	1505.76	1505.78	1505.81	1505.83	1505.85	1505.88	1505.90	1505.93	1505.95
28.7	1505.97	1506.00	1506.02	1506.05	1506.07	1506.09	1506.12	1506.14	1506.17	1506.19
28.8	1506.21	1506.24	1506.26	1506.29	1506.31	1506.33	1506.36	1506.38	1506.40	1506.43
28.9	1506.45	1506.48	1506.50	1506.52	1506.55	1506.57	1506.60	1506.62	1506.64	1506.67
29.0	1506.69	1506.71	1506.74	1506.76	1506.79	1506.81	1506.83	1506.86	1506.88	1506.90
29.1	1506.93	1506.95	1506.98	1507.00	1507.02	1507.05	1507.07	1507.09	1507.12	1507.14
29.2	1507.16	1507.19	1507.21	1507.24	1507.26	1507.28	1507.31	1507.33	1507.35	1507.38
29.3	1507.40	1507.42	1507.45	1507.47	1507.50	1507.52	1507.54	1507.57	1507.59	1507.61
29.4	1507.64	1507.66	1507.68	1507.71	1507.73	1507.75	1507.78	1507.80	1507.82	1507.85
29.5	1507.87	1507.89	1507.92	1507.94	1507.96	1507.99	1508.01	1508.03	1508.06	1508.08
29.6	1508.10	1508.13	1508.15	1508.17	1508.20	1508.22	1508.24	1508.27	1508.29	1508.31
29.7	1508.34	1508.36	1508.38	1508.41	1508.43	1508.45	1508.48	1508.50	1508.52	1508.55
29.8	1508.57	1508.59	1508.62	1508.64	1508.66	1508.69	1508.71	1508.73	1508.76	1508.78
29.9	1508.80	1508.83	1508.85	1508.87	1508.89	1508.92	1508.94	1508.96	1508.99	1509.01
30.0	1509.03	1509.06	1509.08	1509.10	1509.13	1509.15	1509.17	1509.19	1509.22	1509.24
30.1	1509.26	1509.29	1509.31	1509.33	1509.36	1509.38	1509.40	1509.42	1509.45	1509.47
30.2	1509.49	1509.52	1509.54	1509.56	1509.59	1509.61	1509.63	1509.65	1509.68	1509.70
30.3	1509.72	1509.75	1509.77	1509.79	1509.81	1509.84	1509.86	1509.88	1509.91	1509.93
30.4	1509.95	1509.97	1510.00	1510.02	1510.04	1510.06	1510.09	1510.11	1510.13	1510.16
30.5	1510.18	1510.20	1510.22	1510.25	1510.27	1510.29	1510.31	1510.34	1510.36	1510.38
30.6	1510.41	1510.43	1510.45	1510.47	1510.50	1510.52	1510.54	1510.56	1510.59	1510.61
30.7	1510.63	1510.65	1510.68	1510.70	1510.72	1510.74	1510.77	1510.79	1510.81	1510.83
30.8	1510.86	1510.88	1510.90	1510.92	1510.95	1510.97	1510.99	1511.01	1511.04	1511.06
30.9	1511.08	1511.10	1511.13	1511.15	1511.17	1511.19	1511.22	1511.24	1511.26	1511.28
TEMP	0	1	2	3	4	5	6	7	8	9

Table 3 — The Speed of Sound in Water in Feet/Sec as a Function of Temperature in Degrees Fahrenheit 60.00 to 63.99

TEMP	0	1	2	3	4	5	6	7	8	9
60.0	4815.53	4815.59	4815.66	4815.72	4815.78	4815.84	4815.91	4815.97	4816.03	4816.10
60.1	4816.16	4816.22	4816.28	4816.35	4816.41	4816.47	4816.53	4816.60	4816.66	4816.72
60.2	4816.79	4816.85	4816.91	4816.97	4817.04	4817.10	4817.16	4817.22	4817.29	4817.35
60.3	4817.41	4817.47	4817.54	4817.60	4817.66	4817.73	4817.79	4817.85	4817.91	4817.98
60.4	4818.04	4818.10	4818.16	4818.23	4818.29	4818.35	4818.41	4818.48	4818.54	4818.60
60.5	4818.66	4818.73	4818.79	4818.85	4818.91	4818.98	4819.04	4819.10	4819.16	4819.22
60.6	4819.29	4819.35	4819.41	4819.47	4819.54	4819.60	4819.66	4819.72	4819.79	4819.85
60.7	4819.91	4819.97	4820.03	4820.10	4820.16	4820.22	4820.28	4820.35	4820.41	4820.47
60.8	4820.53	4820.59	4820.66	4820.72	4820.78	4820.84	4820.91	4820.97	4821.03	4821.09
60.9	4821.15	4821.22	4821.28	4821.34	4821.40	4821.46	4821.53	4821.59	4821.65	4821.71
61.0	4821.77	4821.84	4821.90	4821.96	4822.02	4822.08	4822.15	4822.21	4822.27	4822.33
61.1	4822.39	4822.46	4822.52	4822.58	4822.64	4822.70	4822.76	4822.83	4822.89	4822.95
61.2	4823.01	4823.07	4823.14	4823.20	4823.26	4823.32	4823.38	4823.44	4823.51	4823.57
61.3	4823.63	4823.69	4823.75	4823.81	4823.88	4823.94	4824.00	4824.06	4824.12	4824.18
61.4	4824.25	4824.31	4824.37	4824.43	4824.49	4824.55	4824.62	4824.68	4824.74	4824.80
61.5	4824.86	4824.92	4824.99	4825.05	4825.11	4825.17	4825.23	4825.29	4825.35	4825.42
61.6	4825.48	4825.54	4825.60	4825.66	4825.72	4825.78	4825.85	4825.91	4825.97	4826.03
61.7	4826.09	4826.15	4826.21	4826.28	4826.34	4826.40	4826.46	4826.52	4826.58	4826.64
61.8	4826.70	4826.77	4826.83	4826.89	4826.95	4827.01	4827.07	4827.13	4827.19	4827.26
61.9	4827.32	4827.38	4827.44	4827.50	4827.56	4827.62	4827.68	4827.75	4827.81	4827.87
62.0	4827.93	4827.99	4828.05	4828.11	4828.17	4828.23	4828.30	4828.36	4828.42	4828.48
62.1	4828.54	4828.60	4828.66	4828.72	4828.78	4828.84	4828.91	4828.97	4829.03	4829.09
62.2	4829.15	4829.21	4829.27	4829.33	4829.39	4829.45	4829.51	4829.58	4829.64	4829.70
62.3	4829.76	4829.82	4829.88	4829.94	4830.01	4830.06	4830.12	4830.18	4830.24	4830.31
62.4	4830.37	4830.43	4830.49	4830.55	4830.61	4830.67	4830.73	4830.79	4830.85	4830.91
62.5	4830.97	4831.03	4831.09	4831.16	4831.22	4831.28	4831.34	4831.40	4831.46	4831.52
62.6	4831.58	4831.64	4831.70	4831.76	4831.82	4831.88	4831.94	4832.00	4832.06	4832.12
62.7	4832.18	4832.25	4832.31	4832.37	4832.43	4832.49	4832.55	4832.61	4832.67	4832.73
62.8	4832.79	4832.85	4832.91	4832.97	4833.03	4833.09	4833.15	4833.21	4833.27	4833.33
62.9	4833.39	4833.45	4833.51	4833.57	4833.63	4833.69	4833.75	4833.81	4833.87	4833.94
63.0	4834.00	4834.06	4834.12	4834.18	4834.24	4834.30	4834.36	4834.42	4834.48	4834.54
63.1	4834.60	4834.66	4834.72	4834.78	4834.84	4834.90	4834.96	4835.02	4835.08	4835.14
63.2	4835.20	4835.26	4835.32	4835.38	4835.44	4835.50	4835.56	4835.62	4835.68	4835.74
63.3	4835.80	4835.86	4835.92	4835.98	4836.04	4836.10	4836.16	4836.22	4836.28	4836.34
63.4	4836.40	4836.46	4836.52	4836.58	4836.64	4836.70	4836.76	4836.82	4836.88	4836.94
63.5	4837.00	4837.06	4837.12	4837.18	4837.23	4837.29	4837.35	4837.41	4837.47	4837.53
63.6	4837.59	4837.65	4837.71	4837.77	4837.83	4837.89	4837.95	4838.01	4838.07	4838.13
63.7	4838.19	4838.25	4838.31	4838.37	4838.43	4838.49	4838.55	4838.61	4838.67	4838.73
63.8	4838.79	4838.85	4838.90	4838.96	4839.02	4839.08	4839.14	4839.20	4839.26	4839.32
63.9	4839.38	4839.44	4839.50	4839.56	4839.62	4839.68	4839.74	4839.80	4839.86	4839.92

Table 3 (Continued) — The Speed of Sound in Water in Feet/Sec as a Function of Temperature in Degrees Fahrenheit 64.00 to 67.99

TEMP	0	1	2	3	4	5	6	7	8	9
64.0	4839.97	4840.03	4840.09	4840.15	4840.21	4840.27	4840.33	4840.39	4840.45	4840.51
64.1	4840.57	4840.63	4840.69	4840.75	4840.80	4840.86	4840.92	4840.98	4841.04	4841.10
64.2	4841.16	4841.22	4841.28	4841.34	4841.40	4841.46	4841.51	4841.57	4841.63	4841.69
64.3	4841.75	4841.81	4841.87	4841.93	4841.99	4842.05	4842.11	4842.16	4842.22	4842.28
64.4	4842.34	4842.40	4842.46	4842.52	4842.58	4842.64	4842.70	4842.75	4842.81	4842.87
64.5	4842.93	4842.99	4843.05	4843.11	4843.17	4843.23	4843.28	4843.34	4843.40	4843.46
64.6	4843.52	4843.58	4843.64	4843.70	4843.76	4843.81	4843.87	4843.93	4843.99	4844.05
64.7	4844.11	4844.17	4844.23	4844.28	4844.34	4844.40	4844.46	4844.52	4844.58	4844.64
64.8	4844.70	4844.75	4844.81	4844.87	4844.93	4844.99	4845.05	4845.11	4845.16	4845.22
64.9	4845.28	4845.34	4845.40	4845.46	4845.52	4845.57	4845.63	4845.69	4845.75	4845.81
65.0	4845.87	4845.93	4845.98	4846.04	4846.10	4846.16	4846.22	4846.28	4846.33	4846.39
65.1	4846.45	4846.51	4846.57	4846.63	4846.68	4846.74	4846.80	4846.86	4846.92	4846.98
65.2	4847.03	4847.09	4847.15	4847.21	4847.27	4847.33	4847.38	4847.44	4847.50	4847.56
65.3	4847.62	4847.68	4847.73	4847.79	4847.85	4847.91	4847.97	4848.02	4848.08	4848.14
65.4	4848.20	4848.26	4848.32	4848.37	4848.43	4848.49	4848.55	4848.61	4848.66	4848.72
65.5	4848.78	4848.84	4848.90	4848.95	4849.01	4849.07	4849.13	4849.19	4849.24	4849.30
65.6	4849.36	4849.42	4849.48	4849.53	4849.59	4849.65	4849.71	4849.77	4849.82	4849.88
65.7	4849.94	4850.00	4850.06	4850.11	4850.17	4850.23	4850.29	4850.34	4850.40	4850.46
65.8	4850.52	4850.58	4850.63	4850.69	4850.75	4850.81	4850.87	4850.92	4850.98	4851.04
65.9	4851.10	4851.15	4851.21	4851.27	4851.33	4851.38	4851.44	4851.50	4851.56	4851.62
66.0	4851.67	4851.73	4851.79	4851.85	4851.90	4851.96	4852.02	4852.08	4852.13	4852.19
66.1	4852.25	4852.31	4852.36	4852.42	4852.48	4852.54	4852.59	4852.65	4852.71	4852.77
66.2	4852.82	4852.88	4852.94	4853.00	4853.05	4853.11	4853.17	4853.23	4853.28	4853.34
66.3	4853.40	4853.46	4853.51	4853.57	4853.63	4853.68	4853.74	4853.80	4853.86	4853.91
66.4	4853.97	4854.03	4854.09	4854.14	4854.20	4854.26	4854.32	4854.37	4854.43	4854.49
66.5	4854.54	4854.60	4854.66	4854.72	4854.77	4854.83	4854.89	4854.94	4855.00	4855.06
66.6	4855.12	4855.17	4855.23	4855.29	4855.34	4855.40	4855.46	4855.52	4855.57	4855.63
66.7	4855.69	4855.74	4855.80	4855.86	4855.91	4855.97	4856.03	4856.09	4856.14	4856.20
66.8	4856.26	4856.31	4856.37	4856.43	4856.48	4856.54	4856.60	4856.65	4856.71	4856.77
66.9	4856.83	4856.88	4856.94	4857.00	4857.05	4857.11	4857.17	4857.22	4857.28	4857.34
67.0	4857.39	4857.45	4857.51	4857.56	4857.62	4857.68	4857.73	4857.79	4857.85	4857.90
67.1	4857.96	4858.02	4858.07	4858.13	4858.19	4858.24	4858.30	4858.36	4858.41	4858.47
67.2	4858.53	4858.58	4858.64	4858.70	4858.75	4858.81	4858.87	4858.92	4858.98	4859.04
67.3	4859.09	4859.15	4859.21	4859.26	4859.32	4859.38	4859.43	4859.49	4859.55	4859.60
67.4	4859.66	4859.72	4859.77	4859.83	4859.88	4859.94	4860.00	4860.05	4860.11	4860.17
67.5	4860.22	4860.28	4860.34	4860.39	4860.45	4860.50	4860.56	4860.62	4860.67	4860.73
67.6	4860.79	4860.84	4860.90	4860.95	4861.01	4861.07	4861.12	4861.18	4861.24	4861.29
67.7	4861.35	4861.40	4861.46	4861.52	4861.57	4861.63	4861.69	4861.74	4861.80	4861.85
67.8	4861.91	4861.97	4862.02	4862.08	4862.13	4862.19	4862.25	4862.30	4862.36	4862.41
67.9	4862.47	4862.53	4862.58	4862.64	4862.69	4862.75	4862.81	4862.86	4862.92	4862.97
TEMP	0	1	2	3	4	5	6	7	8	9

Table 3 (Continued) — The Speed of Sound in Water in Feet/Sec as a Function of Temperature in Degrees Fahrenheit 68.00 to 71.99

TEMP	0	1	2	3	4	5	6	7	8	9
68.0	4863.03	4863.09	4863.14	4863.20	4863.25	4863.31	4863.37	4863.42	4863.48	4863.53
68.1	4863.59	4863.65	4863.70	4863.76	4863.81	4863.87	4863.92	4863.98	4864.04	4864.09
68.2	4864.15	4864.20	4864.26	4864.32	4864.37	4864.43	4864.48	4864.54	4864.59	4864.65
68.3	4864.71	4864.76	4864.82	4864.87	4864.93	4864.98	4865.04	4865.09	4865.15	4865.21
68.4	4865.26	4865.32	4865.37	4865.43	4865.48	4865.54	4865.60	4865.65	4865.71	4865.76
68.5	4865.82	4865.87	4865.93	4865.98	4866.04	4866.09	4866.15	4866.21	4866.26	4866.32
68.6	4866.37	4866.43	4866.48	4866.54	4866.59	4866.65	4866.70	4866.76	4866.82	4866.87
68.7	4866.93	4866.98	4867.04	4867.09	4867.15	4867.20	4867.26	4867.31	4867.37	4867.42
68.8	4867.48	4867.53	4867.59	4867.65	4867.70	4867.76	4867.81	4867.87	4867.92	4867.98
68.9	4868.03	4868.09	4868.14	4868.20	4868.25	4868.31	4868.36	4868.42	4868.47	4868.53
69.0	4868.58	4868.64	4868.69	4868.75	4868.80	4868.86	4868.91	4868.97	4869.02	4869.08
69.1	4869.13	4869.19	4869.24	4869.30	4869.35	4869.41	4869.46	4869.52	4869.57	4869.63
69.2	4869.68	4869.74	4869.79	4869.85	4869.90	4869.96	4870.01	4870.07	4870.13	4870.18
69.3	4870.23	4870.29	4870.34	4870.40	4870.45	4870.51	4870.56	4870.62	4870.67	4870.73
69.4	4870.78	4870.84	4870.89	4870.95	4871.00	4871.06	4871.11	4871.16	4871.22	4871.27
69.5	4871.33	4871.38	4871.44	4871.49	4871.55	4871.60	4871.66	4871.71	4871.77	4871.82
69.6	4871.88	4871.93	4871.98	4872.04	4872.09	4872.15	4872.20	4872.26	4872.31	4872.37
69.7	4872.42	4872.48	4872.53	4872.58	4872.64	4872.69	4872.75	4872.80	4872.86	4872.91
69.8	4872.97	4873.02	4873.08	4873.13	4873.18	4873.24	4873.29	4873.35	4873.40	4873.46
69.9	4873.51	4873.56	4873.62	4873.67	4873.73	4873.78	4873.84	4873.89	4873.95	4874.00
70.0	4874.05	4874.11	4874.16	4874.22	4874.27	4874.32	4874.38	4874.43	4874.49	4874.54
70.1	4874.60	4874.65	4874.70	4874.76	4874.81	4874.87	4874.92	4874.98	4875.03	4875.08
70.2	4875.14	4875.19	4875.25	4875.30	4875.35	4875.41	4875.46	4875.52	4875.57	4875.62
70.3	4875.68	4875.73	4875.79	4875.84	4875.89	4875.95	4876.00	4876.06	4876.11	4876.16
70.4	4876.22	4876.27	4876.33	4876.38	4876.43	4876.49	4876.54	4876.60	4876.65	4876.70
70.5	4876.76	4876.81	4876.87	4876.92	4876.97	4877.03	4877.08	4877.13	4877.19	4877.24
70.6	4877.30	4877.35	4877.40	4877.46	4877.51	4877.57	4877.62	4877.67	4877.73	4877.78
70.7	4877.83	4877.89	4877.94	4877.99	4878.05	4878.10	4878.16	4878.21	4878.26	4878.32
70.8	4878.37	4878.42	4878.48	4878.53	4878.59	4878.64	4878.69	4878.75	4878.80	4878.85
70.9	4878.91	4878.96	4879.01	4879.07	4879.12	4879.17	4879.23	4879.28	4879.33	4879.39
71.0	4879.44	4879.50	4879.55	4879.60	4879.66	4879.71	4879.76	4879.82	4879.87	4879.92
71.1	4879.98	4880.03	4880.08	4880.14	4880.19	4880.24	4880.30	4880.35	4880.40	4880.46
71.2	4880.51	4880.56	4880.62	4880.67	4880.72	4880.78	4880.83	4880.88	4880.94	4880.99
71.3	4881.04	4881.10	4881.15	4881.21	4881.25	4881.31	4881.36	4881.41	4881.47	4881.52
71.4	4881.57	4881.63	4881.68	4881.73	4881.79	4881.84	4881.89	4881.95	4882.00	4882.05
71.5	4882.11	4882.16	4882.21	4882.26	4882.32	4882.37	4882.42	4882.48	4882.53	4882.58
71.6	4882.64	4882.69	4882.74	4882.79	4882.85	4882.90	4882.95	4883.01	4883.06	4883.11
71.7	4883.16	4883.22	4883.27	4883.32	4883.38	4883.43	4883.48	4883.53	4883.59	4883.64
71.8	4883.69	4883.75	4883.80	4883.85	4883.90	4883.96	4884.01	4884.06	4884.12	4884.17
71.9	4884.22	4884.27	4884.33	4884.38	4884.43	4884.48	4884.54	4884.59	4884.64	4884.70
TEMP	0	1	2	3	4	5	6	7	8	9

Table 3 (Continued) — The Speed of Sound in Water in Feet/Sec as a Function of Temperature in Degrees Fahrenheit 72.00 to 75.99

TEMP	0	1	2	3	4	5	6	7	8	9
72.0	4884.75	4884.80	4884.85	4884.91	4884.96	4885.01	4885.06	4885.12	4885.17	4885.22
72.1	4885.27	4885.33	4885.38	4885.43	4885.48	4885.54	4885.59	4885.64	4885.69	4885.75
72.2	4885.80	4885.85	4885.90	4885.96	4886.01	4886.06	4886.11	4886.69	4886.22	4886.27
72.3	4886.32	4886.38	4886.43	4886.48	4886.53	4886.59	4886.64	4886.69	4886.74	4886.80
72.4	4886.85	4886.90	4886.95	4887.01	4887.06	4887.11	4887.16	4887.21	4887.27	4887.32
72.5	4887.37	4887.42	4887.48	4887.53	4887.58	4887.63	4887.68	4887.74	4887.79	4887.84
72.6	4887.89	4887.95	4888.00	4888.05	4888.10	4888.15	4888.21	4888.26	4888.31	4888.36
72.7	4888.41	4888.47	4888.52	4888.57	4888.62	4888.68	4888.73	4888.78	4888.83	4888.88
72.8	4888.94	4888.99	4889.04	4889.09	4889.14	4889.20	4889.25	4889.30	4889.35	4889.40
72.9	4889.46	4889.51	4889.56	4889.61	4889.66	4889.71	4889.77	4889.82	4889.87	4889.92
73.0	4889.97	4890.03	4890.08	4890.13	4890.18	4890.23	4890.29	4890.34	4890.39	4890.44
73.1	4890.49	4890.54	4890.60	4890.65	4890.70	4890.75	4890.80	4890.85	4890.91	4890.96
73.2	4891.01	4891.06	4891.11	4891.16	4891.22	4891.27	4891.32	4891.37	4891.42	4891.47
73.3	4891.53	4891.58	4891.63	4891.68	4891.73	4891.78	4891.84	4891.89	4891.94	4891.99
73.4	4892.04	4892.09	4892.15	4892.20	4892.25	4892.30	4892.35	4892.40	4892.45	4892.51
73.5	4892.56	4892.61	4892.66	4892.71	4892.76	4892.81	4892.87	4892.92	4892.97	4893.02
73.6	4893.07	4893.12	4893.17	4893.23	4893.28	4893.33	4893.38	4893.43	4893.48	4893.53
73.7	4893.58	4893.64	4893.69	4893.74	4893.79	4893.84	4893.89	4893.94	4894.00	4894.05
73.8	4894.10	4894.15	4894.20	4894.25	4894.30	4894.35	4894.40	4894.46	4894.51	4894.56
73.9	4894.61	4894.66	4894.71	4894.76	4894.81	4894.87	4894.92	4894.97	4895.02	4895.07
74.0	4895.12	4895.17	4895.22	4895.27	4895.32	4895.38	4895.43	4895.48	4895.53	4895.58
74.1	4895.63	4895.68	4895.73	4895.78	4895.83	4895.89	4895.94	4895.99	4896.04	4896.09
74.2	4896.14	4896.19	4896.24	4896.29	4896.34	4896.39	4896.45	4896.50	4896.55	4896.60
74.3	4896.65	4896.70	4896.75	4896.80	4896.85	4896.90	4896.95	4897.00	4897.06	4897.11
74.4	4897.16	4897.21	4897.26	4897.31	4897.36	4897.41	4897.46	4897.51	4897.56	4897.61
74.5	4897.66	4897.71	4897.77	4897.82	4897.87	4897.92	4897.97	4898.02	4898.07	4898.12
74.6	4898.17	4898.22	4898.27	4898.32	4898.37	4898.42	4898.47	4898.52	4898.57	4898.63
74.7	4898.68	4898.73	4898.78	4898.83	4898.88	4898.93	4898.98	4899.03	4899.08	4899.13
74.8	4899.18	4899.23	4899.28	4899.33	4899.38	4899.43	4899.48	4899.53	4899.58	4899.63
74.9	4899.68	4899.73	4899.78	4899.84	4899.89	4899.94	4899.99	4900.04	4900.09	4900.14
75.0	4900.19	4900.24	4900.29	4900.34	4900.39	4900.44	4900.49	4900.54	4900.59	4900.64
75.1	4900.69	4900.74	4900.79	4900.84	4900.89	4900.94	4900.99	4901.04	4901.09	4901.14
75.2	4901.19	4901.24	4901.29	4901.34	4901.39	4901.44	4901.49	4901.54	4901.59	4901.64
75.3	4901.69	4901.74	4901.79	4901.84	4901.89	4901.94	4901.99	4902.04	4902.09	4902.14
75.4	4902.19	4902.24	4902.29	4902.34	4902.39	4902.44	4902.49	4902.54	4902.59	4902.64
75.5	4902.69	4902.74	4902.79	4902.84	4902.89	4902.94	4902.99	4903.04	4903.09	4903.14
75.6	4903.19	4903.24	4903.29	4903.34	4903.39	4903.44	4903.49	4903.54	4903.59	4903.64
75.7	4903.69	4903.74	4903.79	4903.84	4903.89	4903.94	4903.99	4904.04	4904.09	4904.13
75.8	4904.18	4904.23	4904.28	4904.33	4904.38	4904.43	4904.48	4904.53	4904.58	4904.63
75.9	4904.68	4904.73	4904.78	4904.83	4904.88	4904.93	4904.98	4905.03	4905.08	4905.13
TEMP	0	1	2	3	4	5	6	7	8	9

Table 3 (Continued) — The Speed of Sound in Water in Feet/Sec as a Function of Temperature in Degrees Fahrenheit 76.00 to 79.99

TEMP	0	1	2	3	4	5	6	7	8	9
76.0	4905.18	4905.23	4905.27	4905.32	4905.37	4905.42	4905.47	4905.52	4905.57	4905.62
76.1	4905.67	4905.72	4905.77	4905.82	4905.87	4905.92	4905.97	4906.02	4906.07	4906.11
76.2	4906.16	4906.21	4906.26	4906.31	4906.36	4906.41	4906.46	4906.51	4906.56	4906.61
76.3	4906.66	4906.71	4906.76	4906.81	4906.85	4906.90	4906.95	4907.00	4907.05	4907.10
76.4	4907.15	4907.20	4907.25	4907.30	4907.35	4907.40	4907.44	4907.49	4907.54	4907.59
76.5	4907.64	4907.69	4907.74	4907.79	4907.84	4907.89	4907.94	4907.98	4908.03	4908.08
76.6	4908.13	4908.18	4908.23	4908.28	4908.33	4908.38	4908.43	4908.47	4908.52	4908.57
76.7	4908.62	4908.67	4908.72	4908.77	4908.82	4908.87	4908.92	4908.96	4909.01	4909.06
76.8	4909.11	4909.16	4909.21	4909.26	4909.31	4909.35	4909.40	4909.45	4909.50	4909.55
76.9	4909.60	4909.65	4909.70	4909.75	4909.79	4909.84	4909.89	4909.94	4909.99	4910.04
77.0	4910.09	4910.14	4910.18	4910.23	4910.28	4910.33	4910.38	4910.43	4910.48	4910.52
77.1	4910.57	4910.62	4910.67	4910.72	4910.77	4910.82	4910.87	4910.91	4910.96	4911.01
77.2	4911.06	4911.11	4911.16	4911.21	4911.25	4911.30	4911.35	4911.40	4911.45	4911.50
77.3	4911.54	4911.59	4911.64	4911.69	4911.74	4911.79	4911.84	4911.88	4911.93	4911.98
77.4	4912.03	4912.08	4912.13	4912.17	4912.22	4912.27	4912.32	4912.37	4912.42	4912.46
77.5	4912.51	4912.56	4912.61	4912.66	4912.71	4912.75	4912.80	4912.85	4912.90	4912.95
77.6	4913.00	4913.04	4913.09	4913.14	4913.19	4913.24	4913.29	4913.33	4913.38	4913.43
77.7	4913.48	4913.53	4913.57	4913.62	4913.67	4913.72	4913.77	4913.82	4913.86	4913.91
77.8	4913.96	4914.01	4914.06	4914.10	4914.15	4914.20	4914.25	4914.30	4914.34	4914.39
77.9	4914.44	4914.49	4914.54	4914.58	4914.63	4914.68	4914.73	4914.78	4914.82	4914.87
78.0	4914.92	4914.97	4915.02	4915.06	4915.11	4915.16	4915.21	4915.26	4915.30	4915.35
78.1	4915.40	4915.45	4915.50	4915.54	4915.59	4915.64	4915.69	4915.73	4915.78	4915.83
78.2	4915.88	4915.93	4915.97	4916.02	4916.07	4916.12	4916.16	4916.21	4916.26	4916.31
78.3	4916.36	4916.40	4916.45	4916.50	4916.55	4916.59	4916.64	4916.69	4916.74	4916.78
78.4	4916.83	4916.88	4916.93	4916.98	4917.02	4917.07	4917.12	4917.17	4917.21	4917.26
78.5	4917.31	4917.36	4917.40	4917.45	4917.50	4917.55	4917.59	4917.64	4917.69	4917.74
78.6	4917.78	4917.83	4917.88	4917.93	4917.97	4918.02	4918.07	4918.12	4918.16	4918.21
78.7	4918.26	4918.31	4918.35	4918.40	4918.45	4918.50	4918.54	4918.59	4918.64	4918.68
78.8	4918.73	4918.78	4918.83	4918.87	4918.92	4918.97	4919.02	4919.06	4919.11	4919.16
78.9	4919.21	4919.25	4919.30	4919.35	4919.39	4919.44	4919.49	4919.54	4919.58	4919.63
79.0	4919.68	4919.72	4919.77	4919.82	4919.87	4919.91	4919.96	4920.01	4920.05	4920.10
79.1	4920.15	4920.20	4920.24	4920.29	4920.34	4920.38	4920.43	4920.48	4920.53	4920.57
79.2	4920.62	4920.67	4920.71	4920.76	4920.81	4920.86	4920.90	4920.95	4921.00	4921.04
79.3	4921.09	4921.14	4921.18	4921.23	4921.28	4921.32	4921.37	4921.42	4921.47	4921.51
79.4	4921.56	4921.61	4921.65	4921.70	4921.75	4921.79	4921.84	4921.89	4921.93	4921.98
79.5	4922.03	4922.07	4922.12	4922.17	4922.22	4922.26	4922.31	4922.36	4922.40	4922.45
79.6	4922.50	4922.54	4922.59	4922.64	4922.68	4922.73	4922.78	4922.82	4922.87	4922.92
79.7	4922.96	4923.01	4923.06	4923.11	4923.15	4923.20	4923.24	4923.29	4923.34	4923.38
79.8	4923.43	4923.48	4923.52	4923.57	4923.62	4923.66	4923.71	4923.76	4923.80	4923.85
79.9	4923.89	4923.94	4923.99	4924.03	4924.08	4924.13	4924.17	4924.22	4924.27	4924.31
TEMP	0	1	2	3	4	5	6	7	8	9

Table 3 (Continued) — The Speed of Sound in Water in Feet/Sec as a Function of Temperature in Degrees Fahrenheit 80.00 to 83.99

TEMP	0	1	2	3	4	5	6	7	8	9
80.0	4924.36	4924.41	4924.45	4924.50	4924.54	4924.59	4924.64	4924.68	4924.73	4924.78
80.1	4924.82	4924.87	4924.92	4924.96	4925.01	4925.06	4925.10	4925.15	4925.19	4925.24
80.2	4925.29	4925.33	4925.38	4925.43	4925.47	4925.52	4925.56	4925.61	4925.66	4925.70
80.3	4925.75	4925.80	4925.84	4925.89	4925.93	4925.98	4926.03	4926.07	4926.12	4926.16
80.4	4926.21	4926.26	4926.30	4926.35	4926.40	4926.44	4926.49	4926.53	4926.58	4926.63
80.5	4926.67	4926.72	4926.76	4926.81	4926.86	4926.90	4926.95	4926.99	4927.04	4927.09
80.6	4927.13	4927.18	4927.22	4927.27	4927.32	4927.36	4927.41	4927.45	4927.50	4927.55
80.7	4927.59	4927.64	4927.68	4927.73	4927.78	4927.82	4927.87	4927.91	4927.96	4928.00
80.8	4928.05	4928.10	4928.14	4928.19	4928.23	4928.28	4928.33	4928.37	4928.42	4928.46
80.9	4928.51	4928.55	4928.60	4928.65	4928.69	4928.74	4928.78	4928.83	4928.87	4928.92
81.0	4928.97	4929.01	4929.06	4929.10	4929.15	4929.19	4929.24	4929.29	4929.33	4929.38
81.1	4929.42	4929.47	4929.51	4929.56	4929.60	4929.65	4929.70	4929.74	4929.79	4929.83
81.2	4929.88	4929.92	4929.97	4930.01	4930.06	4930.11	4930.15	4930.20	4930.24	4930.29
81.3	4930.33	4930.38	4930.42	4930.47	4930.52	4930.56	4930.61	4930.65	4930.70	4930.74
81.4	4930.79	4930.83	4930.88	4930.92	4930.97	4931.01	4931.06	4931.11	4931.15	4931.20
81.5	4931.24	4931.29	4931.33	4931.38	4931.42	4931.47	4931.51	4931.56	4931.60	4931.65
81.6	4931.69	4931.74	4931.78	4931.83	4931.88	4931.92	4931.97	4932.01	4932.06	4932.10
81.7	4932.15	4932.19	4932.24	4932.28	4932.33	4932.37	4932.42	4932.46	4932.51	4932.55
81.8	4932.60	4932.64	4932.69	4932.73	4932.78	4932.82	4932.87	4932.91	4932.96	4933.00
81.9	4933.05	4933.09	4933.14	4933.18	4933.23	4933.27	4933.32	4933.36	4933.41	4933.45
82.0	4933.50	4933.54	4933.59	4933.63	4933.68	4933.72	4933.77	4933.81	4933.86	4933.90
82.1	4933.95	4933.99	4934.04	4934.08	4934.13	4934.17	4934.22	4934.26	4934.31	4934.35
82.2	4934.40	4934.44	4934.49	4934.53	4934.57	4934.62	4934.66	4934.71	4934.75	4934.80
82.3	4934.84	4934.89	4934.93	4934.98	4935.02	4935.07	4935.11	4935.16	4935.20	4935.25
82.4	4935.29	4935.34	4935.38	4935.42	4935.47	4935.51	4935.56	4935.60	4935.65	4935.69
82.5	4935.74	4935.78	4935.83	4935.87	4935.91	4935.96	4936.00	4936.05	4936.09	4936.14
82.6	4936.18	4936.23	4936.27	4936.32	4936.36	4936.40	4936.45	4936.49	4936.54	4936.58
82.7	4936.63	4936.67	4936.72	4936.76	4936.80	4936.85	4936.89	4936.94	4936.98	4937.03
82.8	4937.07	4937.12	4937.16	4937.20	4937.25	4937.29	4937.34	4937.38	4937.43	4937.47
82.9	4937.51	4937.56	4937.60	4937.65	4937.69	4937.74	4937.78	4937.82	4937.87	4937.91
83.0	4937.96	4938.00	4938.05	4938.09	4938.13	4938.18	4938.22	4938.27	4938.31	4938.35
83.1	4938.40	4938.44	4938.49	4938.53	4938.58	4938.62	4938.66	4938.71	4938.75	4938.80
83.2	4938.84	4938.88	4938.93	4938.97	4939.02	4939.06	4939.10	4939.15	4939.19	4939.24
83.3	4939.28	4939.32	4939.37	4939.41	4939.46	4939.50	4939.54	4939.59	4939.63	4939.68
83.4	4939.72	4939.76	4939.81	4939.85	4939.90	4939.94	4939.98	4940.03	4940.07	4940.11
83.5	4940.16	4940.20	4940.25	4940.29	4940.33	4940.38	4940.42	4940.47	4940.51	4940.55
83.6	4940.60	4940.64	4940.68	4940.73	4940.77	4940.82	4940.86	4940.90	4940.95	4940.99
83.7	4941.03	4941.08	4941.12	4941.17	4941.21	4941.25	4941.30	4941.34	4941.38	4941.43
83.8	4941.47	4941.51	4941.56	4941.60	4941.65	4941.69	4941.73	4941.78	4941.82	4941.86
83.9	4941.91	4941.95	4941.99	4942.04	4942.08	4942.12	4942.17	4942.21	4942.26	4942.30
TEMP	0	1	2	3	4	5	6	7	8	9

Table 3 (Continued) — The Speed of Sound in Water in Feet/Sec as a Function of Temperature in Degrees Fahrenheit 84.00 to 87.99

TEMP	0	1	2	3	4	5	6	7	8	9
84.0	4942.34	4942.39	4942.43	4942.47	4942.52	4942.56	4942.60	4942.65	4942.69	4942.73
84.1	4942.78	4942.82	4942.86	4942.91	4942.95	4942.99	4943.04	4943.08	4943.12	4943.17
84.2	4943.21	4943.25	4943.30	4943.34	4943.38	4943.43	4943.47	4943.51	4943.56	4943.60
84.3	4943.64	4943.69	4943.73	4943.77	4943.82	4943.86	4943.90	4943.95	4943.99	4944.03
84.4	4944.08	4944.12	4944.16	4944.21	4944.25	4944.29	4944.34	4944.38	4944.42	4944.46
84.5	4944.51	4944.55	4944.59	4944.64	4944.68	4944.72	4944.77	4944.81	4944.85	4944.90
84.6	4944.94	4944.98	4945.02	4945.07	4945.11	4945.15	4945.20	4945.24	4945.28	4945.33
84.7	4945.37	4945.41	4945.45	4945.50	4945.54	4945.58	4945.63	4945.67	4945.71	4945.76
84.8	4945.80	4945.84	4945.88	4945.93	4945.97	4946.01	4946.06	4946.10	4946.14	4946.18
84.9	4946.23	4946.27	4946.31	4946.36	4946.40	4946.44	4946.48	4946.53	4946.57	4946.61
85.0	4946.66	4946.70	4946.74	4946.78	4946.83	4946.87	4946.91	4946.95	4947.00	4947.04
85.1	4947.08	4947.13	4947.17	4947.21	4947.25	4947.30	4947.34	4947.38	4947.42	4947.47
85.2	4947.51	4947.55	4947.59	4947.64	4947.68	4947.72	4947.76	4947.81	4947.85	4947.89
85.3	4947.94	4947.98	4948.02	4948.06	4948.11	4948.15	4948.19	4948.23	4948.28	4948.32
85.4	4948.36	4948.40	4948.45	4948.49	4948.53	4948.57	4948.62	4948.66	4948.70	4948.74
85.5	4948.78	4948.83	4948.87	4948.91	4948.95	4949.00	4949.04	4949.08	4949.12	4949.17
85.6	4949.21	4949.25	4949.29	4949.34	4949.38	4949.42	4949.46	4949.50	4949.55	4949.59
85.7	4949.63	4949.67	4949.72	4949.76	4949.80	4949.84	4949.89	4949.93	4949.97	4950.01
85.8	4950.05	4950.10	4950.14	4950.18	4950.22	4950.27	4950.31	4950.35	4950.39	4950.43
85.9	4950.48	4950.52	4950.56	4950.60	4950.64	4950.69	4950.73	4950.77	4950.81	4950.85
86.0	4950.90	4950.94	4950.98	4951.02	4951.06	4951.11	4951.15	4951.19	4951.23	4951.27
86.1	4951.32	4951.36	4951.40	4951.44	4951.48	4951.53	4951.57	4951.61	4951.65	4951.69
86.2	4951.74	4951.78	4951.82	4951.86	4951.90	4951.95	4951.99	4952.03	4952.07	4952.11
86.3	4952.16	4952.20	4952.24	4952.28	4952.32	4952.36	4952.41	4952.45	4952.49	4952.53
86.4	4952.57	4952.62	4952.66	4952.70	4952.74	4952.78	4952.82	4952.87	4952.91	4952.95
86.5	4952.99	4953.03	4953.07	4953.12	4953.16	4953.20	4953.24	4953.28	4953.32	4953.37
86.6	4953.41	4953.45	4953.49	4953.53	4953.57	4953.62	4953.66	4953.70	4953.74	4953.78
86.7	4953.82	4953.86	4953.91	4953.95	4953.99	4954.03	4954.07	4954.11	4954.16	4954.20
86.8	4954.24	4954.28	4954.32	4954.36	4954.40	4954.45	4954.49	4954.53	4954.57	4954.61
86.9	4954.65	4954.69	4954.74	4954.78	4954.82	4954.86	4954.90	4954.94	4954.98	4955.03
87.0	4955.07	4955.11	4955.15	4955.19	4955.23	4955.27	4955.31	4955.36	4955.40	4955.44
87.1	4955.48	4955.52	4955.56	4955.60	4955.65	4955.69	4955.73	4955.77	4955.81	4955.85
87.2	4955.89	4955.93	4955.97	4956.02	4956.06	4956.10	4956.14	4956.18	4956.22	4956.26
87.3	4956.30	4956.35	4956.39	4956.43	4956.47	4956.51	4956.55	4956.59	4956.63	4956.67
87.4	4956.72	4956.76	4956.80	4956.84	4956.88	4956.92	4956.96	4957.00	4957.04	4957.08
87.5	4957.13	4957.17	4957.21	4957.25	4957.29	4957.33	4957.37	4957.41	4957.45	4957.49
87.6	4957.54	4957.58	4957.62	4957.66	4957.70	4957.74	4957.78	4957.82	4957.86	4957.90
87.7	4957.94	4957.98	4958.03	4958.07	4958.11	4958.15	4958.19	4958.23	4958.27	4958.31
87.8	4958.35	4958.39	4958.43	4958.47	4958.52	4958.56	4958.60	4958.64	4958.68	4958.72
87.9	4958.76	4958.80	4958.84	4958.88	4958.92	4958.96	4959.00	4959.04	4959.09	4959.13
TEMP	0	1	2	3	4	5	6	7	8	9

Table 3 (Continued) — The Speed of Sound in Water in Feet/Sec as a Function of Temperature in Degrees Fahrenheit 88.00 to 91.99

TEMP	0	1	2	3	4	5	6	7	8	9
88.0	4959.17	4959.21	4959.25	4959.29	4959.33	4959.37	4959.41	4959.45	4959.49	4959.53
88.1	4959.57	4959.61	4959.65	4959.69	4959.74	4959.78	4959.82	4959.86	4959.90	4959.94
88.2	4959.98	4960.02	4960.06	4960.10	4960.14	4960.18	4960.22	4960.26	4960.30	4960.34
88.3	4960.38	4960.42	4960.46	4960.50	4960.54	4960.59	4960.63	4960.67	4960.71	4960.75
88.4	4960.79	4960.83	4960.87	4960.91	4960.95	4960.99	4961.03	4961.07	4961.11	4961.15
88.5	4961.19	4961.23	4961.27	4961.31	4961.35	4961.39	4961.43	4961.47	4961.51	4961.55
88.6	4961.59	4961.63	4961.67	4961.71	4961.75	4961.79	4961.83	4961.87	4961.91	4961.95
88.7	4961.99	4962.03	4962.08	4962.12	4962.16	4962.20	4962.24	4962.28	4962.32	4962.36
88.8	4962.40	4962.44	4962.48	4962.52	4962.56	4962.60	4962.64	4962.68	4962.72	4962.76
88.9	4962.80	4962.84	4962.88	4962.92	4962.96	4963.00	4963.04	4963.08	4963.12	4963.16
89.0	4963.20	4963.24	4963.28	4963.32	4963.36	4963.40	4963.44	4963.48	4963.52	4963.56
89.1	4963.60	4963.64	4963.68	4963.72	4963.76	4963.79	4963.83	4963.87	4963.91	4963.95
89.2	4963.99	4964.03	4964.07	4964.11	4964.15	4964.19	4964.23	4964.27	4964.31	4964.35
89.3	4964.39	4964.43	4964.47	4964.51	4964.55	4964.59	4964.63	4964.67	4964.71	4964.75
89.4	4964.79	4964.83	4964.87	4964.91	4964.95	4964.99	4965.03	4965.07	4965.11	4965.15
89.5	4965.19	4965.22	4965.26	4965.30	4965.34	4965.38	4965.42	4965.46	4965.50	4965.54
89.6	4965.58	4965.62	4965.66	4965.70	4965.74	4965.78	4965.82	4965.86	4965.90	4965.94
89.7	4965.98	4966.02	4966.05	4966.09	4966.13	4966.17	4966.21	4966.25	4966.29	4966.33
89.8	4966.37	4966.41	4966.45	4966.49	4966.53	4966.57	4966.61	4966.65	4966.69	4966.72
89.9	4966.76	4966.80	4966.84	4966.88	4966.92	4966.96	4967.00	4967.04	4967.08	4967.12
90.0	4967.16	4967.20	4967.24	4967.27	4967.31	4967.35	4967.39	4967.43	4967.47	4967.51
90.1	4967.55	4967.59	4967.63	4967.67	4967.71	4967.75	4967.78	4967.82	4967.86	4967.90
90.2	4967.94	4967.98	4968.02	4968.06	4968.10	4968.14	4968.18	4968.21	4968.25	4968.29
90.3	4968.33	4968.37	4968.41	4968.46	4968.49	4968.53	4968.57	4968.60	4968.64	4968.68
90.4	4968.72	4968.76	4968.80	4968.84	4968.88	4968.92	4968.96	4968.99	4969.03	4969.07
90.5	4969.11	4969.15	4969.19	4969.23	4969.27	4969.31	4969.34	4969.38	4969.42	4969.46
90.6	4969.50	4969.54	4969.58	4969.62	4969.66	4969.69	4969.73	4969.77	4969.81	4969.85
90.7	4969.89	4969.93	4969.97	4970.00	4970.04	4970.08	4970.12	4970.16	4970.20	4970.24
90.8	4970.28	4970.31	4970.35	4970.39	4970.43	4970.47	4970.51	4970.55	4970.59	4970.62
90.9	4970.66	4970.70	4970.74	4970.78	4970.82	4970.86	4970.89	4970.93	4970.97	4971.01
91.0	4971.05	4971.09	4971.13	4971.16	4971.20	4971.24	4971.28	4971.32	4971.36	4971.40
91.1	4971.43	4971.47	4971.51	4971.55	4971.59	4971.63	4971.67	4971.70	4971.74	4971.78
91.2	4971.82	4971.86	4971.90	4971.93	4971.97	4972.01	4972.05	4972.09	4972.13	4972.16
91.3	4972.20	4972.24	4972.28	4972.32	4972.36	4972.39	4972.43	4972.47	4972.51	4972.55
91.4	4972.59	4972.62	4972.66	4972.70	4972.74	4972.78	4972.82	4972.85	4972.89	4972.93
91.5	4972.97	4973.01	4973.05	4973.08	4973.12	4973.16	4973.20	4973.24	4973.27	4973.31
91.6	4973.35	4973.39	4973.43	4973.47	4973.50	4973.54	4973.58	4973.62	4973.66	4973.69
91.7	4973.73	4973.77	4973.81	4973.85	4973.88	4973.92	4973.96	4974.00	4974.04	4974.08
91.8	4974.11	4974.15	4974.19	4974.23	4974.27	4974.30	4974.34	4974.38	4974.42	4974.46
91.9	4974.49	4974.53	4974.57	4974.61	4974.64	4974.68	4974.72	4974.76	4974.80	4974.83
TEMP	0	1	2	3	4	5	6	7	8	9

INDEX

absorbing cylinders, 139
absorbing spheres, 195, 199
absorption, 56, 125, 140, 146, 148,
  158, 159, 217, 220
acoustic impedance, 22, 27
air facility, 365
angular frequency, 2
antisymmetric modes, 105, 108-115,
  337
attenuation, 95-97, 151
attenuation measurements, 316,
  321-323
background contribution, 262
"background" integral, 40
beam-displacement, 297, 307
bistatic form function, 140
bistatic reflection, 143, 144, 146,
  275, 276
boundary condition, 3, 127
bounded beam, 297
bulk properties determined from
  reflection, 289
calibration using a sphere, 279
characteristic frequencies, 254
circumferential waves, 35, 57, 58, 75,
  92, 98, 109, 117, 230
circumferential wave speed, 50
compressional waves, 95
creeping wave attenuation, 49, 51, 64
creeping wave paths, 185
creeping waves ("Regge poles"), 35, 41,
  43, 44, 47-51, 57, 62, 180, 257, 265,
  268
curved plates, 342-344
curved surface, 11
cylinder, 12, 40, 55, 59, 62, 142, 239,
  258, 290
cylinder at high ka, 239
cylindrical cavity, 151
cylindrical shells, 125, 342
differential scattering section, 37, 43,
  47, 207

eigen frequency, 73
elastic sphere in solid, 205
elastic spheres, 161, 177
facilities, 236
farfield, 139
finite cylinders, 140
finite plane, 5, 18
finite wedge, 8
flaws in plates, 247
flexural modes, 56, 92
fluid filled shell, 129
form function, 56, 60, 102, 139, 177,
  178, 196, 208, 225, 238, 242,
  259, 266
"form-function", 163
Fourier theorem, 178
Fourier transform, 60-62
"Franz-type" waves, 36, 57, 70, 73, 75
Franz-type zeros, 39, 44
Franz wave, 45, 47, 62, 71, 117
Franz wave speed, 65
frequency of least reflection, 306, 309
Fresnel, 1
Fresnel integral, 19, 20, 23, 27, 29
gaussian-shaped beam, 301
general shape, 11
graphical solution for finite plane, 7
grounding, 357
group velocity, 94, 99, 102, 332, 334,
  339, 348
group wave-speed, 332, 334
Hankel asymptotic form, 59, 259
headwave, 315, 331, 333, 347
Huygens' construction, 30-32
hydrophone experiments, 49-51
impedance, 290
incident shear wave, 153
interference, 102
Kirchhoff approximation, 1, 12, 14,
  17, 21
Lamb modes (table), 338
Lamb theory, 336-339

Lamb waves, 93, 105, 117, 325, 336-339
Lamé constants, 36
lateral displacement beam, 299, 304
lateral waves, 75
layered cylinders, 125, 290
leaky Rayleigh wave, 58, 75, 80
lenses, 225
longitudinal critical angle, 311, 325
longitudinal mode, 56, 92
longitudinal waves, 36
material constants, 36, 102, 130, 146 153, 193, 208, 237, 305,
mechanical impedance, 264
modal impedance, 243
mode conversion, 158,
monostatic reflection, 129
near-field measurements, 21, 24
nonstatic reflection, 2,
normal mode, 35, 55, 56, 73, 100, 116, 126, 130, 242, 244
normal plane, 3, 5, 12
partial-wave scattering, 257, 265, 267
particle velocity, 4,
phase jump, 273
phase velocity, 105, 107, 337, 348
plates, 325, 327
plates and shells, 342
"potential scattering", 262
quadric surfaces, 12
quality factor (Q), 253
radiation frequencies, 254
radiation impedance, 243, 264
Rayleigh angle, 297
Rayleigh phase velocity, 315
Rayleigh radiation, 48
Rayleigh resonances, 272
Rayleigh series, 55, 57, 58
"Rayleigh-type" waves, 36, 57, 73, 75
Rayleigh-type zeros, 39, 45, 73
Rayleigh wave, 47, 76, 268, 297
Rayleigh wave speed, 315
rectangular piston, 28
rectangular plane, 1
reflection coefficient, 289
reflection reduction, 293
"Regge poles", 269
resonance, 72, 75, 80, 91, 257, 267, 283

resonance decay time, 257
resonance frequencies, 260
resonance scattering from cylinders, 2, 58
resonance width, 261
resonant modes, 167
retarded potential, 17
rigid sphere, 177, 180
rubber cylinders, 225
schileren, 12, 46-48, 140, 141, 247, 325, 354, 367-372
Schoch displacement distance, 298, 304
shadow region, 47
shear critical angle, 309, 325
Shear wave speed, 315, 319, 320
shells, 56, 91, 94, 104
silicone rubber, 225
slow waveguides, 225
solid cylinder, 130
Sommerfeld-Watson, 35, 38, 39, 57, 58, 63, 75
sound speed measurement, 373-383
sound speed tables, 383-396,
source strength, 3,
specular reflection, 331
sphere, 12, 161, 238, 265, 281
sphere as a standard reflector, 283
sphere at high ka, 238
spheroid, 12, 183-185
steady-state reflection, 161, 194, 280
summation formula, 2, 9, 15
surface wave radiation, 298
symmetric mode, 105, 108-115, 337
temperature effect, 172
test tank size, 359
thin plate, 336
tone burst, 194
total impedance, 290
total scattering cross section, 151
transient signals, 177
transmission, 327
transverse waves, 36
velocity potential, 1-3, 12, 17
wave speed, 22, 56
"whispering gallery" modes, 268
"whispering gallery" resonances, 272
"whispering gallery" waves, 75